Advances in Molecular Mechanisms of Gastrointestinal Tumors

Advances in Molecular Mechanisms of Gastrointestinal Tumors

Editor

Shihori Tanabe

Basel • Beijing • Wuhan • Barcelona • Belgrade • Novi Sad • Cluj • Manchester

Editor
Shihori Tanabe
National Institute of Health Sciences
Kawasaki
Japan

Editorial Office
MDPI AG
Grosspeteranlage 5
4052 Basel, Switzerland

This is a reprint of articles from the Special Issue published online in the open access journal *Cancers* (ISSN 2072-6694) (available at: https://www.mdpi.com/journal/cancers/special_issues/molecular_gastrointestinal).

For citation purposes, cite each article independently as indicated on the article page online and as indicated below:

Lastname, A.A.; Lastname, B.B. Article Title. *Journal Name* **Year**, *Volume Number*, Page Range.

ISBN 978-3-7258-1681-1 (Hbk)
ISBN 978-3-7258-1682-8 (PDF)
doi.org/10.3390/books978-3-7258-1682-8

© 2024 by the authors. Articles in this book are Open Access and distributed under the Creative Commons Attribution (CC BY) license. The book as a whole is distributed by MDPI under the terms and conditions of the Creative Commons Attribution-NonCommercial-NoDerivs (CC BY-NC-ND) license.

Contents

About the Editor vii

Shihori Tanabe
Advances in Molecular Mechanisms of Gastrointestinal Tumors
Reprinted from: *Cancers* **2024**, *16*, 1603, doi:10.3390/cancers16081603 1

Xiyang Tang, Jie Yang, Anping Shi, Yanlu Xiong, Miaomiao Wen, Zhonglin Luo, et al.
CD155 Cooperates with PD-1/PD-L1 to Promote Proliferation of Esophageal Squamous Cancer Cells via PI3K/Akt and MAPK Signaling Pathways
Reprinted from: *Cancers* **2022**, *14*, 5610, doi:10.3390/cancers14225610 5

Elizabeth Proaño-Pérez, Eva Serrano-Candelas, Cindy Mancia, Arnau Navinés-Ferrer, Mario Guerrero and Margarita Martin
SH3BP2 Silencing Increases miRNAs Targeting ETV1 and Microphthalmia-Associated Transcription Factor, Decreasing the Proliferation of Gastrointestinal Stromal Tumors
Reprinted from: *Cancers* **2022**, *14*, 6198, doi:10.3390/cancers14246198 23

Eman A. Abdul Razzaq, Khuloud Bajbouj, Amal Bouzid, Noura Alkhayyal, Rifat Hamoudi and Riyad Bendardaf
Transcriptomic Changes Associated with *ERBB2* Overexpression in Colorectal Cancer Implicate a Potential Role of the Wnt Signaling Pathway in Tumorigenesis
Reprinted from: *Cancers* **2023**, *15*, 130, doi:10.3390/cancers15010130 38

Wuhan Yu, Ning Liu, Xiaogang Song, Lang Chen, Mancai Wang, Guohui Xiao, et al.
EZH2: An Accomplice of Gastric Cancer
Reprinted from: *Cancers* **2023**, *15*, 425, doi:10.3390/cancers15020425 55

Huan Yan, Jing-Ling Zhang, Kam-Tong Leung, Kwok-Wai Lo, Jun Yu, Ka-Fai To and Wei Kang
An Update of G-Protein-Coupled Receptor Signaling and Its Deregulation in Gastric Carcinogenesis
Reprinted from: *Cancers* **2023**, *15*, 736, doi:10.3390/cancers15030736 74

John M. Macharia, Timea Varjas, Ruth W. Mwangi, Zsolt Káposztás, Nóra Rozmann, Márton Pintér, et al.
Modulatory Properties of *Aloe secundiflora's* Methanolic Extracts on Targeted Genes in Colorectal Cancer Management
Reprinted from: *Cancers* **2023**, *15*, 5002, doi:10.3390/cancers15205002 104

Joanna Kamińska, Olga Martyna Koper-Lenkiewicz, Donata Ponikwicka-Tyszko, Weronika Lebiedzińska, Ewelina Palak, Maria Sztachelska, et al.
New Insights on the Progesterone (P4) and PGRMC1/NENF Complex Interactions in Colorectal Cancer Progression
Reprinted from: *Cancers* **2023**, *15*, 5074, doi:10.3390/cancers15205074 120

Xiaofei Cheng, Feng Zhao, Bingxin Ke, Dong Chen and Fanlong Liu
Harnessing Ferroptosis to Overcome Drug Resistance in Colorectal Cancer: Promising Therapeutic Approaches
Reprinted from: *Cancers* **2023**, *15*, 5209, doi:10.3390/cancers15215209 141

Jianyun Shi, Wenjing Li, Zhenhua Jia, Ying Peng, Jiayi Hou, Ning Li, et al.
Synaptotagmin 1 Suppresses Colorectal Cancer Metastasis by Inhibiting ERK/MAPK
Signaling-Mediated Tumor Cell Pseudopodial Formation and Migration
Reprinted from: *Cancers* **2023**, *15*, 5282, doi:10.3390/cancers15215282 . . . . **159**

Milica Mitrovic Jovanovic, Aleksandra Djuric Stefanovic, Dimitrije Sarac, Jelena Kovac, Aleksandra Jankovic, Dusan J. Saponjski, et al.
Possibility of Using Conventional Computed Tomography Features and Histogram Texture
Analysis Parameters as Imaging Biomarkers for Preoperative Prediction of High-Risk
Gastrointestinal Stromal Tumors of the Stomach
Reprinted from: *Cancers* **2023**, *15*, 5840, doi:10.3390/cancers15245840 . . . . **179**

David Aebisher, Paweł Woźnicki, Klaudia Dynarowicz, Aleksandra Kawczyk-Krupka, Grzegorz Cieślar and Dorota Bartusik-Aebisher
Photodynamic Therapy and Immunological View in Gastrointestinal Tumors
Reprinted from: *Cancers* **2024**, *16*, 66, doi:10.3390/cancers16010066 . . . . **194**

About the Editor

Shihori Tanabe

Shihori Tanabe, PhD, is a senior researcher at the Division of Risk Assessment, Center for Biological Safety and Research in the National Institute of Health Sciences (NIHS). She completed her doctorate in pharmaceutical science within the field of molecular biology and pharmacology at the University of Tokyo. She also performed research mainly focused on biochemistry and cellular biology at the University of Illinois College of Medicine in Chicago. Part of her research is focused on molecular networks, gene expression profiling, and finding important factors of various cells, including cancer and stem cells, using new bioinformatics approaches and methodologies. She has also conducted research on cellular signaling, cardiac cells and protein regulation mechanisms, and made contributions in the field of molecular biology, pharmacology and biochemistry. She is also interested in finding essential factors to distinguish cancer cell types for revealing the mechanism of cancer development and safely applying cells as cellular therapeutics in regenerative medicine, as well as in assessing the long-term influences of alterations in gene expression and cellular phenotypes from the point of view of regulatory science. She has been a delegate for the Japanese Society for Regenerative Medicine, a councillor for the Society for Regulatory Science of Medical Products, a delegate for the Pharmaceutical Society of Japan, an international associate for the Japanese Pharmacological Society, an Editorial Board Member for several journals, a Book Editor for several books. She leads the Adverse Outcome Pathway (AOP) Coach Team and has been a co-lead of the External Review subgroup in OECD. She has published more than 90 papers and presented more than 140 presentations. She has organized several international conferences and was awarded an Overseas Study Grant by the Foundation for Research on Metabolic Disorders in 2002.

Editorial

Advances in Molecular Mechanisms of Gastrointestinal Tumors

Shihori Tanabe

Division of Risk Assessment, Center for Biological Safety and Research, National Institute of Health Sciences, Kawasaki 210-9501, Japan; stanabe@nihs.go.jp; Tel.: +81-44-270-6686

Citation: Tanabe, S. Advances in Molecular Mechanisms of Gastrointestinal Tumors. *Cancers* **2024**, *16*, 1603. https://doi.org/10.3390/cancers16081603

Received: 17 April 2024
Accepted: 19 April 2024
Published: 22 April 2024

Copyright: © 2024 by the author. Licensee MDPI, Basel, Switzerland. This article is an open access article distributed under the terms and conditions of the Creative Commons Attribution (CC BY) license (https://creativecommons.org/licenses/by/4.0/).

1. Introduction

Gastrointestinal cancer is one of the most common malignancies worldwide. The molecular mechanisms of gastrointestinal cancer, particularly several types that are resistant to treatment, have not been fully elucidated. The Special Issue entitled "Advances in Molecular Mechanisms of Gastrointestinal Tumors" includes a collection of a variety of articles on gastrointestinal stromal tumors, colorectal cancer, esophageal squamous cancer, gastrointestinal tumors, gastric carcinogenesis, and gastric cancer. This editorial aims to summarize recent perspectives on the mechanisms of gastrointestinal tumors, where molecular pathway networks are involved.

Epithelial–mesenchymal transition (EMT) is essential to the development of drug resistance in cancer, metastasis, and recurrence of cancer [1]. The microenvironment and EMT are involved in gastrointestinal tumor progression such as metastatic colorectal cancer [2]. Recent findings highlight the importance of molecular mechanisms in terms of microenvironmental and immune regulations in gastrointestinal tumors [3–7]. Chronic inflammation and the gut microbiota, in relation to immune response, have been closely investigated in gastrointestinal tumors [8,9].

Furthermore, phytochemicals have been found to be effective in gastrointestinal cancer, which underscores the importance of understanding the molecular pathway mechanisms regulated by phytochemicals as anti-gastrointestinal tumor agents [10]. The modes of action of phytochemicals include inhibiting pathways related to either wingless-type MMTV integration site family (Wnt)/β-catenin, apoptosis, phosphoinositide 3-kinase (PI3K)/protein kinase B (PKB, AKT)/mammalian target of rapamycin (mTOR), mitogen-activated protein kinase (MAPK), or NF-κB, or otherwise detoxification enzymes or adenosine monophosphate (AMP)-activated protein kinase [10]. It is crucial to reveal the molecular mechanisms of gastrointestinal tumors to develop novel therapeutics to overcome drug resistance.

2. An Overview of Published Articles in the Special Issue

The Special Issue "Advances in Molecular Mechanisms of Gastrointestinal Tumors" (https://www.mdpi.com/journal/cancers/special_issues/molecular_gastrointestinal) (accessed on 20 April 2024) was created on 23 November 2021 and the call for submissions of manuscripts was closed on 15 September 2023. Twenty-eight manuscripts were submitted for consideration for this Special Issue, and all of them were subject to the rigorous *Cancers* review process. In total, eleven papers were finally accepted for publication in this Special Issue, including seven articles and four reviews (as of 17 January 2024).

Tan X. et al. focused on the role of CD155 in relation to immunotherapies such as anti-PD-1 and anti-PD-L1 antigens in esophageal squamous cell cancer (ESCA). CD155 is highly expressed in ESCA tissues and is associated with poor patient prognosis. The expression of *CD155* is positively associated with *PD1*, *PDL1*, *CD4*, *IL2RA*, and *S100A9* expression in ESCA. CD155 may be involved in ESCA proliferation.

Proaño-Pérez, E. et al. investigated that the silencing of SH3 Binding Protein 2 (SH3BP2) downregulated KIT, platelet-derived growth factor receptor alpha (PDGFRA), and microphthalmia-associated transcription factor (MITF). It was revealed that SH3BP2 silencing decreased the ETV1 level through miR-1246 and miR-5100, which led to the reduced tumor growth of gastrointestinal stromal tumors (GISTs). The KIT-SH3BP2-MITF/ETV1 pathway may play a role in GIST growth.

Abdul Razzaq E. et al. revealed that overexpression of erb-b2 receptor tyrosine kinase 2 (*ERBB2*) (human epidermal growth factor receptor 2 (HER2)) in colorectal cancer (CRC) is associated with the Wnt signaling pathway in tumorigenesis. HER2 is suggested to be a target for revealing the CRC pathogenesis.

Yu W. et al. highlighted the importance of the enhancer of zeste homolog 2 (EZH2), a catalytic subunit of polycomb repressor complex 2 (PRC2), in gastric cancer (GC). The correlation between the EZH2 gene and gastric carcinogenesis was described, concluding that high expression of EZH2 leads to poor prognosis in GC.

Yan H. et al. focused on G-protein-coupled receptor (GPCR) signaling in GC initiation and progression. GPCR-mediated metastasis and tumor microenvironment remodeling were summarized in terms of their influence on the extracellular matrix, immune cells, stromal cells, sphingosine-1 phosphate receptors, thrombin receptors, and chemokine-chemokine receptors.

Macharia J. et al. revealed that *Aloe secundiflora* extracts have some potential in CRC treatment. The *Aloe secundiflora* methanolic extracts regulated the gene expression of the specific genes in CRC and the rate of apoptosis in Caco-2 colorectal cancer cell lines.

Kamińska, J. et al. focused on the progesterone (P4) and P4 receptor membrane component 1 (PGRMC1)/neuron-derived neurotrophic factor (NENF) complex interactions in CRC. The PGRMC1 and NENF in non-classical P4 signaling may interact as a complex that induces tumor proliferation and invasion.

Cheng, X. et al. investigated the mechanism related to ferroptosis to overcome drug resistance in CRC. Ferroptosis is a unique form of cell death, which is characterized by the iron-dependent accumulation of lipid peroxides. Targeting ferroptosis is a potential therapeutic strategy for CRC.

Shi, J. et al. revealed that synaptotagmin 1 (SYT1) inhibits EMT by negatively regulating ERK/MAPK signaling to suppress CRC cell migration and invasion. It is suggested that SYT1 represses CRC metastasis through blood vessels.

Jovanovic, M. et al. identified the morphological computed tomography features of tumors and the texture analysis parameters. These features represent imaging biomarkers that may be useful for the preoperative prediction of high-risk GISTs.

Aebisher, D. et al. summarized cancer treatment using photodynamic therapy and associated immunological anti-tumor mechanisms in gastrointestinal tumors. Photodynamic therapy is based on oxygen, photosensitizers, and light to induce tumor cell death through the production of reactive oxygen species (ROS).

3. Conclusions

In conclusion, the elucidation of the mechanisms of gastrointestinal tumors leads to the progression of advanced therapeutics for cancer. Targeting the components essential in the signaling pathways of gastrointestinal tumors has high potential as therapeutics and diagnostic markers in gastrointestinal tumors.

Funding: This research was funded by the Japan Agency for Medical Research and Development (AMED), grant numbers JP21mk0101216, JP22mk0101216, and JP23mk0101216, and the Japan Society for the Promotion of Science (JSPS) KAKENHI, grant number 21K12133.

Acknowledgments: As Guest Editor of the Special Issue "Advances in Molecular Mechanisms of Gastrointestinal Tumors", I would like to express my deep appreciation to all authors whose valuable work was published under this issue and thus contributed to the success of the edition. The author would like to thank members of National Institute of Health Sciences, Japan and the collaborators involved.

Conflicts of Interest: The author declares no conflicts of interest.

List of Contributions:

1. Tang, X.; Yang, J.; Shi, A.; Xiong, Y.; Wen, M.; Luo, Z.; Tian, H.; Zheng, K.; Liu, Y.; Shu, C.; et al. CD155 Cooperates with PD-1/PD-L1 to Promote Proliferation of Esophageal Squamous Cancer Cells via PI3K/Akt and MAPK Signaling Pathways. *Cancers* **2022**, *14*, 5610. https://doi.org/10.3390/cancers14225610.
2. Proaño-Pérez, E.; Serrano-Candelas, E.; Mancia, C.; Navinés-Ferrer, A.; Guerrero, M.; Martin, M. SH3BP2 Silencing Increases miRNAs Targeting ETV1 and Microphthalmia-Associated Transcription Factor, Decreasing the Proliferation of Gastrointestinal Stromal Tumors. *Cancers* **2022**, *14*, 6198. https://doi.org/10.3390/cancers14246198.
3. Abdul Razzaq, E.; Bajbouj, K.; Bouzid, A.; Alkhayyal, N.; Hamoudi, R.; Bendardaf, R. Transcriptomic Changes Associated with ERBB2 Overexpression in Colorectal Cancer Implicate a Potential Role of the Wnt Signaling Pathway in Tumorigenesis. *Cancers* **2023**, *15*, 130. https://doi.org/10.3390/cancers15010130.
4. Yu, W.; Liu, N.; Song, X.; Chen, L.; Wang, M.; Xiao, G.; Li, T.; Wang, Z.; Zhang, Y. EZH2: An Accomplice of Gastric Cancer. *Cancers* **2023**, *15*, 425. https://doi.org/10.3390/cancers15020425.
5. Yan, H.; Zhang, J.; Leung, K.; Lo, K.; Yu, J.; To, K.; Kang, W. An Update of G-Protein-Coupled Receptor Signaling and Its Deregulation in Gastric Carcinogenesis. *Cancers* **2023**, *15*, 736. https://doi.org/10.3390/cancers15030736.
6. Macharia, J.; Varjas, T.; Mwangi, R.; Káposztás, Z.; Rozmann, N.; Pintér, M.; Wagara, I.; Raposa, B. Modulatory Properties of *Aloe secundiflora*'s Methanolic Extracts on Targeted Genes in Colorectal Cancer Management. *Cancers* **2023**, *15*, 5002. https://doi.org/10.3390/cancers15205002.
7. Kamińska, J.; Koper-Lenkiewicz, O.; Ponikwicka-Tyszko, D.; Lebiedzińska, W.; Palak, E.; Sztachelska, M.; Bernaczyk, P.; Dorf, J.; Guzińska-Ustymowicz, K.; Zaręba, K.; et al. New Insights on the Progesterone (P4) and PGRMC1/NENF Complex Interactions in Colorectal Cancer Progression. *Cancers* **2023**, *15*, 5074. https://doi.org/10.3390/cancers15205074.
8. Cheng, X.; Zhao, F.; Ke, B.; Chen, D.; Liu, F. Harnessing Ferroptosis to Overcome Drug Resistance in Colorectal Cancer: Promising Therapeutic Approaches. *Cancers* **2023**, *15*, 5209. https://doi.org/10.3390/cancers15215209.
9. Shi, J.; Li, W.; Jia, P.; Peng, Y.; Hou, J.; Li, N.; Meng, R.; Fu, W.; Feng, Y.; Wu, L.; et al. Synaptotagmin 1 Suppresses Colorectal Cancer Metastasis by Inhibiting ERK/MAPK Signaling-Mediated Tumor Cell Pseudopodial Formation and Migration. *Cancers* **2023**, *15*, 5282. https://doi.org/10.3390/cancers15215282.
10. Jovanovic, M.; Stefanovic, A.; Sarac, D.; Kovac, J.; Jankovic, A.; Saponjski, D.; Tadic, B.; Kostadinovic, M.; Veselinovic, M.; Sljukic, V.; et al. Possibility of Using Conventional Computed Tomography Features and Histogram Texture Analysis Parameters as Imaging Biomarkers for Preoperative Prediction of High-Risk Gastrointestinal Stromal Tumors of the Stomach. *Cancers* **2023**, *15*, 5840. https://doi.org/10.3390/cancers15245840.
11. Aebisher, D.; Woźnicki, P.; Dynarowicz, K.; Kawczyk-Krupka, A.; Cieślar, G.; Bartusik-Aebisher, D. Photodynamic Therapy and Immunological View in Gastrointestinal Tumors. *Cancers* **2024**, *16*, 66. https://doi.org/10.3390/cancers16010066.

References

1. Zhang, Y.; Weinberg, R.A. Epithelial-to-mesenchymal transition in cancer: Complexity and opportunities. *Front. Med.* **2018**, *12*, 361–373. [CrossRef] [PubMed]
2. Shin, A.E.; Giancotti, F.G.; Rustgi, A.K. Metastatic colorectal cancer: Mechanisms and emerging therapeutics. *Trends Pharmacol. Sci.* **2023**, *44*, 222–236. [CrossRef] [PubMed]
3. Fang, P.; Zhou, J.; Liang, Z.; Yang, Y.; Luan, S.; Xiao, X.; Li, X.; Zhang, H.; Shang, Q.; Zeng, X.; et al. Immunotherapy resistance in esophageal cancer: Possible mechanisms and clinical implications. *Front. Immunol.* **2022**, *13*, 975986. [CrossRef] [PubMed]
4. Kim, S.M. Cellular and Molecular Mechanisms of 3,3′-Diindolylmethane in Gastrointestinal Cancer. *Int. J. Mol. Sci.* **2016**, *17*, 1155. [CrossRef] [PubMed]
5. Li, S.; Yuan, L.; Xu, Z.Y.; Xu, J.L.; Chen, G.P.; Guan, X.; Pan, G.Z.; Hu, C.; Dong, J.; Du, Y.A.; et al. Integrative proteomic characterization of adenocarcinoma of esophagogastric junction. *Nat. Commun.* **2023**, *14*, 778. [CrossRef] [PubMed]
6. Peng, C.; Ouyang, Y.; Lu, N.; Li, N. The NF-κB Signaling Pathway, the Microbiota, and Gastrointestinal Tumorigenesis: Recent Advances. *Front. Immunol.* **2020**, *11*, 1387. [CrossRef] [PubMed]
7. Shah, S.C.; Itzkowitz, S.H. Colorectal Cancer in Inflammatory Bowel Disease: Mechanisms and Management. *Gastroenterology* **2022**, *162*, 715–730.e713. [CrossRef] [PubMed]
8. Waldum, H.; Fossmark, R. Inflammation and Digestive Cancer. *Int. J. Mol. Sci.* **2023**, *24*, 3503. [CrossRef] [PubMed]

9. Weng, M.T.; Chiu, Y.T.; Wei, P.Y.; Chiang, C.W.; Fang, H.L.; Wei, S.C. Microbiota and gastrointestinal cancer. *J. Formos. Med. Assoc.* **2019**, *118* (Suppl. 1), S32–S41. [CrossRef] [PubMed]
10. Al-Ishaq, R.K.; Overy, A.J.; Büsselberg, D. Phytochemicals and Gastrointestinal Cancer: Cellular Mechanisms and Effects to Change Cancer Progression. *Biomolecules* **2020**, *10*, 105. [CrossRef] [PubMed]

Disclaimer/Publisher's Note: The statements, opinions and data contained in all publications are solely those of the individual author(s) and contributor(s) and not of MDPI and/or the editor(s). MDPI and/or the editor(s) disclaim responsibility for any injury to people or property resulting from any ideas, methods, instructions or products referred to in the content.

Article

CD155 Cooperates with PD-1/PD-L1 to Promote Proliferation of Esophageal Squamous Cancer Cells via PI3K/Akt and MAPK Signaling Pathways

Xiyang Tang [1,†], Jie Yang [1,†], Anping Shi [2], Yanlu Xiong [1], Miaomiao Wen [1], Zhonglin Luo [3], Huanhuan Tian [3], Kaifu Zheng [1], Yujian Liu [1], Chen Shu [1], Nan Ma [4], Rui Wang [5,*] and Jinbo Zhao [1,*]

[1] Department of Thoracic Surgery, Tangdu Hospital, Air Force Medical University, 569 Xinsi Road, Xi'an 710038, China
[2] Department of Radiology, Functional and Molecular Imaging Key Lab of Shaanxi Province, Tangdu Hospital, Fourth Military Medical University (Air Force Medical University), 569 Xinsi Road, Xi'an 710038, China
[3] Department of Cardiothoracic Surgery, Peace Hospital, Changzhi Medical College, 161 Jiefang East Street, Changzhi 046000, China
[4] Department of Ophthalmology, Tangdu Hospital, Air Force Medical University, 569 Xinsi Road, Xi'an 710038, China
[5] Medical Department, Tangdu Hospital, Air Force Medical University, 569 Xinsi Road, Xi'an 710038, China
* Correspondence: tdjrjz@fmmu.edu.cn (R.W.); zhaojinbo@aliyun.com (J.Z.)
† These authors contributed equally to this work.

Simple Summary: Some immunotherapies, such as anti-PD1 and anti-PD-L1 treatments, have been used to treat various tumors. However, they are less efficient against esophageal cancer, partially owing to a lack of research on the cellular and molecular mechanisms of this cancer. Therefore, various emerging immune checkpoints have been discovered in this post-PD-1 era. One such immune checkpoint is CD155, a protein belonging to the Nectin-like family and expressed on the surface of cancer and immune cells. Exploring the mechanisms and therapeutic applications of these immune checkpoints may effectively improve cellular responses to immunotherapies. In this study, we aimed to explore the role of CD155 in esophageal squamous cell cancer (ESCA) and its underlying molecular mechanism. CD155 was positively associated with PD-1/PD-L1 expression and could support ESCA proliferation. The downregulation of CD155 expression inhibited ESCA cell proliferation by impairing the cell cycle and inducing cell apoptosis. This occurred via the inhibition of PI3K/Akt and MAPK signaling pathways. In addition, Nectin3 may be the ligand of CD155 and may be involved in ESCA proliferation. Thus, our study suggests novel targets for tumor therapy, especially for ESCA treatment.

Abstract: Background: Esophageal cancer is still a leading cause of death among all tumors in males, with unsatisfactory responses to novel immunotherapies such as anti-PD-1 agents. Herein, we explored the role of CD155 in esophageal squamous cell cancer (ESCA) and its underlying molecular mechanisms. Methods: Publicly available datasets were used for differential gene expression and immune infiltration analyses, and their correlation with patient survival. A total of 322 ESCA and 161 paracancer samples were collected and evaluated by performing immunohistochemistry and the H score was obtained by performing semiquantitative analysis. In vitro transfection of ESCA cell lines with lentivirus vectors targeting CD155 was performed to knockdown the protein. These cells were analyzed by conducting RNA sequencing, and the effects of CD155 knockdown on cell cycle and apoptosis were verified with flow cytometry and Western blotting. In addition, in vivo experiments using these engineered cell lines were performed to determine the role of CD155 in tumor formation. A small interfering RNA-mediated knockdown of Nectin3 was used to determine whether it phenocopied the profile of CD155 knockdown. Results: CD155 is highly expressed in ESCA tissues and is positively associated with *PD1*, *PDL1*, *CD4*, *IL2RA*, and *S100A9* expression. Furthermore, CD155 knockdown inhibited ESCA cells' proliferation by impairing the cell cycle and inducing cell apoptosis. Bioinformatics analysis of the gene expression profile of these engineered cells showed that CD155 mainly contributed to the regulation of PI3K/Akt and MAPK signals. The

Citation: Tang, X.; Yang, J.; Shi, A.; Xiong, Y.; Wen, M.; Luo, Z.; Tian, H.; Zheng, K.; Liu, Y.; Shu, C.; et al. CD155 Cooperates with PD-1/PD-L1 to Promote Proliferation of Esophageal Squamous Cancer Cells via PI3K/Akt and MAPK Signaling Pathways. *Cancers* **2022**, *14*, 5610. https://doi.org/10.3390/cancers14225610

Academic Editor: Shihori Tanabe

Received: 14 October 2022
Accepted: 9 November 2022
Published: 15 November 2022

Publisher's Note: MDPI stays neutral with regard to jurisdictional claims in published maps and institutional affiliations.

Copyright: © 2022 by the authors. Licensee MDPI, Basel, Switzerland. This article is an open access article distributed under the terms and conditions of the Creative Commons Attribution (CC BY) license (https://creativecommons.org/licenses/by/4.0/).

downregulation of Nectin3 expression phenocopied the profile of CD155 knockdown. Discussion: CD155 may cooperate with PD-1/PD-L1 to support ESCA proliferation in ways other than regulating its underlying immune mechanisms. Indeed, CD155 downregulation can impair ESCA cell pro-cancerous behavior via the inhibition of the PI3K/Akt and MAPK signaling pathways. Moreover, Nectin3 may be a ligand of CD155 and participate in the regulation of ESCA cells' proliferation. Hence, the inhibition of CD155 may enhance the therapeutic effect of anti-PD-1 immunotherapies in ESCA.

Keywords: CD155; PD-1/PD-L1; immunotherapy; esophageal squamous cell cancer

1. Introduction

Esophageal cancer is the sixth leading cause of death among all tumors [1]. In particular, China has the largest number of patients with esophageal cancer worldwide, with about 193,000 deaths per year, of which 90% are due to esophageal squamous cell cancer (ESCA) [2,3]. Currently, immunotherapy is the most widely adopted anticancer treatment option, especially for lung cancer, melanoma, and hepatocellular carcinoma [4–6]. Nevertheless, the immune checkpoint therapy anti-programmed death protein 1 (PD-1) has failed to induce a satisfactory therapeutic effect in advanced and refractory esophageal cancer. Indeed, the overall response rate in refractory esophageal cancer reaches only 8% [7]; however, more manageable toxicity occurs in advanced patients [8]. Thus, the application of immunotherapies, such as using anti-PD-1 agents, in esophageal cancer warrants further research.

PD-1 is expressed on the surface of multiple immune cells, including macrophages and B, T, natural killer, and natural killer T cells [9], whereas its ligand PD-L1 is mainly expressed by cancer cells, such as gastric, non-small cell lung, and breast cancer cells [10]. The interaction between PD-1 and PD-L1 promotes strong inhibitory immune signals in the tumor microenvironment, in particular toward T cell inhibition [11]. Moreover, both PD-1 and PD-L1 were shown to promote tumor proliferation by regulating the cell cycle of several cancer cell types [12–16], including esophageal cancer cells [17], further demonstrating the pro-proliferative role of PD-1/PD-L1 signals in cancer.

A preliminary study of the differential expression of 35 immune checkpoint genes in ESCA using publicly available data suggested that CD155 could mediate important regulatory effects in this cancer type (Figure S1). CD155 (also named poliovirus receptor or PVR) is an immune checkpoint protein that belongs to the Nectin-like family and is expressed on the surface of cancer and immune cells [18,19]. CD155 mainly regulates the immune activity of natural killer [20,21] and T [22] cells in the tumor microenvironment of various cancers. Nectin3 (also called PVRL3) was proven to be one of the ligands of CD155 and may bind with CD155 in regulating cell movement and adhesion [23,24]. Based on the unsatisfactory therapeutic effect of anti-PD-1 therapies in ESCA, CD155 may represent a good therapeutic candidate to achieve improved anticancer responses with fewer immune-related adverse events. Therefore, herein, we explored the role of CD155 in ESCA for the first time, including its relationship with PD-1/PD-L1 signals in the underlying mechanisms of this life-threatening disease.

2. Materials and Methods

2.1. Analysis of Gene Expression and Immune Infiltration

Data from the Gene Expression Profiling Interactive Analysis database (GEPIA, http://gepia.cancer-pku.cn/, accessed on 4 July 2022), which comprise RNA-sequencing data from the Genotypic-Tissue Expression project and The Cancer Genome Atlas, were used for gene expression and gene–gene correlation analyses [25]. Pancancer analysis was performed using the University of Alabama at Birmingham Cancer data analysis platform (UALCAN, http://ualcan.path.uab.edu/, accessed on 4 July 2022) [26] and the Assistant for Clinical Bioinformatics database (ACLBI, www.aclbi.com, accessed on 4 July 2022), which also facilitated simultaneous immune infiltration analysis. The relationship between gene expression profiles, tumor immune infiltration, and patient survival was evaluated using data from the GEPIA and UALCAN databases.

2.2. Analysis of Gene and Protein Interaction

Data from the GeneMANIA (https://genemania.org/, accessed on 7 July 2022) database were used for gene prioritization network integration and gene–gene interaction prediction [27], and protein–protein interaction prediction was performed in the STRING database (www.string-db.org, accessed on 8 July 2022) [28].

2.3. ESCA Sample Collection

A total of 322 ESCA and 161 paracancer samples were collected from patients who were surgically treated in Tangdu Hospital (Xi'an, China). The inclusion criteria and exclusion criteria were formulated as previously reported [29]. This study was approved by the Ethics Committee of the Air Force Medical University (No. 202108-05).

2.4. Immunohistochemistry

All tissue samples were embedded in paraffin and cut into 3 μm slices. Immunohistochemistry was conducted as previously reported [29]. The tissue slices were incubated overnight at 4 °C with anti-CD155, anti-PD1, anti-PD-L1, anti-CD4, anti-IL-2RA, anti-S100A9, anti-CD3D, anti-CD8, anti-FOXP3, anti-TPSB2, anti-CD79A, anti-GNLY, and anti-Nectin3 (Signalway Antibody, Greenbelt, MD, USA), respectively.

A semiquantitative analysis of the H score was performed using the Aipathwell.v2 software as follows:

H score = $\sum (pi \times i)$ = (percentage of weak intensity × 1) + (percentage of moderate intensity × 2) + (percentage of strong intensity × 3) [30–32].

To assess the correlation between CD155 and PD-1/PD-L1, H scores of <25% were considered as negative expression. The semiquantitative analysis was supported by Wuhan Servicebio Technology Co. (Wuhan, China).

2.5. Immunofluorescence Analysis

TE1 and KYSE-520 (K520) human esophageal squamous cancer cells were cultured in 6-well plates until they reached 70% density. A membrane breaking solution (100 μL; Servicebio) was added to the wells and incubated at 25 °C for 20 min. The cells were then washed three times with PBS (phosphate-buffered saline solution) and were incubated overnight at 4 °C with anti-CD155 polyclonal antibody (1:50 dilution; Signalway Antibody, Greenbelt, MD, USA); the respective secondary antibody was incubated at 25 °C for 50 min. To stain the nucleus of the cells, 4′,6-diamidino-2-phenylindole (DAPI) was used.

2.6. Cell Culture

The cultures of TE1 and K520 human esophageal squamous cancer cells (iCell Bioscience Inc., Shanghai, China) were the same as previously reported [29].

2.7. CD155 and Nectin3 Knockdown

CD155 was knocked down (KD) in TE1 and K520 cells using a lentiviral vector synthesized by CytoBiotech (Guangzhou, China) that harbored a luciferase tag and a short hairpin RNA which specifically targets *CD155* through the sequence 5′–CTGTGAACCTCACCGTGTA–3′. Nectin3 was knocked down in TE1 and K520 cells using siRNA (RiboBio, Guangzhou, China). The sequence of the siRNA used was 5′-GACATCCGATACTCTTTCA-3′. A non-targeting vector was used as the control (NC).

2.8. Xenograft Tumor Experiment

Male mice (BALB/cJGpt-Foxn1nu/Gpt; 5 weeks old; 20–25 g) were purchased from GemPharmatech (Beijing, China). The mice were randomly divided into two groups ($n = 5$ in each group) and injected subcutaneously in the back with CD155_NC or CD155_KD TE1 cells (1×10^6 cells/animal). Within 7–17 days post-injection, the volume of the tumors was measured daily based on the two largest perpendicular dimensions. The tumor volume (mm^3) was calculated as (tumor length [mm] × square of tumor width [mm]^2)/2. All in vivo experiments were reviewed and approved by the Ethics Committee of the Air Force Medical University (No. 202203-145).

2.9. Flow Cytometry

For apoptosis analysis, the cells were washed once with phosphate-buffered saline solution and resuspended in a 100 µL (1×) binding buffer. Fluorochrome-conjugated Annexin V (5 µL) was added and incubated for 10–15 min, protected from light at room temperature, and immediately analyzed using a Beckman Coulter (Brea, CA, USA) flow cytometer.

2.10. Western Blotting

Western blotting was conducted as previously reported [29]. The membranes were incubated overnight at 4 °C with the following specific antibodies: anti-β-actin, anti-CD155, anti-PI3K, anti-phosphorylated Akt (Ser473), anti-P38, anti-P38 MAPK, anti-ERK1/2, anti-phosphorylated ERK1/2 (Thr202/Tyr204), anti-JNK1/2/3, anti-phosphorylated JNK1/2/3 (Thr183/Tyr185), anti-cyclin A2, anti-cyclin B1, anti-cyclin D1, anti-cyclin E1, anti-CDK2, anti-CDK4, anti-CDK6, anti-caspase 3, anti-caspase 7, anti-caspase 9, anti-cleaved caspase 9, anti-PARP 1, anti-cleaved PARP, and anti-Nectin3 (Signalway Antibody, Greenbelt, MD, USA), respectively.

2.11. Cell Proliferation Analysis

CD155_NC and CD155_KD ESCA cells were collected; the analyses in a real-time cell analyzer (Agilent Technologies, Santa Clara, CA, USA) and cell proliferation were conducted as previously reported [29]. Similarly, 1000 cells from each of these two groups were seeded onto 6-well plates, cultured for the next 10 days, and washed three times with phosphate-buffered saline solution; then, all of the cells were fixed in methanol for 25 min and stained with 5% crystal violet for 40 min.

2.12. mRNA Sequencing and Analysis

RNA from CD155_NC and CD155_KD TE1 cells was isolated using TRIzol Reagent (Thermo Fisher Scientific, Carlsbad, USA) and the samples were submitted to Genergy Bio-Technology Co. (Shanghai, China) for mRNA sequencing and analysis.

2.13. Statistical Analysis

Prism 8.2.1 (GraphPad Software, San Diego, CA, USA) and SPSS 26 (IBM Corp, Armonk, NY, USA) were used for statistical analysis. Statistical data are expressed as mean ± standard deviation. Differences between the two groups were evaluated using the Student's *t* test. H scores without equal standard deviations were analyzed using the Mann–Whitney test for rank comparison. $p \leq 0.05$ was considered statistically significant.

3. Results

3.1. CD155 Is Highly Expressed in ESCA and Is Associated with Poor Patient Prognosis

The analysis of publicly available data in different databases showed that CD155 was highly expressed in ESCA (Figure 1A,B). In agreement with this finding, a pancancer analysis confirmed the high expression of CD155 in ESCA, as well as in other cancers, including cholangiocarcinoma, colon adenocarcinoma, and pancreatic adenocarcinoma (Figure 1C) (Supplementary Figure S2). To further validate these results, 322 primary ESCA and 161 paracancerous tissues were evaluated by performing immunohistochemistry, which revealed that CD155 was significantly more expressed in ESCA tissues than in healthy cells ($p < 0.0001$) (Figure 1D). Altogether, these results demonstrate that CD155 is highly expressed in ESCA, thus potentially playing an important role in ESCA development.

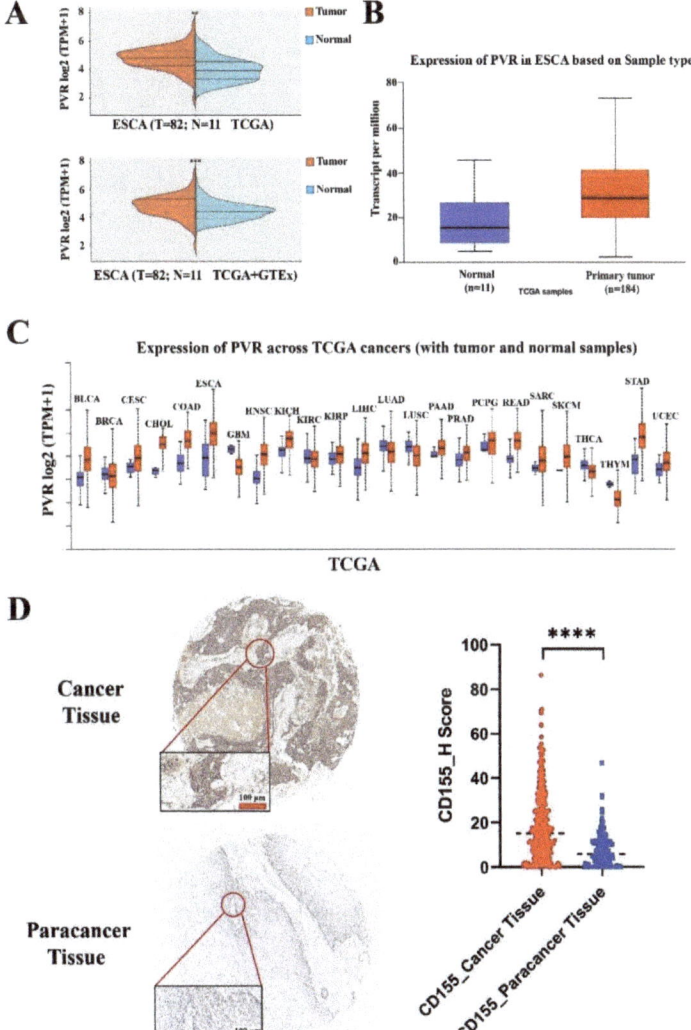

Figure 1. CD155 is highly expressed in ESCA. Data from (**A**) ACLBI and (**B**) UALCAN databases all showed higher expression of CD155 in ESCA compared to normal samples. Pancancer analysis from the (**C**) UALCAN databases showed higher CD155 expression could also be found in CHOL, COAD, and PAAD, as well as other tumors. (**D**) Further validation for the protein level of CD155 was evaluated by immunohistochemistry. ESCA: esophageal squamous cancer; CHOL: cholangio carcinoma; COAD: colon adenocarcinoma; PAAD: pancreatic adenocarcinoma. ** $p < 0.01$, *** $p < 0.001$, **** $p < 0.0001$.

CD155 expression was also evaluated concerning the different stages of ESCA. Noteworthily, CD155 was expressed in all cancer stages (Figure 2A,B). Furthermore, the patient survival analysis showed that a high expression of CD155 predicted poor ESCA prognosis (Figure 2C,D).

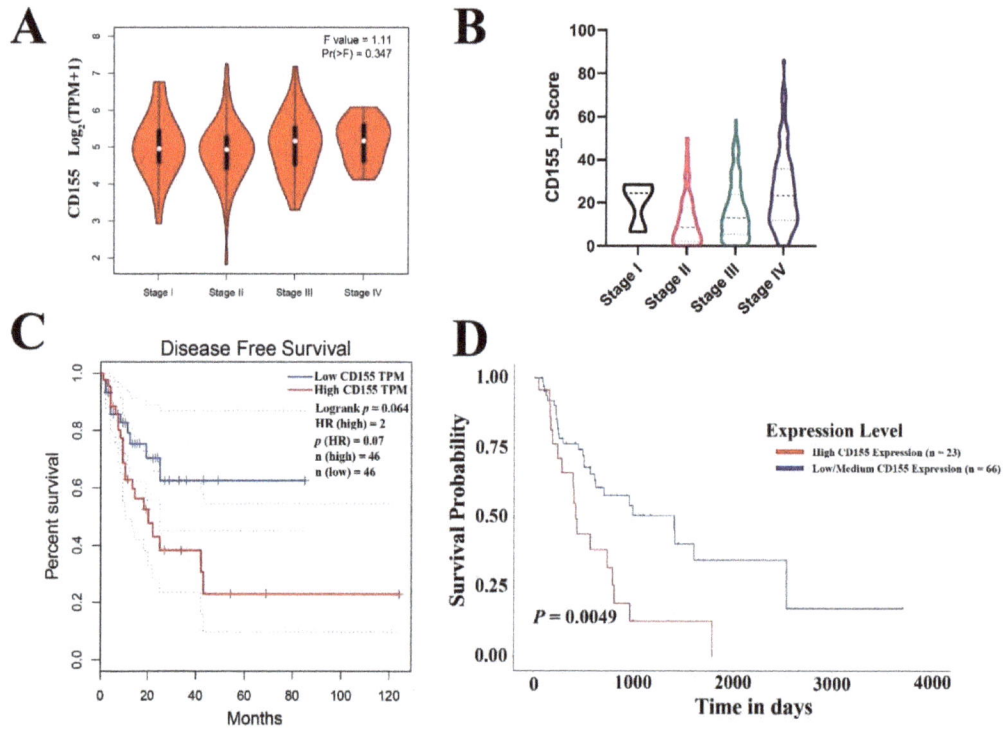

Figure 2. CD155 is expressed in all cancer stages and associated with poor patient prognosis. (**A**) The expression of CD155 in the different stages of ESCA was evaluated using the GEPIA database, the numbers in the Y axis indicate the transcription level of CD155. (**B**) The expression of CD155 in the different stages of ESCA was validated by performing immunohistochemistry, the numbers in the Y axis indicate the protein level of CD155. Survival analyses related to CD155 were performed in the (**C**) GEPIA and (**D**) UALCAN databases.

3.2. CD155 Is Positively Associated with the Expression of CD4, IL-2Rα and S100A9 in ESCA

The gene–gene interaction analysis revealed that *CD155* may interact with *CD96*, *CD226*, and *NECTIN3* via physical interactions (Figure S3A). In addition, the protein–protein interaction analysis further confirmed that CD155 can interact with CD96 and CD226 (Figure S3B).

To further explore the potential role of CD155 in ESCA, an immune infiltration analysis was performed. Interestingly, *CD155* expression was found to be positively associated with the levels of M1 macrophages ($r = 0.11$, $p = 0.011$), myeloid dendritic cells ($r = 0.11$, $p = 0.01$), and neutrophils ($r = 0.09$, $p = 0.051$), but negatively associated with the levels of B cells ($r = -0.23$, $p = 9.46 \times 10^{-8}$), M2 macrophages ($r = -0.23$, $p = 7.73 \times 10^{-8}$), natural killer cells ($r = -0.10$, $p = 0.03$), CD8$^+$ T cells ($r = -0.09$, $p = 0.038$), and T regulatory cells ($r = -0.12$, $p = 0.006$) (Figure 3). A further protein analysis confirmed that both CD8 and CD79A were decreased in ESCA tissues compared with the controls ($p < 0.05$), as well as TPSB2 ($p < 0.01$); however, the expressions of interleukin (IL)-2 receptor α (IL-2Rα) and S100A9 were all significantly elevated in ESCA ($p < 0.0001$). These results were confirmed by performing a correlation analysis, which showed significantly positive correlations between *CD155* and *CD4* (CD4$^+$ T cell marker, $r = 0.1655$, $p = 0.0033$), and *IL2RA* (B cell marker, $r = 0.2850$, $p < 0.0001$) and *S100A9* (neutrophil marker, $r = 0.2425$, $p < 0.0001$) (Figure 4); however, there was no correlation with other immune markers (Figure S4). Thus, CD155 may contribute to regulating immune infiltration in ESCA, especially the regulation of CD4+ T cells, B cells, and neutrophils.

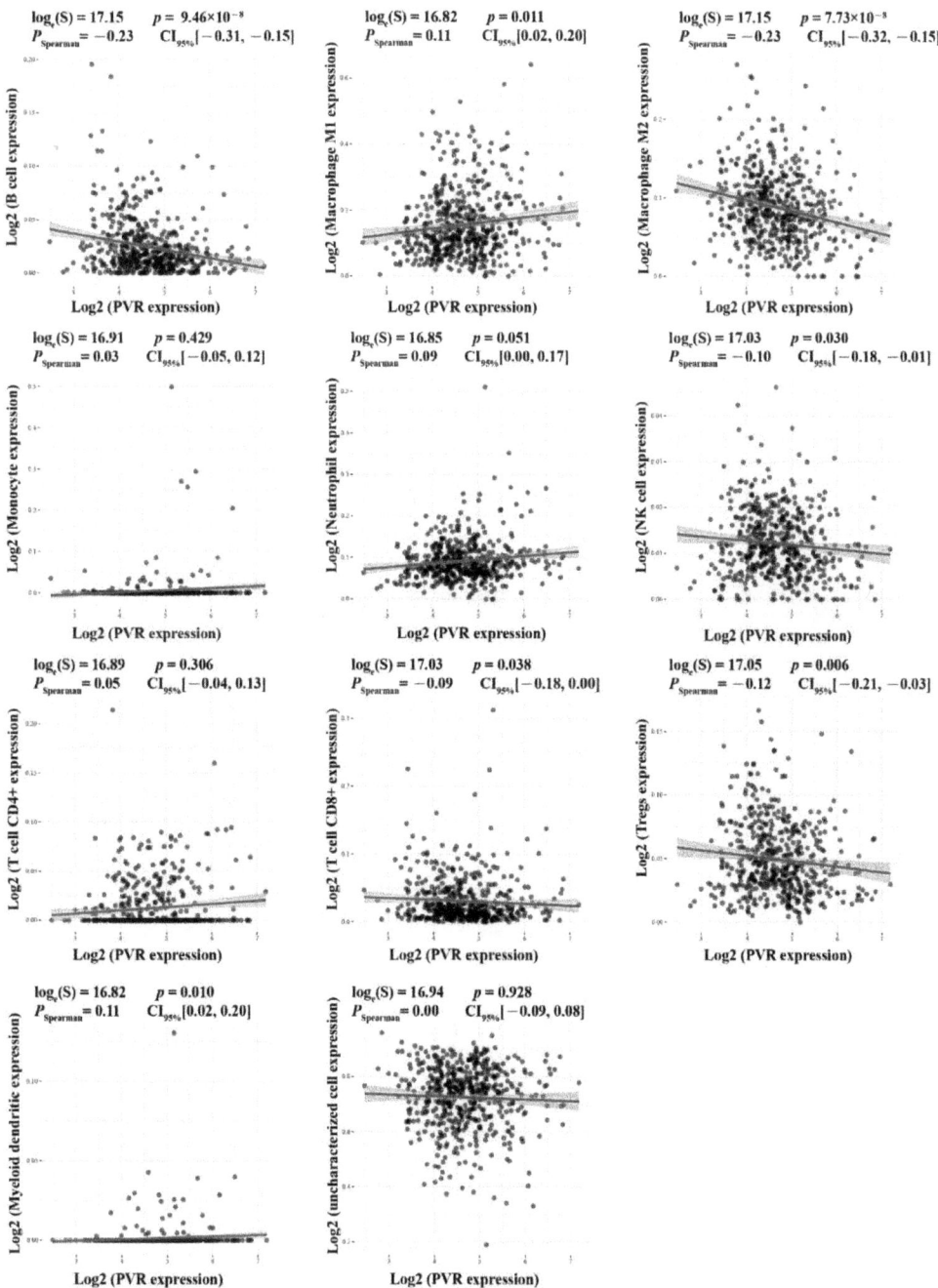

Figure 3. CD155 can be involved in immune infiltration. Immune infiltration analysis was performed in the ACLBI databases.

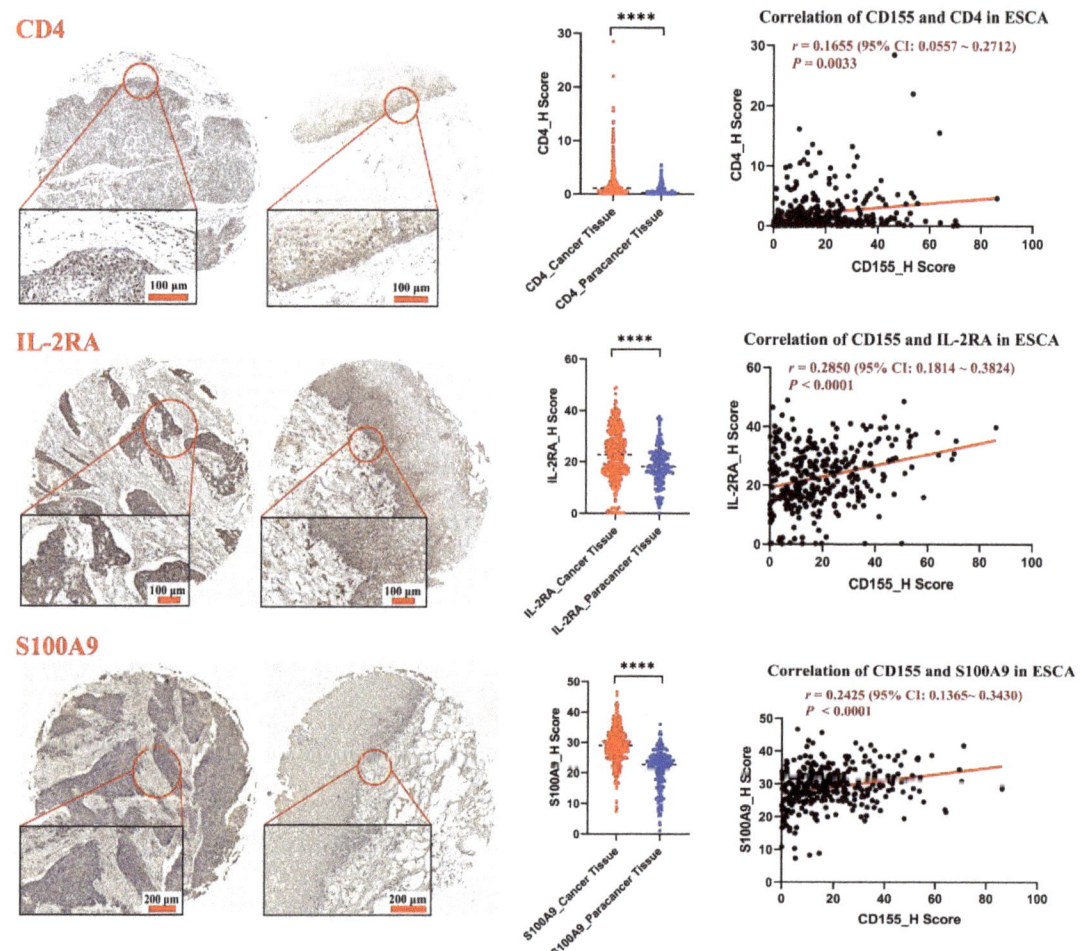

Figure 4. CD155 is positively associated with the expression of CD4, IL-2RA, and S100A9. Three immune-related markers were stained by performing immunohistochemistry. A correlation analysis between CD155 and immune markers was performed, and positive correlations were observed between CD155 and CD4 ($r = 0.1655$, $p = 0.0033$), IL-2RA ($r = 0.2850$, $p < 0.0001$), and S100A9 ($r = 0.2425$, $p < 0.0001$). **** $p < 0.0001$.

3.3. CD155 Can Cooperate with PD-1/PD-L1

To further analyze the relationship between CD155 and PD-1/PD-L1, 322 ESCA samples were divided into eight categories according to their phenotype: PD-1$^+$PD-L1$^+$CD155$^+$, PD-1$^-$PD-L1$^-$CD155$^-$, PD-1$^+$PD-L1$^+$CD155$^-$, PD-1$^+$PD-L1$^-$CD155$^+$, and PD-1$^-$PD-L1$^+$CD155$^+$, PD-1$^+$PD-L1$^-$CD155$^-$, PD-1$^-$PD-L1$^+$CD155$^-$, and PD-1$^-$PD-L1$^-$CD155$^+$. For this classification, quartiles were considered as cutoff values and H scores of less than 25% were considered as negative expressions. Among all samples, triple and double positive samples (PD-1$^+$CD155$^+$ or PD-L1$^+$CD155$^+$) accounted for 58.1% and 10.9%, respectively, further suggesting that CD155 is highly associated with PD1 and PD-L1 expression (Figures 5 and 6). Hence, CD155 may cooperate with PD1/PD-L1 and may have an impact on the therapeutic effect of anti-PD1/PD-L1 treatments in ESCA.

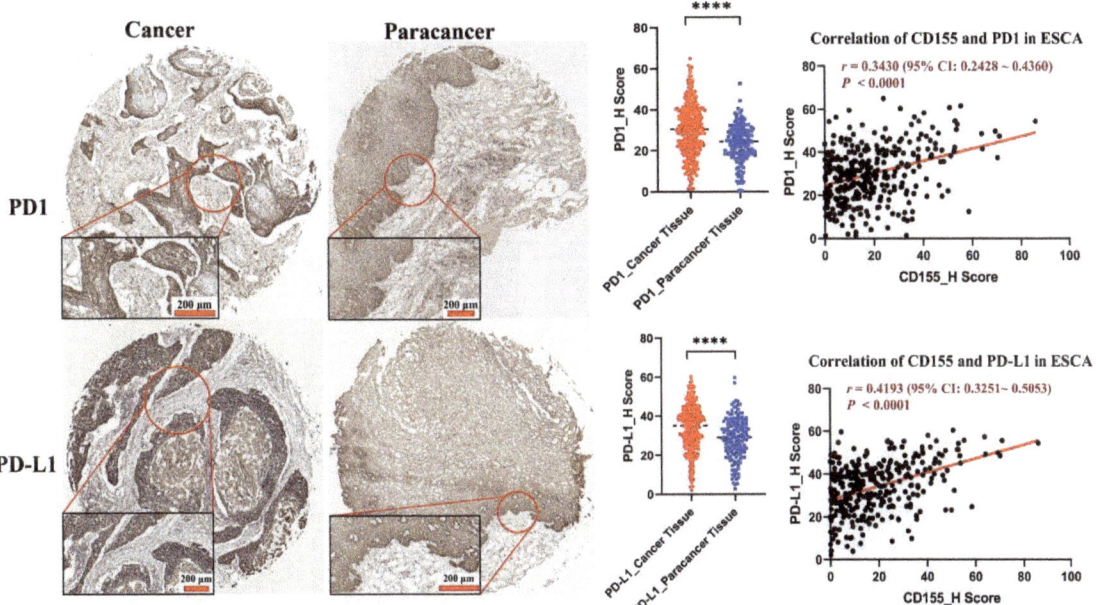

Figure 5. [CD155 is positively associated with PD1/PD-L1]. The correlation analysis between CD155 and PD1 or PD-L1 was performed, and CD155 was positively associated with PD1 ($r = 0.343$, $p < 0.0001$) and PD-L1 ($r = 0.4193$, $p < 0.0001$). A total of 322 ESCA samples was divided into 8 categories based on the expression type of CD155, PD1, and PD-L1; triple positive samples accounted for 58.1%, and double positive samples, PD1$^+$CD155$^+$, or PD-L1$^+$CD155$^+$ accounted for 10.9%. **** $p < 0.0001$.

3.4. CD155 Can Regulate the PI3K/Akt and MAPK Pathways in ESCA

To further explore the potential cellular and molecular effects of CD155 in ESCA, we analyzed TE1 and K520 ESCA cells that were genetically engineered to lack CD155 expression. Western blot and immunofluorescence data confirmed the decreased expression of CD155 in ESCA cells, especially in the cytoplasm (Figure 7A,B). The RNA sequencing of these cells and respective gene ontology analyses showed that CD155 mainly contributed to developmental processes and signaling receptor binding, which could be associated with tumor growth. In addition, KEGG (Kyoto Encyclopedia of Genes and Genomes) and GSEA (Gene Set Enrichment Analyses) indicated that the CD155-related differentially expressed genes were mainly associated with the PI3K/Akt and MAPK signaling pathways (Figure 7C–F). To verify the signaling pathways that CD155 was involved with, the levels of major signaling proteins were evaluated. Interestingly, all the analyzed markers were decreased in the cells lacking CD155 (Figure 7G), further supporting that CD155 exerts an effect in ESCA cells via the PI3K/Akt and MAPK pathways.

3.5. CD155 Downregulation Inhibits ESCA Cell Proliferation by Impairing Cell Cycle and Inducing Cell Apoptosis

Since the gene ontology results suggested that CD155 could be involved in cell developmental processes, we next explored the potential role of CD155 in cell cycle and apoptosis. The proliferation of TE1 and K520 ESCA cells was significantly inhibited for the CD155_KD group compared with the CD155_NC group, as determined using a real-time cell analyzer, colony formation assays, and in vivo experiments (Figure 8A–C). Moreover, these cells showed increased levels of cell apoptosis (Figure 8D,E), with elevated caspase 3 and cleaved PARP (Figure 8F). Noteworthily, ESCA cells lacking CD155 had an impaired cell cycle, with decreased levels of cyclins B1, D1, E1, and CDK6 (Figure 8G). In summary, CD155 downregulation can effectively inhibit the proliferation of ESCA cells by preventing their proliferation and promoting cell apoptosis.

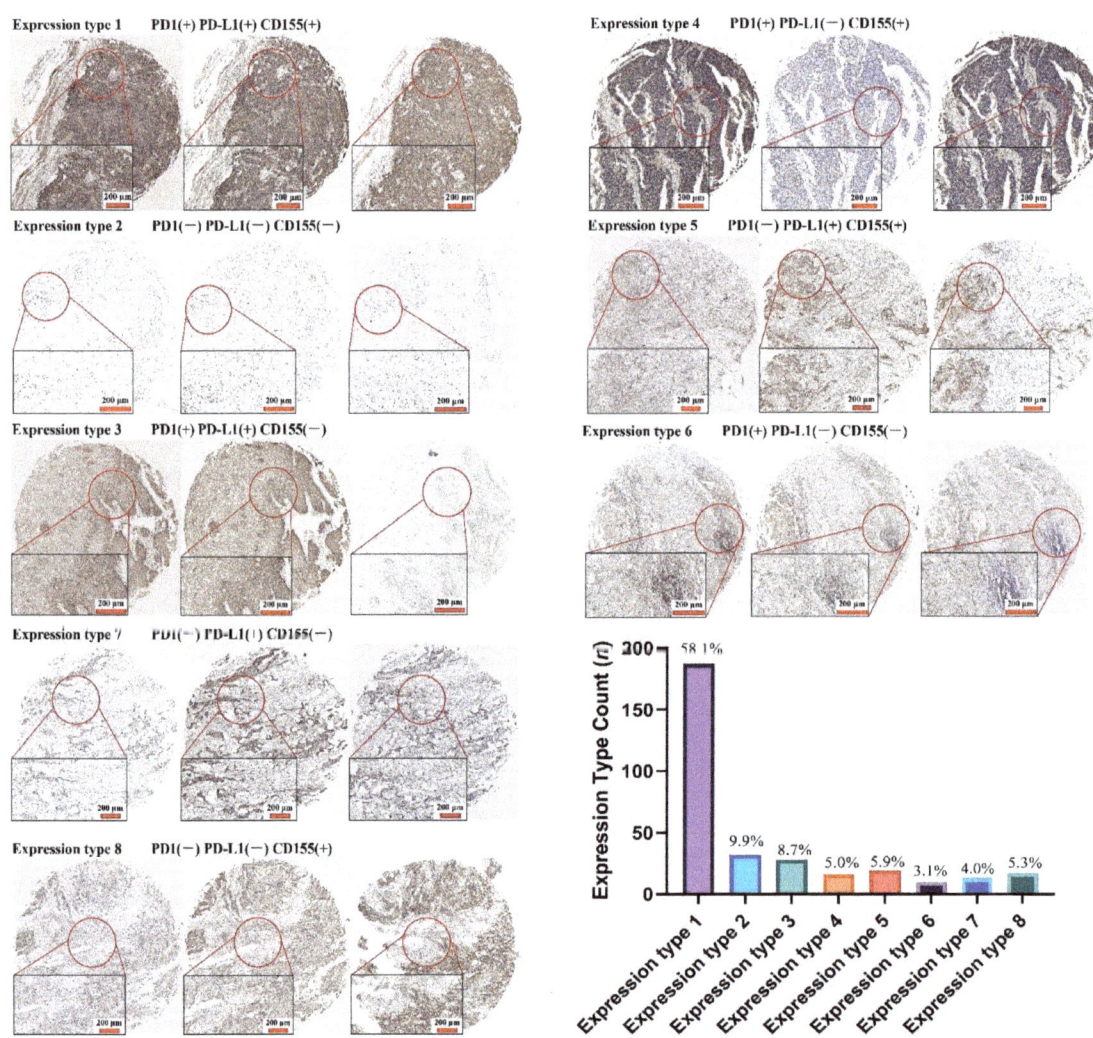

Figure 6. CD155 is tightly associated with PD1/PD-L1. A total of 322 ESCA samples were divided into 8 categories based on the expression type of CD155, PD1, and PD-L1; triple positive samples accounted for 58.1%, and double positive samples, PD1$^+$CD155$^+$, or PD-L1$^+$CD155$^+$ accounted for 10.9%.

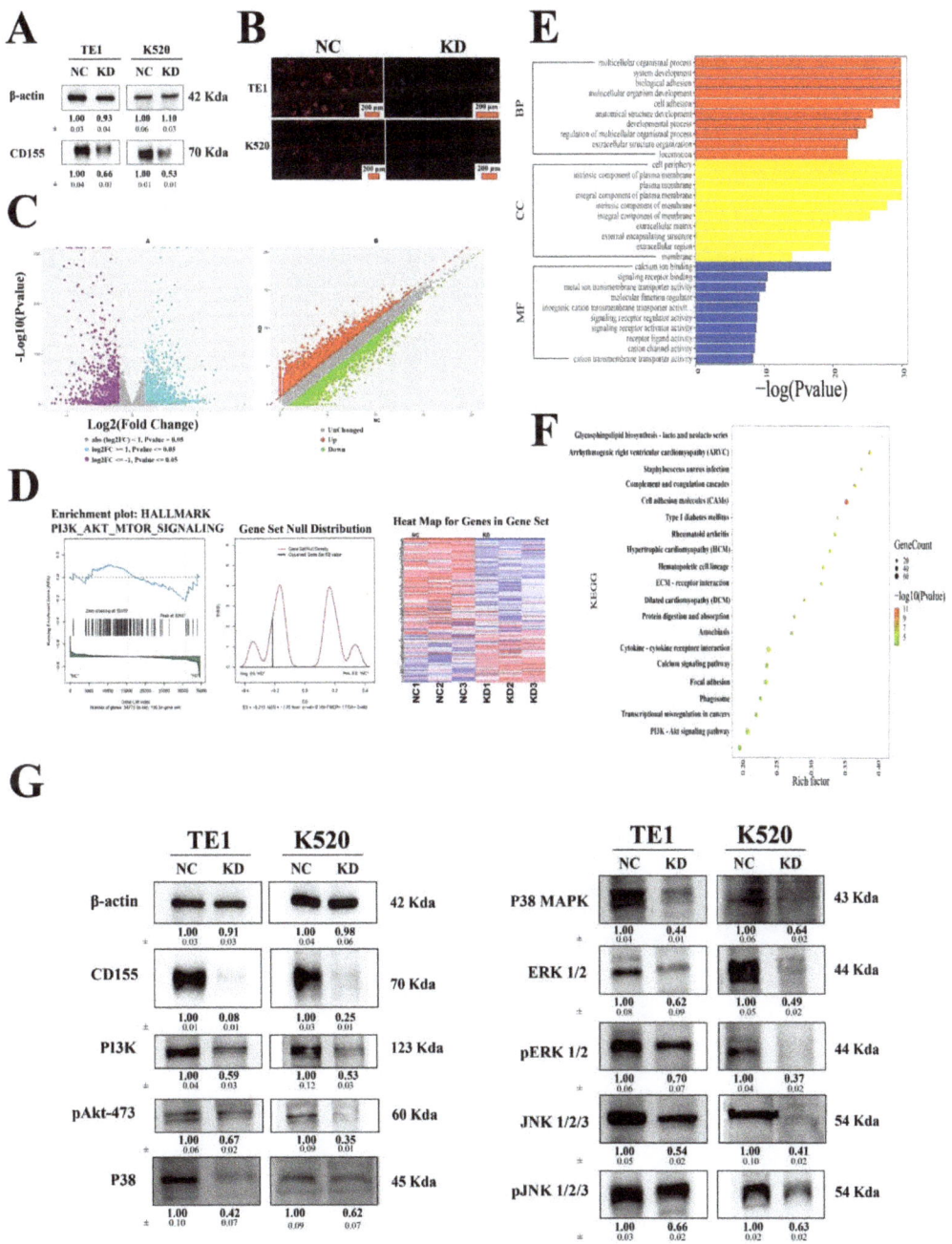

Figure 7. CD155 can regulate the PI3K/Akt and MAPK pathways in ESCA. (**A**) Western blot and (**B**) immunofluorescence data confirmed the decreased expression of CD155 in ESCA cells. (**C**) RNA sequencing of these cells and a respective (**D**) GSEA analyses were conducted, followed by (**E**) GO and (**F**) KEGG. KEGG and GSEA analyses both indicated that the CD155-related differentially expressed genes were mainly associated with the PI3K/Akt and MAPK signaling pathways. (**G**) The two signaling pathways related to CD155 were verified with Western blotting. The original WB Blots of subfigure (**A**) is Figure S5. The original WB Blots of subfigure (**G**) are Figure S6.

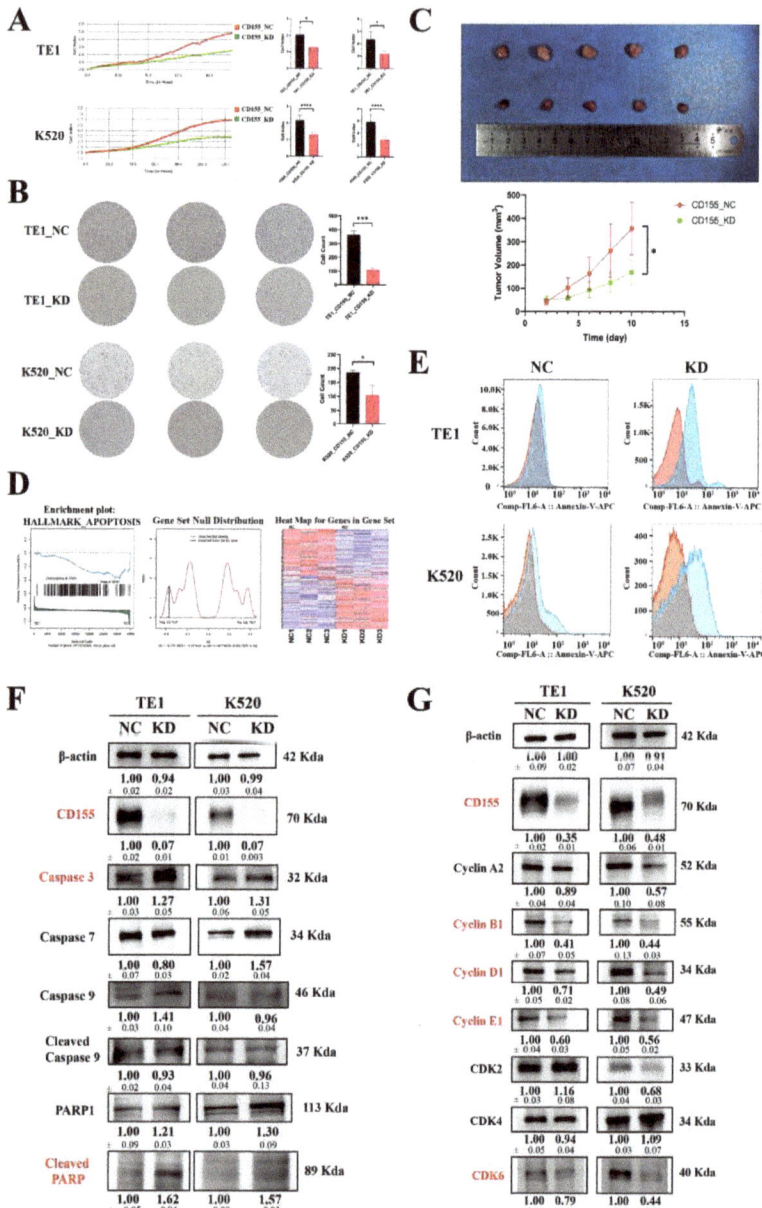

Figure 8. The downregulation of CD155 inhibits ESCA cells' proliferation by inducing cell cycle S phase arrest and cell apoptosis. (**A**) Real-time cell analyzer, (**B**) colony formation assays, and (**C**) in vivo experiments were performed to validate the proliferation between CD155_NC and CD155_KD groups in TE1 cells and K520 cells. These cells showed increased levels of cell apoptosis, supported by (**D**) GSEA analysis from RNA-seq (**E**) cell apoptosis flow analysis, and (**F**) Western blotting revealed a higher level of caspase 3 and cleaved PARP in the CD155_KD group of TE1 and K520 cells. Moreover, (**G**) Western blot indicated that the downregulation of CD155 may induce a decreased level of Cyclin B1, D1, E1, and CDK6 in TE1 and K520 cells. * $p < 0.05$; *** $p < 0.001$; **** $p < 0.0001$. The original WB Blots of subfigure (**F**) is Figure S7. The original WB Blots of subfigure (**G**) are Figure S8.

3.6. CD155 May Interact with Nectin3 and Regulate ESCA Proliferation

The knockdown of Nectin3 using siRNA phenocopies the profile of CD155 knockdown. Nectin3 was highly expressed in ESCA tissues compared to the adjacent tissues. In addition, it was positively associated with the expression of CD155, which was predicted in the GEPIA analysis (Figure 9A) and validated through tissue microarray (Figure 9B). Nectin3 was knocked down by siRNA, and its protein level was confirmed through Western blot analysis. The levels of PI3K, pAKT-473, P38, P38 MAPK, pERK1/2, pJNK1/2/3, Cyclin B1, Cyclin D1, and CDK6 were all decreased in TE1 and K520 cells. Furthermore, these cells showed increased levels of caspase 3 and cleaved PARP (Figure 9C). Similar results were obtained for the Western blot analysis of CD155 (Figure 8F,G), indicating that the knockdown of Nectin3 phenocopies the profile of CD155 knockdown. These results suggest that Nectin3 could be a ligand of CD155 and that their interaction could promote ESCA proliferation.

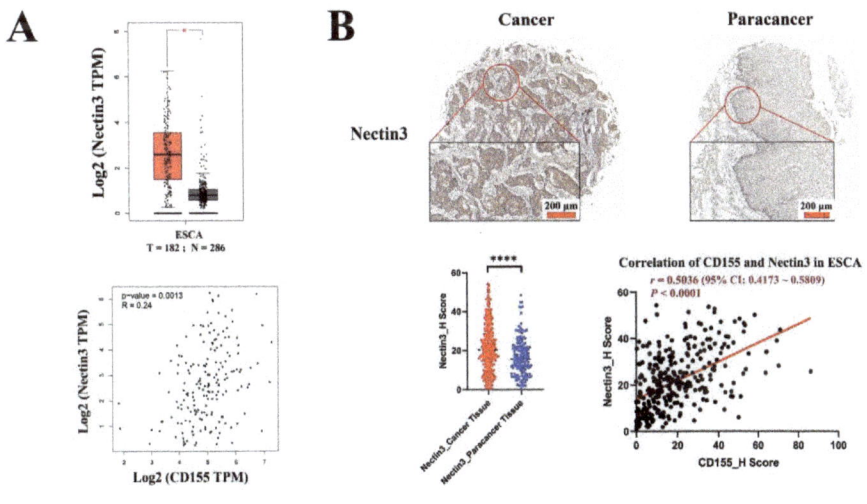

Figure 9. *Cont.*

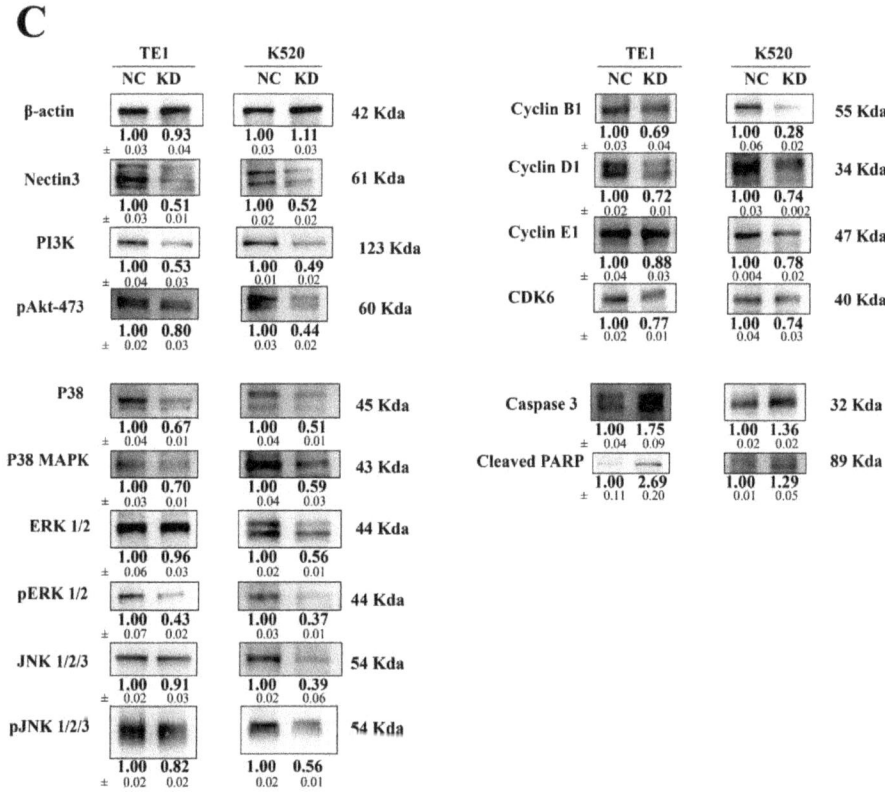

Figure 9. Downregulation of Nectin3 expression phenocopies the profile of CD155 knockdown. (**A**) Nectin3 expression and the correlation analysis between CD155 and Nectin3 expression were predicted using the GEPIA database and (**B**) validated at the protein level through tissue microarray. (**C**) Nectin3 is knocked down using siRNA. Changes similar to those of CD155 knockdown were seen in TE1 and K520 cells. * $p < 0.05$; **** $p < 0.0001$. The original WB Blots of subfigure (**C**) are Figure S9.

4. Discussion

Currently, the available immunotherapies lack efficiency in treating esophageal cancer [7,8], which can be, at least in part, explained by our lack of knowledge on its cellular and molecular mechanisms. Herein, we demonstrate that CD155 can be a valuable therapeutic candidate to treat these patients.

Overall, we found that CD155 is highly associated with PD-1/PD-L1 in ESCA. In particular, ESCA cells express high CD155 levels, which are positively associated with PD-1, PD-L1, CD4, IL-2Rα, and S100A9 levels, indicating that CD155 may exert immune regulatory effects in ESCA, especially toward CD4[+] T cells, B cells, and neutrophils. Moreover, given the strong correlation between CD155 and PD1/PD-L1, CD155 may be related to the efficacy of anti-PD1/PD-L1 treatment in ESCA. However, we suggest that CD155 may cooperate with PD-1/PD-L1 in this cancer via a regulatory role other than immune regulation. In agreement with this hypothesis, PD-1 and PD-L1 were previously shown to promote tumor proliferation in multiple types of tumors, including esophageal cancer, with the inhibition of PD-1 or PD-L1 effectively preventing the proliferation and inducing the apoptosis of tumor cells [17]. Therefore, further studies on whether the co-expression of CD155 and PD-1/PD-L1 may further enhance tumor growth and progression, and their underlying regulatory mechanisms, are warranted.

In vitro and in vivo experimental assays showed that the downregulation of CD155 inhibits ESCA cells' proliferation by preventing the expression of cell cycle-related proteins and by inducing apoptosis. Moreover, bioinformatics and experimental analyses of the expression profile of ESCA cells lacking CD155 suggested that CD155 downstream effects are mediated by the PI3K/Akt and MAPK signaling pathways, which are known to be involved in tumor proliferation [33–35]. Indeed, CD155 was previously shown to be able to directly bind to signaling proteins [36]. Thus, similar to PD-1, CD155 may play a positive regulatory role in the proliferation and progression of ESCA.

Herein, we determined that CD155 can interact with Nectin3 to promote ESCA proliferation. Nectin3 is a ligand of CD155. The interaction between these molecules can regulate cell behavior [24]. The protein level of Nectin3 is positively associated with CD155 expression in ESCA; both levels are higher in ESCA compared to those in adjacent tissues. Moreover, the downregulation of Nectin3 expression by siRNA phenocopies the profile of CD155 knockdown, especially with respect to the PI3K/Akt and MAPK signaling pathways, cell cycle-related proteins, and apoptosis-related proteins. Thus, we postulate that CD155 interacts with Nectin3 to promote ESCA proliferation. Similar results have been reported for multiple myeloma [36]. The binding of CD155 and Nectin3 can induce the adhesion of multiple myeloma cells to bone marrow stromal cells. Furthermore, CD155 can play multiple roles when binding with different ligands. When binding with the activating receptor DNAM-1, the cytotoxicity of NK and CD8+ T cells is promoted [37]. In contrast, binding with the inhibitory receptors TIGIT and CD96 can result in the inhibition of IFN-γ production and NK and T cell activity [21,38]. Similar immune regulation of CD155-mediated production can be observed in ESCA. However, further research is required.

In summary, the expression of CD155 is significantly positively associated with PD-1/PD-L1 signals, thereby regulating ESCA behavior by regulating its proliferation and viability via the PI3K/Akt and MAPK signaling pathways. Finding new strategies to block CD155, in addition to PD-1/PD-L1, may promote enhanced anticancer responses in patients with ESCA and help to achieve improved therapeutic outcomes.

5. Conclusions

In this study, we aimed to explore the role of CD155 in esophageal squamous cell cancer (ESCA) and its underlying molecular mechanism. CD155 was positively associated with PD-1/PD-L1 expression and could support ESCA proliferation. The downregulation of CD155 expression inhibited ESCA cell proliferation by impairing the cell cycle and inducing cell apoptosis. This occurred via the inhibition of PI3K/Akt and MAPK signaling pathways. In addition, Nectin3 may be the ligand of CD155 and may be involved in ESCA proliferation. Thus, our study suggests novel targets for tumor therapy, especially for ESCA treatment.

Supplementary Materials: The following supporting information can be downloaded at: https://www.mdpi.com/article/10.3390/cancers14225610/s1, Figure S1: Differential expression of 35 immune checkpoint genes in ESCA; Figure S2: CD155 is also highly expressed in various tumors in addition to ESCA; Figure S3: The gene and protein interaction network of CD155; Figure S4: CD155 is not correlated with the expression of various immune markers; Figure S5: The original WB Blots of Figure 7A; Figure S6: The original WB Blots of Figure 7G; Figure S7: The original WB Blots of Figure 8F; Figure S8: The original WB Blots of Figure 8G; Figure S9: The original WB Blots of Figure 9C.

Author Contributions: X.T., Y.X., J.Y. and M.W. collected the related papers and drafted the manuscript; Z.L. and H.T. contributed to manuscript revision; A.S., K.Z., Y.L. and C.S. participated in the design of the research; J.Z., N.M. and R.W. initiated the study and revised and finalized the manuscript. All authors have read and agreed to the published version of the manuscript.

Funding: This work was supported by the National Natural Science Foundation of China (No. 82070101; No. 82002421); 2021 Tangdu Hospital Discipline Innovation Development Plan (No. 2021LCYJ042); and Tangdu Hospital Technological innovation and development fund (No. 2019QYTS004).

Institutional Review Board Statement: The authors are accountable for all aspects of the work in ensuring that questions related to the accuracy or integrity of any part of the work are appropriately investigated and resolved. The study was conducted in accordance with the Declaration of Helsinki (as revised in 2013). The study was approved by the Ethics Committee of the Air Force Medical University (Ethics Approval number: 202108-05, 202203-145 on 30 August 2022 and 9 March 2022, respectively). Written informed consent, which included agreement to use personal clinical data and the collection of tissue and plasma samples, was provided by all patients before any study-related procedures began.

Informed Consent Statement: Informed consent was obtained from all subjects involved in the study to publish this paper.

Data Availability Statement: The data that support the findings of this study are available from the corresponding author upon reasonable request. The RNA sequencing datasets generated in this study can be found in the data depository Figshare, and are publicly accessible at DOI: 10.6084/m9.figshare.20419200.

Acknowledgments: We thank all those involved and their families, and we are grateful for the funding received from the National Natural Science Foundation of China.

Conflicts of Interest: The authors declare that the research was conducted in the absence of any commercial or financial relationships that could be construed as potential conflicts of interest.

Abbreviations

ESCA	Esophageal squamous cell cancer
IL	Interleukin
K520	KYSE-520
KD	Knockdown
NC	Non-targeting control
PD-1	Programmed death protein 1
PD-L1	Programmed death-ligand 1

References

1. Siegel, R.L.; Miller, K.D.; Fuchs, H.E.; Jemal, A. Cancer statistics, 2022. *CA Cancer J. Clin.* **2022**, *72*, 7–33. [CrossRef] [PubMed]
2. Liang, H.; Fan, J.H.; Qiao, Y.L. Epidemiology, etiology, and prevention of esophageal squamous cell carcinoma in China. *Cancer Biol. Med.* **2017**, *14*, 33–41. [CrossRef] [PubMed]
3. Chen, W.; Zheng, R.; Baade, P.D.; Zhang, S.; Zeng, H.; Bray, F.; Jemal, A.; Yu, X.Q.; He, J. Cancer statistics in China, 2015. *CA Cancer J. Clin.* **2016**, *66*, 115–132. [CrossRef] [PubMed]
4. Forde, P.M.; Chaft, J.E.; Smith, K.N.; Anagnostou, V.; Cottrell, T.R.; Hellmann, M.D.; Zahurak, M.; Yang, S.C.; Jones, D.R.; Broderick, S.; et al. Neoadjuvant PD-1 Blockade in Resectable Lung Cancer. *N. Engl. J. Med.* **2018**, *378*, 1976–1986. [CrossRef]
5. El-Khoueiry, A.B.; Sangro, B.; Yau, T.; Crocenzi, T.S.; Kudo, M.; Hsu, C.; Kim, T.Y.; Choo, S.P.; Trojan, J.; Welling, T.H.R.; et al. Nivolumab in patients with advanced hepatocellular carcinoma (CheckMate 040): An open-label, non-comparative, phase 1/2 dose escalation and expansion trial. *Lancet (Lond. Engl.)* **2017**, *389*, 2492–2502. [CrossRef]
6. Tawbi, H.A.; Schadendorf, D.; Lipson, E.J.; Ascierto, P.A.; Matamala, L.; Castillo Gutiérrez, E.; Rutkowski, P.; Gogas, H.J.; Lao, C.D.; De Menezes, J.J.; et al. Relatlimab Nivolumab Versus Nivolumab Untreated Advanced. *Melanoma. N. Engl. J. Med.* **2022**, *386*, 24–34. [CrossRef]
7. de Klerk, L.K.; Patel, A.K.; Derks, S.; Pectasides, E.; Augustin, J.; Uduman, M.; Raman, N.; Akarca, F.G.; McCleary, N.J.; Cleary, J.M.; et al. Phase II study of pembrolizumab in refractory esophageal cancer with correlates of response and survival. *J. Immunother. Cancer* **2021**, *9*, e002472. [CrossRef]
8. Zhang, W.; Yan, C.; Gao, X.; Li, X.; Cao, F.; Zhao, G.; Zhao, J.; Er, P.; Zhang, T.; Chen, X.; et al. Safety and Feasibility of Radiotherapy Plus Camrelizumab for Locally Advanced Esophageal Squamous Cell Carcinoma. *Oncologist* **2021**, *26*, e1110–e1124. [CrossRef]
9. Agata, Y.; Kawasaki, A.; Nishimura, H.; Ishida, Y.; Tsubata, T.; Yagita, H.; Honjo, T. Expression of the PD-1 antigen on the surface of stimulated mouse T and B lymphocytes. *Int. Immunol.* **1996**, *8*, 765–772. [CrossRef]
10. Cha, J.H.; Chan, L.C.; Li, C.W.; Hsu, J.L.; Hung, M.C. Mechanisms Controlling PD-L1 Expression in Cancer. *Mol. Cell* **2019**, *76*, 359–370. [CrossRef]

11. Yi, M.; Jiao, D.; Xu, H.; Liu, Q.; Zhao, W.; Han, X.; Wu, K. Biomarkers for predicting efficacy of PD-1/PD-L1 inhibitors. *Mol. Cancer* **2018**, *17*, 129. [CrossRef] [PubMed]
12. Jiang, Z.; Yan, Y.; Dong, J.; Duan, L. PD-1 expression on uveal melanoma induces tumor proliferation and predicts poor patient survival. *Int. J. Biol. Markers* **2020**, *35*, 50–58. [CrossRef] [PubMed]
13. Pawelczyk, K.; Piotrowska, A.; Ciesielska, U.; Jablonska, K.; Gletzel-Plucinska, N.; Grzegrzolka, J.; Podhorska-Okolow, M.; Dziegiel, P.; Nowinska, K. Role of PD-L1 Expression in Non-Small Cell Lung Cancer and Their Prognostic Significance according to Clinicopathological Factors and Diagnostic Markers. *Int. J. Mol. Sci.* **2019**, *20*, 824. [CrossRef] [PubMed]
14. Lee, S.H.; Koo, B.S.; Kim, J.M.; Huang, S.; Rho, Y.S.; Bae, W.J.; Kang, H.J.; Kim, Y.S.; Moon, J.H.; Lim, Y.C. Wnt/β-catenin signalling maintains self-renewal and tumourigenicity of head and neck squamous cell carcinoma stem-like cells by activating Oct4. *J. Pathol.* **2014**, *234*, 99–107. [CrossRef]
15. Tufano, M.; D'Arrigo, P.; D'Agostino, M.; Giordano, C.; Marrone, L.; Cesaro, E.; Romano, M.F.; Romano, S. PD-L1 Expression Fluctuates Concurrently with Cyclin D in Glioblastoma Cells. *Cells* **2021**, *10*, 2366. [CrossRef]
16. Liu, M.Y.; Klement, J.D.; Langan, C.J.; van Riggelen, J.; Liu, K. Expression regulation and function of PD-1 and PD-L1 in T lymphoma cells. *Cell. Immunol.* **2021**, *366*, 104397. [CrossRef]
17. Davern, M.; RM, O.B.; McGrath, J.; Donlon, N.E.; Melo, A.M.; Buckley, C.E.; Sheppard, A.D.; Reynolds, J.V.; Lynam-Lennon, N.; Maher, S.G.; et al. PD-1 blockade enhances chemotherapy toxicity in oesophageal adenocarcinoma. *Sci. Rep.* **2022**, *12*, 3259. [CrossRef]
18. Bronte, V. The expanding constellation of immune checkpoints: A DNAMic control by CD155. *J. Clin. Investig.* **2018**, *128*, 2199–2201. [CrossRef]
19. Freed-Pastor, W.A.; Lambert, L.J.; Ely, Z.A.; Pattada, N.B.; Bhutkar, A.; Eng, G.; Mercer, K.L.; Garcia, A.P.; Lin, L.; Rideout, W.M., 3rd; et al. The CD155/TIGIT axis promotes and maintains immune evasion in neoantigen-expressing pancreatic cancer. *Cancer Cell* **2021**, *39*, 1342–1360. [CrossRef]
20. Liu, S.; Zhang, H.; Li, M.; Hu, D.; Li, C.; Ge, B.; Jin, B.; Fan, Z. Recruitment of Grb2 and SHIP1 by the ITT-like motif of TIGIT suppresses granule polarization and cytotoxicity of NK cells. *Cell Death Differ.* **2013**, *20*, 456–464. [CrossRef]
21. Li, M.; Xia, P.; Du, Y.; Liu, S.; Huang, G.; Chen, J.; Zhang, H.; Hou, N.; Cheng, X.; Zhou, L.; et al. T-cell immunoglobulin and ITIM domain (TIGIT) receptor/poliovirus receptor (PVR) ligand engagement suppresses interferon-γ production of natural killer cells via β-arrestin 2-mediated negative signaling. *J. Biol. Chem.* **2014**, *289*, 17647–17657. [CrossRef] [PubMed]
22. He, W.; Zhang, H.; Han, F.; Chen, X.; Lin, R.; Wang, W.; Qiu, H.; Zhuang, Z.; Liao, Q.; Zhang, W.; et al. CD155T/TIGIT Signaling Regulates CD8(+) T-cell Metabolism and Promotes Tumor Progression in Human Gastric Cancer. *Cancer Res.* **2017**, *77*, 6375–6388. [CrossRef] [PubMed]
23. Fabre, S.; Reymond, N.; Cocchi, F.; Menotti, L.; Dubreuil, P.; Campadelli-Fiume, G.; Lopez, M. Prominent role of the Ig-like V domain in trans-interactions of nectins. Nectin3 and nectin 4 bind to the predicted C-C′-C″-D beta-strands of the nectin1 V domain. *J. Biol. Chem.* **2002**, *277*, 27006–27013. [CrossRef] [PubMed]
24. Fujito, T.; Ikeda, W.; Kakunaga, S.; Minami, Y.; Kajita, M.; Sakamoto, Y.; Monden, M.; Takai, Y. Inhibition of cell movement and proliferation by cell-cell contact-induced interaction of Necl-5 with nectin-3. *J. Cell Biol.* **2005**, *171*, 165–173. [CrossRef] [PubMed]
25. Liu, X.S.; Gao, Y.; Liu, C.; Chen, X.Q.; Zhou, L.M.; Yang, J.W.; Kui, X.Y.; Pei, Z.J. Comprehensive Analysis of Prognostic and Immune Infiltrates for E2F Transcription Factors in Human Pancreatic Adenocarcinoma. *Front. Oncol.* **2020**, *10*, 606735. [CrossRef] [PubMed]
26. Chandrashekar, D.S.; Bashel, B.; Balasubramanya, S.A.H.; Creighton, C.J.; Ponce-Rodriguez, I.; Chakravarthi, B.; Varambally, S. UALCAN: A Portal for Facilitating Tumor Subgroup Gene Expression and Survival Analyses. *Neoplasia* **2017**, *19*, 649–658. [CrossRef]
27. Warde-Farley, D.; Donaldson, S.L.; Comes, O.; Zuberi, K.; Badrawi, R.; Chao, P.; Franz, M.; Grouios, C.; Kazi, F.; Lopes, C.T.; et al. The GeneMANIA prediction server: Biological network integration for gene prioritization and predicting gene function. *Nucleic Acids Res.* **2010**, *38*, W214–W220. [CrossRef]
28. Szklarczyk, D.; Gable, A.L.; Nastou, K.C.; Lyon, D.; Kirsch, R.; Pyysalo, S.; Doncheva, N.T.; Legeay, M.; Fang, T.; Bork, P.; et al. The STRING database in 2021: Customizable protein-protein networks, and functional characterization of user-uploaded gene/measurement sets. *Nucleic Acids Res.* **2021**, *49*, D605–D612. [CrossRef]
29. Tang, X.Y.; Xiong, Y.L.; Shi, A.P.; Sun, Y.; Han, Q.; Lv, Y.; Shi, X.G.; Frattini, M.; Malhotra, J.; Zheng, K.F.; et al. The downregulation of fibrinogen-like protein 1 inhibits the proliferation of lung adenocarcinoma via regulating MYC-target genes. *Transl. Lung Cancer Res.* **2022**, *11*, 404–419. [CrossRef]
30. Maclean, A.; Bunni, E.; Makrydima, S.; Withington, A.; Kamal, A.M.; Valentijn, A.J.; Hapangama, D.K. Fallopian tube epithelial cells express androgen receptor and have a distinct hormonal responsiveness when compared with endometrial epithelium. *Hum. Reprod. (Oxf. Engl.)* **2020**, *35*, 2097–2106. [CrossRef]
31. Dogan, S.; Vasudevaraja, V.; Xu, B.; Serrano, J.; Ptashkin, R.N.; Jung, H.J.; Chiang, S.; Jungbluth, A.A.; Cohen, M.A.; Ganly, I.; et al. DNA methylation-based classification of sinonasal undifferentiated carcinoma. *Mod. Pathol. Off. J. U.S. Can. Acad. Pathol. Inc.* **2019**, *32*, 1447–1459. [CrossRef] [PubMed]
32. Paschalis, A.; Sheehan, B.; Riisnaes, R.; Rodrigues, D.N.; Gurel, B.; Bertan, C.; Ferreira, A.; Lambros, M.B.K.; Seed, G.; Yuan, W.; et al. Prostate-specific Membrane Antigen Heterogeneity and DNA Repair Defects in Prostate Cancer. *Eur. Urol.* **2019**, *76*, 469–478. [CrossRef] [PubMed]

33. Cheng, R.; Wang, B.; Cai, X.R.; Chen, Z.S.; Du, Q.; Zhou, L.Y.; Ye, J.M.; Chen, Y.L. CD276 Promotes Vasculogenic Mimicry Formation in Hepatocellular Carcinoma via the PI3K/AKT/MMPs Pathway. *OncoTargets Ther.* **2020**, *13*, 11485–11498. [CrossRef] [PubMed]
34. Li, Y.; Guo, G.; Song, J.; Cai, Z.; Yang, J.; Chen, Z.; Wang, Y.; Huang, Y.; Gao, Q. B7-H3 Promotes the Migration and Invasion of Human Bladder Cancer Cells via the PI3K/Akt/STAT3 Signaling Pathway. *J. Cancer* **2017**, *8*, 816–824. [CrossRef] [PubMed]
35. Zhang, H.; Wang, C.; Fan, J.; Zhu, Q.; Feng, Y.; Pan, J.; Peng, J.; Shi, J.; Qi, S.; Liu, Y. CD47 promotes the proliferation and migration of adamantinomatous craniopharyngioma cells by activating the MAPK/ERK pathway, and CD47 blockade facilitates microglia-mediated phagocytosis. *Neuropathol. Appl. Neurobiol.* **2022**, *48*, e12795. [CrossRef] [PubMed]
36. Molfetta, R.; Zitti, B.; Lecce, M.; Milito, N.D.; Stabile, H.; Fionda, C.; Cippitelli, M.; Gismondi, A.; Santoni, A.; Paolini, R. CD155: A Multi-Functional Molecule in Tumor Progression. *Int. J. Mol. Sci.* **2020**, *21*, 922. [CrossRef]
37. Bryceson, Y.T.; March, M.E.; Ljunggren, H.G.; Long, E.O. Synergy among receptors on resting NK cells for the activation of natural cytotoxicity and cytokine secretion. *Blood* **2006**, *107*, 159–166. [CrossRef]
38. Lozano, E.; Dominguez-Villar, M.; Kuchroo, V.; Hafler, D.A. The TIGIT/CD226 axis regulates human T cell function. *J. Immunol.* **2012**, *188*, 3869–3875. [CrossRef]

Article

SH3BP2 Silencing Increases miRNAs Targeting ETV1 and Microphthalmia-Associated Transcription Factor, Decreasing the Proliferation of Gastrointestinal Stromal Tumors

Elizabeth Proaño-Pérez [1,2,3], Eva Serrano-Candelas [1,2,†], Cindy Mancia [1], Arnau Navinés-Ferrer [1,2], Mario Guerrero [1] and Margarita Martin [1,2,*]

1. Biochemistry and Molecular Biology Unit, Biomedicine Department, Faculty of Medicine and Health Sciences, University of Barcelona, 08036 Barcelona, Spain
2. Clinical and Experimental Respiratory Immunoallergy (IRCE), Institut d'Investigacions Biomediques August Pi i Sunyer (IDIBAPS), 08036 Barcelona, Spain
3. Faculty of Health Sciences, Technical University of Ambato, Ambato 180105, Ecuador
* Correspondence: martin_andorra@ub.edu
† Current address: ProtoQSAR SL, Centro Europeo de Empresas Innovadoras (CEEI), Parque Tecnológico de Valencia, 46980 Paterna, Valencia, Spain.

Simple Summary: A previous study showed that silencing the adaptor molecule SH3 Binding Protein 2 (SH3BP2) reduced oncogenic KIT and PDGFRA receptor levels and impaired gastrointestinal stromal tumor (GIST) growth. This study tries to get insights into the molecular mechanism underlying this effect. The silencing of SH3BP2 induces miRNAs (miR-1246 and miR-5100), which target microphthalmia-associated transcription factor (MITF) and ETV1, a linage survival factor involved in GIST tumorigenesis. Altogether, this results in decreased tumor cell viability and enhanced apoptosis.

Abstract: Gastrointestinal stromal tumors (GISTs) are the most common mesenchymal tumors of the gastrointestinal tract. Gain of function in receptor tyrosine kinases type III, KIT, or PDGFRA drives the majority of GIST. Previously, our group reported that silencing of the adaptor molecule SH3 Binding Protein 2 (SH3BP2) downregulated KIT and PDGFRA and microphthalmia-associated transcription factor (MITF) levels and reduced tumor growth. This study shows that SH3BP2 silencing also decreases levels of ETV1, a required factor for GIST growth. To dissect the SH3BP2 pathway in GIST cells, we performed a miRNA array in SH3BP2-silenced GIST cell lines. Among the most up-regulated miRNAs, we found miR-1246 and miR-5100 to be predicted to target *MITF* and *ETV1*. Overexpression of these miRNAs led to a decrease in MITF and ETV1 levels. In this context, cell viability and cell cycle progression were affected, and a reduction in BCL2 and CDK2 was observed. Interestingly, overexpression of MITF enhanced cell proliferation and significantly rescued the viability of miRNA-transduced cells. Altogether, the KIT-SH3BP2-MITF/ETV1 pathway deserves to be considered in GIST cell survival and proliferation.

Keywords: SH3BP2; MITF; ETV1; miRNA; cell survival; cell cycle; gastrointestinal stromal tumors

1. Introduction

Gastrointestinal stromal tumors (GISTs) are the most common type of soft tissue sarcoma in the intestinal tract [1]. They are derived from the interstitial cells of Cajal (ICCs), located in the submucosa and myenteric plexus of the gastrointestinal tract [2]. The pathogenesis of GISTs is defined by mutually exclusive mutations in *KIT* (75–80%) and platelet-derived growth factor receptor α (*PDGFRA*) genes (5–10%). Additionally, 10–15% of GISTs lack *KIT* mutations, the so-called "wild type." They are classified as deficient in succinate dehydrogenase (SDH)-deficient and non-SHD-deficient. Non-SDH-deficient include NF type 1 neurofibromatosis and GISTs with *BRAF*, *KRAS*, and *PIK3CA* mutations [3].

SH3BP2 (cytoplasmic adaptor molecule SH3-binding protein 2) has been described as an active regulator of *KIT* expression and signaling in mast cells [4] and GIST cells [5]. The silencing of SH3BP2 decreases KIT levels and increases the caspase-3/7 activity, which consequently induces apoptosis. Additionally, SH3BP2 regulates KIT at the transcriptional level and MITF (microphthalmia-associated transcription factor) at the post-transcriptional level in mast cells [4]. Overexpression of MITF in GIST cell lines prevented significant cellular apoptosis [5]. MITF is a basic helix-loop-helix leucine zipper, a dimeric transcription factor well-documented in melanocyte differentiation, cell cycle progression, and survival by targeting pigment enzyme genes or CDK2 [6,7], among others. MITF activity is needed for melanocyte development, and deregulation of its activity is reported in melanoma [8]. Besides melanocytes, MITF is essential for mast cell differentiation [9] and binds to the KIT promoter on mast cells [10]. MITF has recently been reported to be involved in GIST cell survival, proliferation, and tumor growth, and MITF silencing leads to an ETV1 reduction [11]. ETV1 is a transcription factor required for the development of interstitial cells of Cajal and the proliferation of GIST cells [12]. ETV1 is regulated by the MEK–MAPK pathway downstream and activated by KIT and PDGFRA [13]. Therefore, KIT inhibition with imatinib reduces ETV1 levels [14]. Likewise, we found that GIST treated with imatinib reduced MITF levels in vitro [5].

Due to the regulation of SH3BP2 over KIT and MITF levels and their mutual regulation, herein, we aimed to study ETV1 involvement in the pathway and tried to dissect the SH3BP2 pathway pursuing the analysis of miRNAs. In this study, a miRNA microarray was performed, comparing the expression levels of several miRNAs in SH3BP2-silenced GIST cell lines with non-silenced cells. We analyzed the highest up-regulated miRNAs that predictively regulate MITF or ETV1 levels. Further, these miRNAs were validated and characterized in GIST cell lines.

2. Materials and Methods

2.1. Antibodies and Reagents

Mouse anti-SH3BP2 (clone C5), mouse anti-KIT (clone Ab81), mouse anti-BCL2, and mouse anti-CDK2 were purchased from Santa Cruz Biotechnology, Inc. (Santa Cruz, CA, USA). Anti-MITF (clone D5G7V) was obtained from Cell Signaling Technology, Inc (Danvers, MA, USA). Mouse anti-β-actin (clone AC-40) was purchased from Sigma (St. Louis, MO, USA). Anti-ETV1 antibody (ER81) (ab81086) was obtained from Abcam technology (Abcam, Cambridge, UK). Anti-mouse and anti-rabbit IgG peroxidase Abs were acquired from DAKO (Carpinteria, CA, USA) and Biorad (Hercules, CA, USA), respectively.

2.2. Cell Culture

Human GIST cell lines GIST882, GIST48, and GIST-T1 were kindly provided by Dr. S. Bauer. GIST cell lines were-cultured as described elsewhere [11]. Transient transfections were carried out using Opti-MEM (Gibco, Carlsbad, CA, USA). The mycoplasma test was performed routinely in all cell lines used.

2.3. RNA Extraction, Retrotranscription, and PCR Assays

Total RNA was extracted with a miRCURY RNA Isolation Kit (Exiqon, Vedbaek, Denmark) from NT control and SH3BP2 knockdown GIST cells. cDNA was generated by reverse transcription using the miRCURY LNA RT Kit. Quantitative, Real-Time PCR for miRNA PCR assay was performed using the miRCURY SYBR Green PCR Kit, and following miRCURY LNA miRNA PCR assay protocol on a LightCycler® 480 Instrument II (LifeScience Roche). miR-30c-5p and miR-335 were used as housekeeping miRNA genes.

2.4. MicroRNA Array Profiling

All experiments were conducted at Exiqon Services, Denmark. The quality of all the total RNA was verified by an Agilent 2100 Bioanalyzer profile. 750 ng total RNA from both sample and reference was labeled with Hy3™ and Hy5™ fluorescent labels, respectively, using the miRCURY LNA™ microRNA Hi-Power Labeling Kit, Hy3™/Hy5™ (Exiqon, Vedbæk, Denmark), following the procedure described by the manufacturer. The Hy3™-labeled samples and a Hy5™-labeled reference RNA sample were mixed pair-wise and hybridized to the miRCURY LNA™ microRNA Array 7th Gen (Exiqon, Denmark), which contains capture probes targeting all microRNAs for humans, mice, or rats registered in the miRBASE 18.0. The hybridization was performed according to the miRCURY LNA™ microRNA Array Instruction manual using a Tecan HS4800™ hybridization station (Tecan, Austria). After hybridization, the microarray slides were scanned and stored in an ozone-free environment (ozone level below 2.0 ppb) to prevent potential bleaching of the fluorescent dyes. The miRCURY LNA™ microRNA Array slides were scanned using the Agilent G2565BA Microarray Scanner System (Agilent Technologies, Inc., Santa Clara, CA, USA), and the image analysis was carried out using the ImaGene® 9 (miRCURY LNA™ microRNA Array Analysis Software, Exiqon, Denmark). The quantified signals were background corrected (Normexp with offset value 10, see [15]) and normalized using the global Lowess (Locally Weighted Scatterplot Smoothing) regression algorithm. Among the 502 human miRNAs detected by the array, we discarded any miRNA with an Average Hy3 signal under 7.5.

2.5. Lentiviral Transduction

Lentiviral particles to silence the *SH3BP2* gene expression were previously described [5]. Lentiviral transduction for NT (non-target) was performed as described in [4] with slight modifications. PLenti-III-mir-GFP-blank was the plasmid used as a control. Plenti-III-miR-GFP miRNAs (miR-1246 and miR-5100) were obtained from Applied Biological Materials Inc (Richmond, BC, Canada). GIST cells were transduced in the presence of 8 µL/mL of Polybrene (Santa Cruz, CA, USA), and puromycin selection (1 µg/mL) was carried out after one day from transduction.

2.6. Cell Viability, Proliferation, and Caspases 3/7 Activity Assays

Cell viability and proliferation were assessed using Crystal violet dye [16], colorimetric assay (WST-1 based) (Roche Diagnostics, Germany), and caspase activity using the Caspase-Glo™ 3/7 Assay (Promega, San Luis Obispo, CA, USA), according to the manufacturer's protocol.

2.7. Western Blotting

Western blotting was performed as described [5,17]. Briefly, transduced cells were lysed at the 5th and 7th days post-lentiviral infection. Electrophoresis and protein blotting was performed using NuPage TM 4–12% Bis-Tris Gel, 1.5 mm × 15 w (Invitrogen, Waltham, MA, USA), and electrotransferred to polyvinylidene difluoride (PVDF) membranes (Millipore, Bedford, MA, USA). Blots were probed with the indicated antibodies. In all blots, proteins were visualized by enhanced chemiluminescence (WesternBright TM ECL, Advansta, San Jose, CA, USA). Original blots see Material S1.

2.8. Cell Cycle Analysis by Flow Cytometry

GIST cells were collected on the 5th and 7th days after transduction. The cells were fixed with 70% ethanol at 4 °C overnight and stained with propidium iodide buffer, as described elsewhere [18]. Data were acquired in FACS Calibur and analyzed using model Dean/Jet/Fox FlowJo 7.6 software.

2.9. MITF Overexpression

MiR-CTL, miR-1246, and miR-5100 were overexpressed by lentiviral transduction in GIST-T1. Cells were selected with puromycin (1 µg/mL) 24 h after infection. MITF overexpression was achieved using MITF A GFPpcDNA 3.1+/C-eGFP or pEGFP N3 (control) plasmid (Genscript). MITF-GFP or GFP plasmids were transfected into GIST-T1 cells by Lipofectamine LTX (Invitrogen) on the 3rd day after transduction following the manufacturer's instructions with slight modifications. Mix plasmids and lipofectamine were incubated overnight with CTS Opti-MEM (Gibco). Cells were maintained in IMDM media (Lonza) with puromycin (1 µg/mL). Cell proliferation was determined on the 7th day using WST-1 (Roche Diagnostics, Mannheim, Germany). MITF and MITF-GFP protein levels were analyzed by Western blotting.

2.10. LNA Anti-miRNA Treatment

GIST-T1 (0.06 × 10^6 cells/w) was cultured in a 96-well plate and treated gymnotically with fluoresceinated LNA oligonucleotides (LNA®, miRCURY®, QIAGEN, Hilden, Germany). LNA anti-miR-1246 (0.1 µM), LNA anti-miR-5100 (0.1 µM), and LNA mixed solution (anti-miR-1246, 0.05 µM + anti-miR-5100, 0.05 µM). Two hours after LNA treatment, sh3BP2 lentiviral particles were added with 8 µg/mL polybrene (Santa Cruz). GIST cells were selected by puromycin (1 µg/mL) after 24 h of lentiviral transduction. We measured cell viability by crystal violet assay [16] on the 4th day post-transduction.

2.11. Statistical Data Analysis

After determining the normal distribution of the samples and variance analysis, an unpaired student's t-test was used to determine significant differences (p-value) between the two experimental groups. A one-way ANOVA test was used to determine significant differences (p-value) between several experimental groups. All results are expressed as mean ± standard error of the mean (SEM).

3. Results

3.1. SH3BP2 Silencing Reduces ETV1 Levels in GIST Cell Lines

In previous work, we showed that silencing of SH3BP2 diminished KIT, PDGFRA, and MITF levels lead to a reduction in tumor growth in vitro and in vivo [5]. To better understand the role of SH3BP2 in GIST survival, we checked whether silencing of SH3BP2 was also affecting ETV1, a master regulator of the normal linage of interstitial Cajal cells, which cooperates with KIT in GIST [14,19]. As shown in Figure 1 and Supplementary Material S1, silencing of SH3BP2 reduces MITF, as previously reported, and ETV1 protein levels in imatinib-sensitive and imatinib-resistant GIST cell lines.

3.2. miRNA Profiling of SH3BP2-Silenced GIST Cells

SH3BP2 silencing reduces MITF at the protein level but not at the mRNA level in GIST [5], suggesting a post-transcriptional regulatory mechanism. KIT can regulate MITF through selective miRNA expression in mast cells [12]. Thus, we next performed a miRNA microarray to identify miRNAs regulated by SH3BP2 in GIST882 and GIST48-silenced cells to get insights into the signaling pathway that leads to apoptosis. Figure 2 shows the heat map representation of the two-way hierarchical clustering of miRNAs and samples. Interestingly, the samples cluster according to their biological group, meaning a very different miRNA profile exists between Non-Target and SH3BP2-silenced cells independently of the cell type.

A p-value < 0.05 was used to define significantly deregulated miRNAs between the different groups. This criterion identified 162 and 130 miRNAs in GIST882 and GIST48, respectively, with 107 in common (Figure 2A). In Figure 2B, a four-way Venn diagram shows that 32 miRNAs are downregulated in both cell lines, and 56 are up-regulated among the significantly changed miRNAs. Among them, a threshold of 1.5-fold change defined the 21 most up-regulated and the 12 most downregulated miRNAs in both cell lines (Figure 2C). Several databases were used to predict miRNA–target interactions with these miRNAs (Tables S1 and S2).

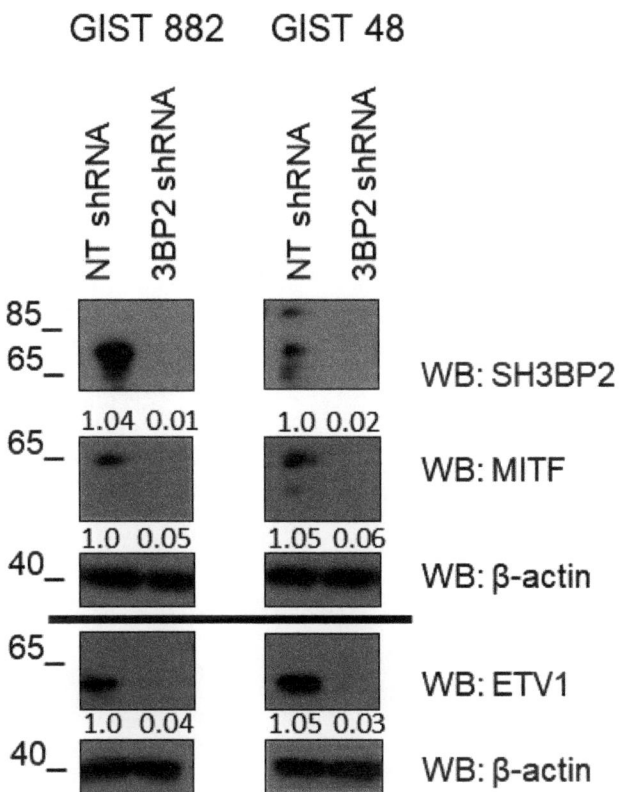

Figure 1. Reduced levels of ETV1 and MITF in SH3BP2-silenced GIST cells. GIST882 and GIST48 were transduced with control NT (Non-target) shRNA and SH3BP2 shRNA. Cell lysates were analyzed on the 7th day post-transduction for MITF, ETV1, and SH3BP2. β-actin was used as a loading control.

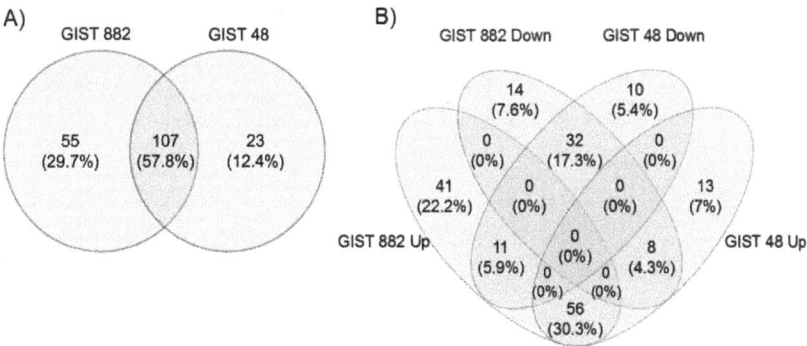

Figure 2. *Cont.*

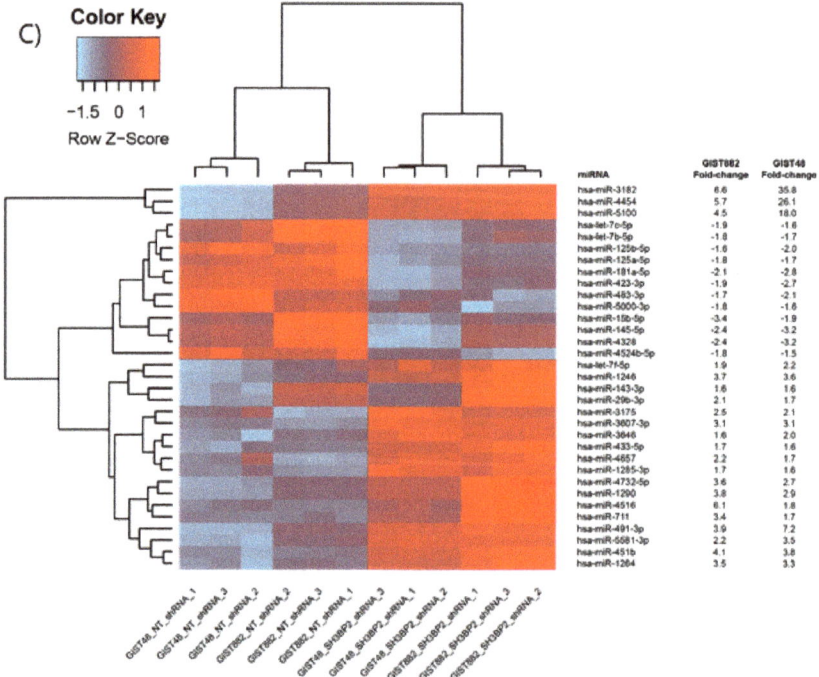

Figure 2. Profile of miRNA expression in SH3BP2-silenced GIST cells. Heat map representation of two-way hierarchical clustering of the miRNAs altered in GIST48 and GIST882 after 3BP2 silencing. (**A**) Venn diagram of miRNAs altered after SH3BP2 silencing. The expression of 107 miRNAs was significantly altered in both cell lines (p-value < 0.05). (**B**) A four-way Venn diagram shows the overlapping of the different miRNAs in both cell lines. The clustering was done using the complete-linkage method and Euclidean distance measure. (**C**) Each column represents a single sample, and each file represents a single miRNA. The red and blue colors represent high and low relative expressions, respectively (p-value < 0.05).

3.3. Validation of Up-Regulated miRNAs That Target MITF and ETV1 in GIST Cell Lines

From the most up-regulated miRNAs, we identified microRNAs that target MITF and ETV1. We used TargetScan [20], miRtar [21], miRwalk 2.0 [22], microT CDS [23], and mirDIP [24]. The different databases identified miR-1246, miR-1264, miR-1290, miR-3182, and miR5100 as putative MITF and ETV1 partners [25]. The results are summarized in Supplementary Tables S1 and S2.

Next, we validated these five miRNAs in various GIST cell lines. Quantitative real-time PCR was carried out in SH3BP2 silenced GIST-T1 (Figure 3A), GIST882 (Figure 3B), and GIST48 cells (Figure 3C). Only two of the five putative miRNAs (miR-1246 and miR-5100) exhibited significant differences between SH3BP2 shRNA and scramble transfection in all GIST cells. In parallel, only miR-1246 and 5100 overexpression in GIST cell lines show a reduction of MITF level by western blot (Supplementary Figure S1). We restricted further studies to these two miRNAs. The miRNAs sequence location on the *MITF-A*, the highest isoform expressed, and ETV1 genes, are shown in Supplementary Figure S2.

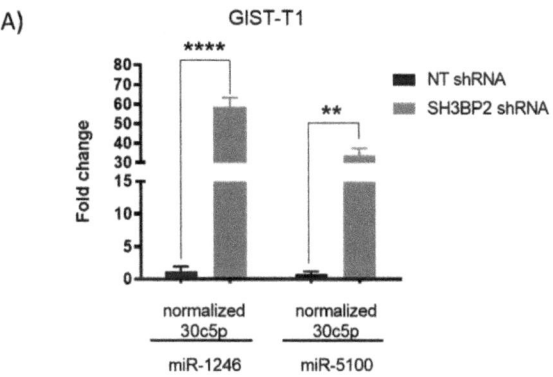
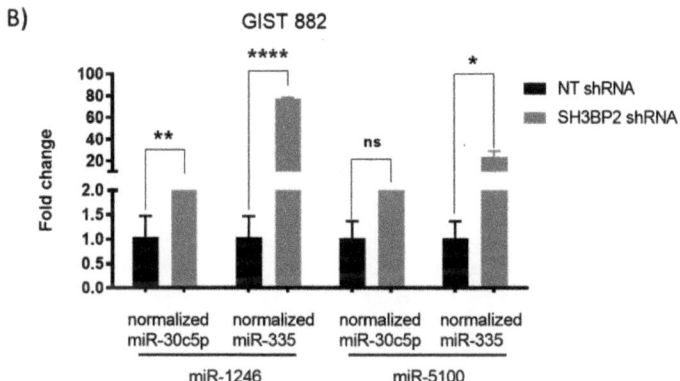
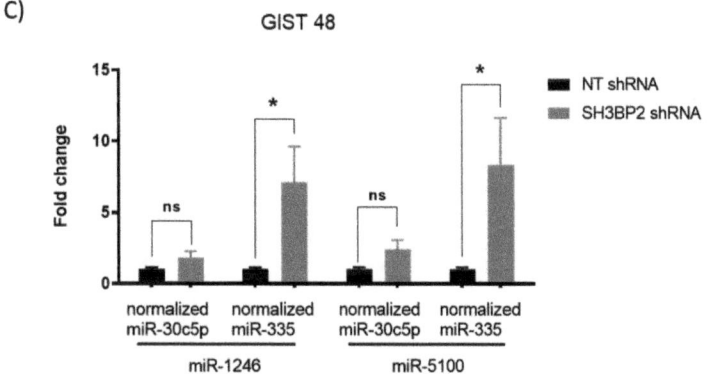

Figure 3. Validation of miRNA upregulation after SH3BP2 silencing by Real-Time PCR in GIST cell lines. GIST-T1, GIST882, and GIST48 cells were transduced with a non-target shRNA sequence or a specific shRNA SH3BP2. MiR-335 and miR-30c5p were used as housekeeping miRNAs. (**A**) GIST-T1 Data represent one biological replicate performed two times. (**B**) GIST 882 Data are representative of two biological replicates performed two times. (**C**) GIST 48 Data represent three biological replicates performed two times. (* $p < 0.05$, ** $p < 0.01$, **** $p < 0.0001$; Unpaired t-test; mean ± SEM).

3.4. MiR-1246 and miR-5100 Target ETV1 and MITF, and Overexpression Significantly Affects Cell Proliferation

As mentioned above, these miRNAs putatively bind to *MITF* or *ETV1* mRNA, so we overexpressed them in the imatinib-sensitive GIST-T1 and imatinib-resistant GIST-48 cell lines to check ETV1 and MITF protein levels. The overexpression of GFP-miR-1246 and GFP-miR-5100 efficiently causes the downregulation of MITF and ETV1 protein levels (Figure 4A,B and Supplementary Material S1). Consistently with this, we reported diminished cell proliferation in GIST cells, Figure 4C. The levels of transfection were similar in all cases (Supplementary Figure S3).

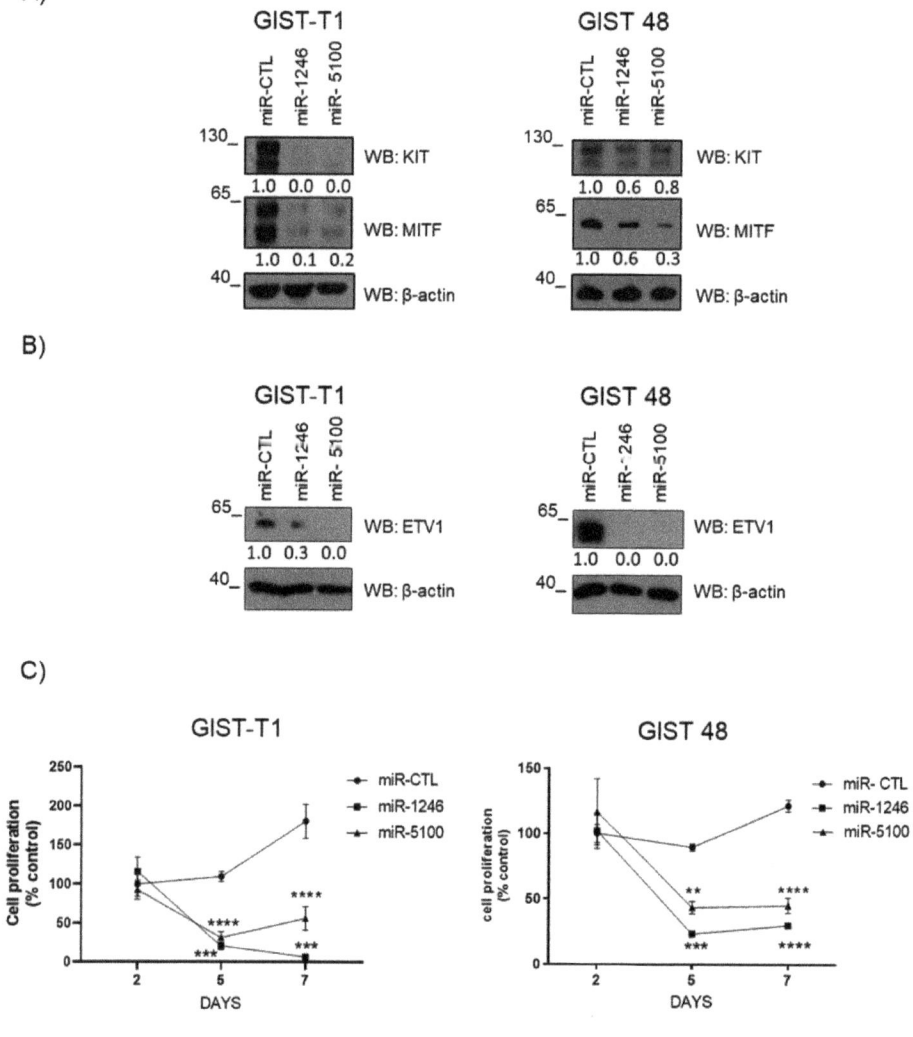

Figure 4. MiR-1246 and miR-5100 reduce cell proliferation in GIST cells. Western blot was performed on the 5th day after lentiviral transduction in GIST-T1 and GIST 48 to determine levels of (**A**) KIT and MITF, (**B**) ETV1. β-actin was used as load control. (**C**) Cell proliferation assay was performed by WST-1 on the 2nd, 5th, and 7th days after lentiviral transduction. (** $p < 0.01$, *** $p < 0.001$, **** $p < 0.0001$; one-way-ANOVA with Bonferroni's post-hoc test) n = 3. GFP-miR-CTL was used as a control.

3.5. MiR-1246 and miR-5100 Promote Apoptosis by Caspases 3/7 in GIST Cells

To analyze how miRNAs affect cell proliferation, we performed a viability assay and measured caspase 3/7 activity on overexpressed miRNAs GIST cells. Our results show a decrease in cell viability that correlates with an increase in caspase 3/7 activity in both cell lines (Figure 5B,C). Previous studies reported that miR-5100 induces apoptosis throughout caspase 3 protein activity [26], and miR-1246 increases apoptosis by promoting caspase 3 and caspase 7 activity [27]; these results are consistent with the anti-apoptotic protein BCL2 (MITF-dependent target) reduction after overexpression of miRNAs (Figure 5A and Supplementary Material S1).

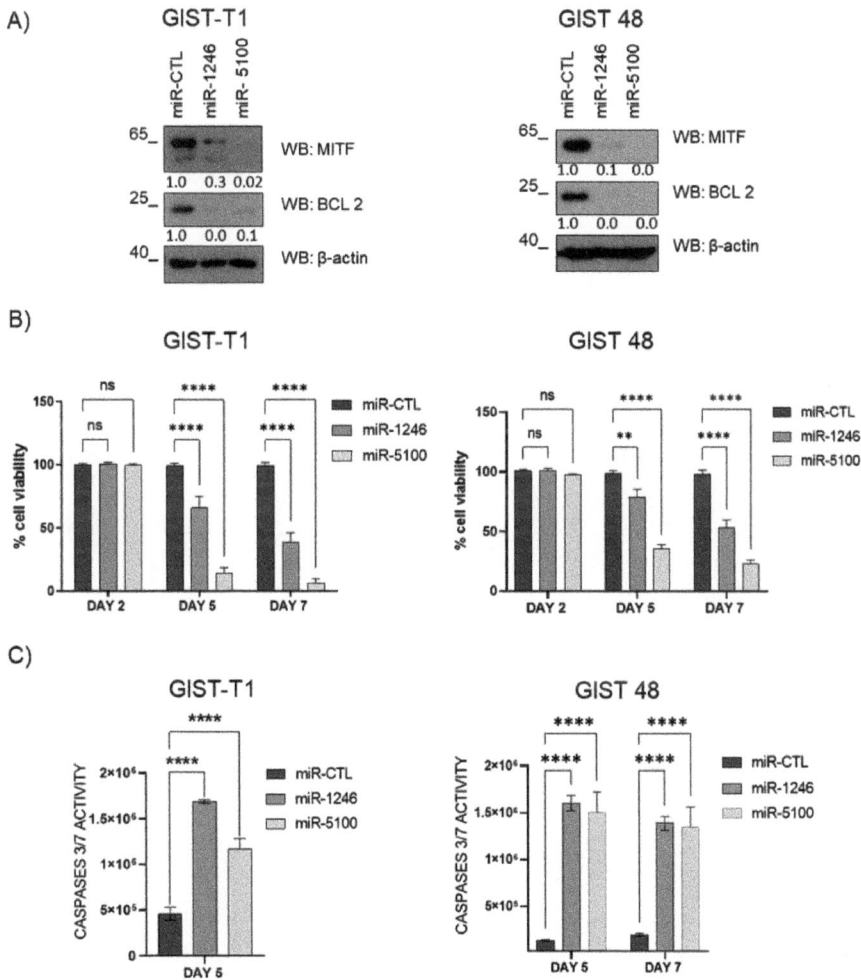

Figure 5. MiR-1246 and miR-5100 induce apoptosis in GIST cells. (**A**) Western blot was carried out on the 5th day after lentiviral transduction in GIST. MITF and BCL2 levels were assessed; β-actin was used as load control. (**B**) Viability was evaluated by crystal violet on the 2nd, 5th, and 7th days after lentiviral transduction. Statistical significance (** $p < 0.01$, **** $p < 0.0001$; one-way-ANOVA with Bonferroni's post-hoc test) GIST-T1 n = 4; GIST 48 n = 3. (**C**) Caspase 3/7 activity was measured on the 5th day on GIST-T1; the 5th and 7th day on GIST 48 post lentiviral transduction (**** $p < 0.0001$; Unpaired *t*-test; mean ± SEM) n = 3. GFPmiR-CTL was used as a control.

3.6. MiR-1246 and miR-5100 Affect Cell Cycle Progression

MITF regulates CDK2 in melanoma, which is critical for tumor cell growth [7,28]. We further analyzed whether CDK2 was altered after miRNA overexpression. MITF reduction was accompanied by decreased CDK2 levels in GIST-T1 and GIST 48 (Figure 6A,B and Supplementary Material S1). The overexpression of both miRNAs had different consequences in the cell cycle in both cell lines. GIST-T1 (Figure 6C) overexpression induced a substantial increase in the G2 phase, while in GIST 48 (Figure 6D), there is an accumulation in the S phase. Altogether, these results indicate that these miRNAs may regulate MITF-dependent targets and cell cycle progression.

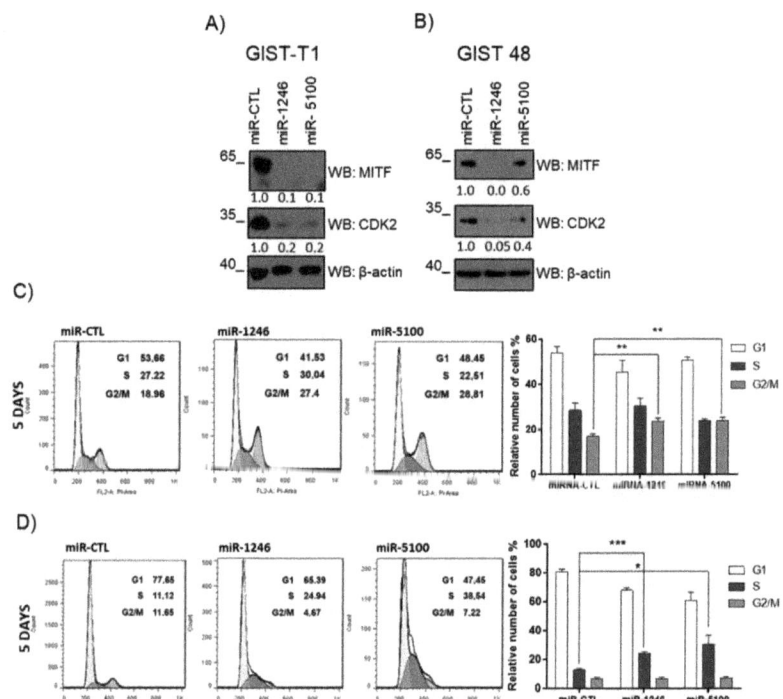

Figure 6. MiR-1246 and miR-5100 arrest cell cycle in GISTs cells. Western blots were performed on GIST cells on the 5th days after lentiviral miRNAs transduction; lysates were analyzed to determine MITF and CDK2 levels in (**A**) GIST-T1 and (**B**) GIST 48. Cell cycle assay was performed by FACS, and miRNA–CTL was used as a control. Results were analyzed by Dean/Jett/Fox model Flow jo 7.0 software (**C**) GIST-T1 n = 4, (**D**) GIST 48 n = 3 (* $p < 0.05$, ** $p < 0.01$, *** $p < 0.001$; Unpaired *t*-test; mean ± SEM).

3.7. MITF Overexpression Significantly Restores Cell Proliferation

Next, we assessed the specificity of the effect of MITF on the proliferation of miRNA-treated GIST cells. For that purpose, after three days of miRNA transduction (when cells were still viable), cells were transfected with MITF-GFP or GFP. Seven days after miRNA transduction, MITF levels and cell proliferation were assessed. Our data show that MITF reconstitution is detected by western blot (Figure 7A and Supplementary Material S1) and significantly increases cell proliferation (Figure 7B).

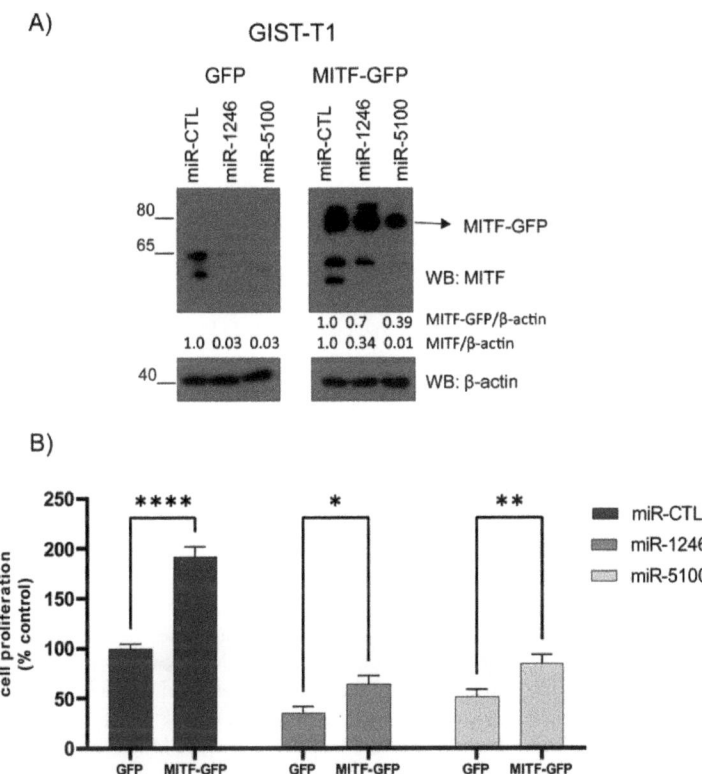

Figure 7. Overexpression of MITF reverses the phenotype produced by miRNAs in GIST-T1. Cells were transiently transduced with GFP or MITF-GFP plasmids on the 3rd day after lentiviral miRNA transduction. (**A**) Western blot shows MITF-GFP and MITF (endogenous) levels in GFP or MITF-GFP overexpressed cells. β-actin was used as a control. (**B**) Cell proliferation was measured by WST-1 on the 7th day after lentiviral miRNAs transduction. (* $p < 0.05$, ** $p < 0.01$, **** $p < 0.0001$. One-way ANOVA with Bonferroni's post-hoc test) n = 3.

3.8. LNA Treatment (Anti-miR-1246 Anti-miR-5100) Was Not Adequate to Revert the Apoptotic Phenotype in GIST-T1 SH3BP2 Silenced Cell

To determine whether miR-1246 and miR-5100 are the main ones responsible for the SH3BP2 silencing apoptotic phenotype, we analyzed the effect of LNA (Locked nucleic acid) miRNA inhibitors treatment on SH3BP2 silenced cells. LNA miRNA inhibitors are antisense oligonucleotides with perfect sequences complementary to their target miRNA that prevent miRNA hybridization with its regular cellular interaction partners. LNAs are taken up naturally by cells by a process known as gymnosis. We checked if LNA treatment reverted the apoptotic phenotype of SH3BP2 silenced cells. For that purpose, GIST-T1 cells were treated with LNA against miR-1246 and miR-5100, and afterward, cells were transduced with lentiviral particles shRNA-SH3BP2. Effective gymnosis was measured by FAN fluorescence microscopy each 24 h in treated cells. Our data show that LNAs treatment cannot block the apoptotic phenotype promoted by SH3BP2 silencing (Supplementary Figure S4). These results suggest that SH3BP2 action on apoptotic phenotype goes beyond miR-1246 and miR-5100.

4. Discussion

GISTs can be successfully treated with imatinib or other TKIs [1,29]. However, the necessity for new therapeutical approaches arose due to clinical resistance. We previously reported that silencing of SH3BP2 leads to a reduction of *KIT* expression at both mRNA and protein levels, as well as MITF at the protein level, resulting in a decrease in GIST tumor growth in vitro and in vivo [5]. In the same study, overexpression of MITF significantly reverses the apoptotic phenotype produced by SH3BP2 silencing, suggesting the involvement of this transcription factor in the regulatory mechanism in which SH3BP2 levels are critical. SH3BP2 silencing did not alter *MITF* mRNA levels but protein levels, suggesting a post-transcriptional mechanism. A miRNA microarray was performed in SH3BP2-silenced GIST882 and GIST48 cell lines (imatinib-sensitive and resistant cells) compared to non-silenced cells to get insights into the KIT-SH3BP2-MITF pathway. This microarray showed a different miRNA pattern when SH3BP2 was silenced. In parallel, we found that SH3BP2 silencing also targets ETV1, a master of ICC-transcription factor whose regulation is dependent on KIT signaling and is directly involved in the tumorigenic phenotype [14,19]. In this study, from the top miRNAs that were up-regulated, we focused on those that putatively target MITF and ETV1. After database analysis and cell validation, the miRNAs: miR-1246 and miR-5100 were selected for further studies. Overexpression assays showed that these miRNAs targeted MITF and ETV1 in GIST48 and GIST-T1. Consequently, the decrease in the levels of these transcription factors leads to a reduction in cell survival.

In this context, miR-1246 has been described as a tumor suppressor miRNA in prostate cancer, as authors showed that miR-1246 overexpression led to the inhibition of xenograft tumor growth over time [30]. They propose the exosomal-mediated release of miR-1246 to serum from tumor cells to evade its tumor suppressor role. Moreover, they suggest exosomal miR-1246 as a good biomarker to discern between benign of aggressive prostate cancer. Interestingly, exosomal miR-1246 has been proposed as a biomarker in gastric cancer (GC), and bioinformatics analysis revealed it as a tumor suppressor in GC [31]. Moreover, miR 1246, which can be induced by tumor suppressor p53, has been described as a tumor suppressor due to its capacity to reduce DYRK1A (a Down syndrome-associate kinase) levels, leading to the nuclear retention of NFATc1 and the induction of apoptosis [32]. Additionally, miR-1246 was downregulated in thyroid cancer, and the overexpression of miR-1246 affects PI3K/AKT pathway by regulating phosphoinositide 3-kinase adapter protein1 (PIK3AP1), resulting in less cell proliferation, diminished migration, and increasing apoptosis [33]. Furthermore, miR-1246 mediates LPS-induced pulmonary endothelial cell apoptosis in vitro and acute lung injury (ALI) in mouse models, which are at least partly attributed to the suppression of angiotensin-converting enzyme 2 (ACE2) [27]. In addition, miRNA-1246 mediates ALI-induced lung inflammation and apoptosis via the NF-κB activation and Wnt/β-catenin suppression [34]. Additionally, miRNA-1246 attenuates renal cell carcinoma's proliferative and migratory abilities by downregulating CXCR4 [35]. Nonetheless, the oncogenic role of miR-1246 has been reported in melanoma by conferring resistance to BRAF inhibitors [36] or enhancing migration and invasion through the adhesion molecule CADM1 in hepatocellular cancer [37].

Regarding miR-5100 activity as a tumor suppressor, our results follow Chijiiwa et al. [38]. The authors show that miR-5100 decreases the aggressiveness of the pancreatic cancer tumor models through the inhibition of PODXL, which promotes anti-adhesive and migratory characteristics of various cancers, and high levels of PODXL correlates with poor prognosis in many of them. Moreover, miR-5100 can increase the apoptosis level of gastric cancer cells and inhibit autophagy by targeting CAAP1 (conserved anti-apoptotic protein 1 or caspase activity and apoptosis inhibitor 1) [26].

However, miR-1246 and miR-5100 have been reported as oncogenic miRNAs in lung cancer [39]. One explanation for these contradictory results could be that miRNAs may vary their affinity to target mRNA depending on the cell lines, the pool of miRNAs that they could be cooperating, and the secondary structures in the 3′UTR of the target mRNA, which can affect the binding of a miRNA [40]. In conclusion, many factors could interfere with the miRNA functional effect in other cell lines.

The proapoptotic role of miR-1246 and miR-5100 in GIST cell lines could result from their ability to affect the cell cycle and regulate cell apoptosis. These actions can be related to a MITF reduction since BCL2 and CDK2 are MITF-dependent targets [7,41–43]. BCL2 is found in most GIST patients [19] and correlates with a poor prognosis before imatinib treatment [44]. These miRNAs also induce cell cycle arrest in a cell line-dependent manner. CDK2 has been reported to regulate both G1/S and G2/M transitions. [45]. As previously noted, high double-negative CDK2-expressing cells were arrested in the mid-S phase. In contrast, low double negative CDK2 expressing cells progressed through early and mid-S phases but were still arrested in the late S/G2 phase [45], suggesting that the active CDK2 can be critical in the different phases. Recent research has shown that CDK2 deficiency slows colorectal cancer's S/G2 progression [46]. It would deserve further consideration to analyze the role and regulation of CDK2 in the different GIST cell lines.

The blockage of miR-5100 and 1246 using LNA did not reduce apoptotic effects due to SH3BP2 silencing, indicating that other miRNAs contribute to this phenotype. However, overexpression of MITF significantly restores cell survival after miR-5100 and 1246 transduction, suggesting that MITF is a crucial target for cell viability.

The role of MITF is well-known in melanoma [47], and recent studies suggest that MITF overexpression in kidney angiomyolipoma cells [48] and clear cell renal cell carcinoma (ccRCC) improve cell growth, proliferation, and invasion in vitro and in vivo [49]. Lately, we have described that the silencing of MITF results in decreased gastrointestinal stromal tumor cell viability in vitro and tumor growth in vivo [11].

5. Conclusions

Our results highlight the KIT-SH3BP2-MITF/ETV1 pathway for GIST cell survival and proliferation. Targeting ETV1 and MITF together will help break the positive feedback loop and indirectly target KIT independently of the mutations in the tyrosine kinase receptor.

Supplementary Materials: The following supporting information can be downloaded at: https://www.mdpi.com/article/10.3390/cancers14246198/s1, Figure S1: MiRNAs validation by western blot; Figure S2: MiRNA sequences target on 3′UTR of MITF-A (A) and ETV1 (B) genes; Figure S3: MiR-GFP expression in GIST cells; Figure S4. Anti-miRNA treatment 1246 and 5100 in GIST T1; Table S1: Prediction of miRNAs that target MITF by different databases; Table S2: Prediction of miRNAs that target ETV1 by different databases; Material S1: Original Western blot images.

Author Contributions: E.P.-P., E.S.-C. and M.M. conceived the experiments and wrote the manuscript. E.P.-P., E.S.-C. and C.M. performed the experiments, A.N.-F. and M.G. provided technical support, and M.M. secured funding. All authors have read and agreed to the published version of the manuscript.

Funding: This study has been funded by grants: RTI2018-096915-B100 from the Spanish Ministry of Science, Innovation and Universities and European Regional Development Fund/European Social Fund "Investing in your future", and grant PID2021-122898OB-100 funded by MCIN/AEI/10.13039/501100011033 and, by "ERDF A way of making Europe".

Institutional Review Board Statement: Not applicable.

Informed Consent Statement: Not applicable.

Data Availability Statement: The datasets used and analyzed during the current study are available in the article and supplementary files or from the corresponding author at reasonable request. GSE213777: Accession number for SH3BP2-silenced GIST microarray.

Acknowledgments: The authors are indebted to the Cytomics core facility of the Institut d'Investigacions Biomèdiques August Pi i Sunyer (IDIBAPS) for their technical support.

Conflicts of Interest: The authors have declared that no conflict of interest exists.

References

1. Kelly, C.M.; Gutierrez Sainz, L.; Chi, P. The management of metastatic GIST: Current standard and investigational therapeutics. *J. Hematol. Oncol.* **2021**, *14*, 2. [CrossRef] [PubMed]
2. Beham, A.W.; Schaefer, I.-M.; Schüler, P.; Cameron, S.; Michael Ghadimi, B. Gastrointestinal stromal tumors. *Int. J. Color. Dis.* **2012**, *27*, 689–700. [CrossRef] [PubMed]
3. Wada, R.; Arai, H.; Kure, S.; Peng, W.X.; Naito, Z. "Wild type" GIST: Clinicopathological features and clinical practice. *Pathol. Int.* **2016**, *66*, 431–437. [CrossRef] [PubMed]
4. Ainsua-Enrich, E.; Serrano-Candelas, E.; Álvarez-Errico, D.; Picado, C.; Sayós, J.; Rivera, J.; Martín, M. The Adaptor 3BP2 Is Required for KIT Receptor Expression and Human Mast Cell Survival. *J. Immunol.* **2015**, *194*, 4309–4318. [CrossRef] [PubMed]
5. Serrano-Candelas, E.; Ainsua-Enrich, E.; Navinés-Ferrer, A.; Rodrigues, P.; García-Valverde, A.; Bazzocco, S.; Macaya, I.; Arribas, J.; Serrano, C.; Sayós, J.; et al. Silencing of adaptor protein SH3BP2 reduces KIT/PDGFRA receptors expression and impairs gastrointestinal stromal tumors growth. *Mol. Oncol.* **2018**, *12*, 1383–1397. [CrossRef] [PubMed]
6. Hoek, K.S.; Schlegel, N.C.; Eichhoff, O.M.; Widmer, D.S.; Praetorius, C.; Einarsson, S.O.; Valgeirsdottir, S.; Bergsteinsdottir, K.; Schepsky, A.; Dummer, R.; et al. Novel MITF targets identified using a two-step DNA microarray strategy. *Pigment Cell Melanoma Res.* **2008**, *21*, 665–676. [CrossRef]
7. Du, J.; Widlund, H.R.; Horstmann, M.A.; Ramaswamy, S.; Ross, K.; Huber, W.E.; Nishimura, E.K.; Golub, T.R.; Fisher, D.E. Critical role of CDK2 for melanoma growth linked to its melanocyte-specific transcriptional regulation by MITF. *Cancer Cell* **2004**, *6*, 565–576. [CrossRef]
8. Levy, C.; Khaled, M.; Fisher, D.E. MITF: Master regulator of melanocyte development and melanoma oncogene. *Trends Mol. Med.* **2006**, *12*, 406–414. [CrossRef]
9. Morii, E.; Ogihara, H.; Kim, D.-K.K.; Ito, A.; Oboki, K.; Lee, Y.-M.M.; Jippo, T.; Nomura, S.; Maeyama, K.; Lamoreux, M.L.; et al. Importance of leucine zipper domain of mitranscription factor (MITF) for differentiation of mast cells demonstrated using mice/mice mutant mice of which MITF lacks the zipper domain. *Blood* **2001**, *97*, 2038–2044. [CrossRef]
10. Tsujimura, T.; Morii, E.; Nozaki, M.; Hashimoto, K.; Moriyama, Y.; Takebayashi, K.; Kondo, T.; Kanakura, Y.; Kitamura, Y. Involvement of Transcription Factor Encoded by the mi Locus in the Expression of c-kit Receptor Tyrosine Kinase in Cultured Mast Cells of Mice. *Blood* **1996**, *88*, 1225–1233. [CrossRef]
11. Proaño-Pérez, E.; Serrano-Candelas, E.; García-Valverde, A.; Rosell, J.; Gómez-Peregrina, D.; Navinés-Ferrer, A.; Guerrero, M.; Serrano, C.; Martín, M.; Elizabeth, P.P.; et al. The microphthalmia-associated transcription factor is involved in gastrointestinal stromal tumor growth. *Cancer Gene Ther.* **2022**, *115*–117. [CrossRef]
12. Lee, Y.-N.; Brandal, S.; Noel, P.; Wentzel, E.; Mendell, J.T.; McDevitt, M.A.; Kapur, R.; Carter, M.; Metcalfe, D.D.; Takemoto, C.M. KIT signaling regulates MITF expression through miRNAs in normal and malignant mast cell proliferation. *Blood* **2011**, *117*, 3629–3640. [CrossRef] [PubMed]
13. Ran, L.; Sirota, I.; Cao, Z.; Murphy, D.; Chen, Y.; Shukla, S.; Xie, Y.; Kaufmann, M.C.; Gao, D.; Zhu, S.; et al. Combined Inhibition of MAP Kinase and KIT Signaling Synergistically Destabilizes ETV1 and Suppresses GIST Tumor Growth. *Cancer Discov.* **2015**, *5*, 304–315. [CrossRef] [PubMed]
14. Chi, P.; Chen, Y.; Zhang, L.; Guo, X.; Wongvipat, J.; Shamu, T.; Fletcher, J.A.; Dewell, S.; Maki, R.G.; Zheng, D.; et al. ETV1 is a lineage survival factor that cooperates with KIT in gastrointestinal stromal tumours. *Nature* **2010**, *467*, 849–853. [CrossRef] [PubMed]
15. Ritchie, M.E.; Silver, J.; Oshlack, A.; Holmes, M.; Diyagama, D.; Holloway, A.; Smyth, G.K. A comparison of background correction methods for two-colour microarrays. *Bioinformatics* **2007**, *23*, 2700–2707. [CrossRef] [PubMed]
16. Feoktistova, M.; Geserick, P.; Leverkus, M. Crystal Violet Assay for Determining Viability of Cultured Cells. *Cold Spring Harb. Protoc.* **2016**, *2016*, pdb.prot087379. [CrossRef] [PubMed]
17. Álvarez-Errico, D.; Oliver-Vila, I.; Ainsua-Enrich, E.; Gilfillan, A.M.; Picado, C.; Sayós, J.; Martín, M. CD84 Negatively Regulates IgE High-Affinity Receptor Signaling in Human Mast Cells. *J. Immunol.* **2011**, *187*, 5577–5586. [CrossRef]
18. Pozarowski, P.; Darzynkiewicz, Z. Analysis of Cell Cycle by Flow Cytometry. In *Checkpoint Controls and Cancer*; Humana Press: Totowa, NJ, USA, 2004; Volume 281, pp. 301–312.
19. Zhang, Y.; Gu, M.-L.; Zhou, X.-X.; Ma, H.; Yao, H.-P.; Ji, F. Altered expression of ETV1 and its contribution to tumorigenic phenotypes in gastrointestinal stromal tumors. *Oncol. Rep.* **2014**, *32*, 927–934. [CrossRef]
20. Agarwal, V.; Bell, G.W.; Nam, J.-W.; Bartel, D.P. Predicting effective microRNA target sites in mammalian mRNAs. *eLife* **2015**, *4*, e05005. [CrossRef]
21. Hsu, J.; Chiu, C.-M.; Hsu, S.-D.; Huang, W.-Y.; Chien, C.-H.; Lee, T.-Y.; Huang, H.-D. miRTar: An integrated system for identifying miRNA-target interactions in human. *BMC Bioinform.* **2011**, *12*, 300. [CrossRef]
22. Dweep, H.; Sticht, C.; Pandey, P.; Gretz, N. miRWalk—Database: Prediction of possible miRNA binding sites by "walking" the genes of three genomes. *J. Biomed. Inform.* **2011**, *44*, 839–847. [CrossRef] [PubMed]
23. Paraskevopoulou, M.D.; Georgakilas, G.; Kostoulas, N.; Vlachos, I.S.; Vergoulis, T.; Reczko, M.; Filippidis, C.; Dalamagas, T.; Hatzigeorgiou, A.G. DIANA-microT web server v5.0: Service integration into miRNA functional analysis workflows. *Nucleic Acids Res.* **2013**, *41*, W169–W173. [CrossRef] [PubMed]
24. Tokar, T.; Pastrello, C.; Rossos, A.E.M.; Abovsky, M.; Hauschild, A.-C.; Tsay, M.; Lu, R.; Jurisica, I. mirDIP 4.1—Integrative database of human microRNA target predictions. *Nucleic Acids Res.* **2018**, *46*, D360–D370. [CrossRef] [PubMed]

25. Rehmsmeier, M.; Steffen, P.; Hochsmann, M.; Giegerich, R. Fast and effective prediction of microRNA/target duplexes. *RNA* **2004**, *10*, 1507–1517. [CrossRef]
26. Zhang, H.M.; Li, H.; Wang, G.X.; Wang, J.; Xiang, Y.; Huang, Y.; Shen, C.; Dai, Z.T.; Li, J.P.; Zhang, T.C.; et al. MKL1/miR-5100/CAAP1 loop regulates autophagy and apoptosis in gastric cancer cells. *Neoplasia* **2020**, *22*, 220–230. [CrossRef]
27. Fang, Y.; Gao, F.; Hao, J.; Liu, Z. microRNA-1246 mediates lipopolysaccharide-induced pulmonary endothelial cell apoptosis and acute lung injury by targeting angiotensin-converting enzyme 2. *Am. J. Transl. Res.* **2017**, *9*, 1287–1296.
28. Kawakami, A.; Fisher, D.E. The master role of microphthalmia-associated transcription factor in melanocyte and melanoma biology. *Lab. Investig.* **2017**, *97*, 649–656. [CrossRef]
29. Poveda, A.; García del Muro, X.; López-Guerrero, J.A.; Cubedo, R.; Martínez, V.; Romero, I.; Serrano, C.; Valverde, C.; Martín-Broto, J. GEIS guidelines for gastrointestinal sarcomas (GIST). *Cancer Treat. Rev.* **2017**, *55*, 107–119. [CrossRef]
30. Bhagirath, D.; Yang, T.L.; Bucay, N.; Sekhon, K.; Majid, S.; Shahryari, V.; Dahiya, R.; Tanaka, Y.; Saini, S. microRNA-1246 is an exosomal biomarker for aggressive prostate cancer. *Cancer Res.* **2018**, *78*, 1833–1844. [CrossRef]
31. Shi, Y.; Wang, Z.; Zhu, X.; Chen, L.; Ma, Y.; Wang, J.; Yang, X.; Liu, Z. Exosomal miR-1246 in serum as a potential biomarker for early diagnosis of gastric cancer. *Int. J. Clin. Oncol.* **2019**, *25*, 89–99. [CrossRef]
32. Zhang, Y.; Liao, J.M.; Zeng, S.X.; Lu, H. p53 downregulates Down syndrome-associated DYRK1A through miR-1246. *EMBO Rep.* **2011**, *12*, 811–817. [CrossRef] [PubMed]
33. Li, J.; Zhang, Z.; Hu, J.; Wan, X.; Huang, W.; Zhang, H.; Jiang, N. MiR-1246 regulates the PI3K/AKT signaling pathway by targeting PIK3AP1 and inhibits thyroid cancer cell proliferation and tumor growth. *Mol. Cell. Biochem.* **2022**, *477*, 649–661. [CrossRef] [PubMed]
34. Suo, T.; Chen, G.; Huang, Y.; Zhao, K.; Wang, T.; Hu, K. miRNA-1246 suppresses acute lung injury-induced inflammation and apoptosis via the NF-κB and Wnt/β-catenin signal pathways. *Biomed. Pharmacother.* **2018**, *108*, 783–791. [CrossRef] [PubMed]
35. Liu, H.T.; Fan, W.X. MiRNA-1246 suppresses the proliferation and migration of renal cell carcinoma through targeting CXCR4. *Eur. Rev. Med. Pharmacol. Sci.* **2020**, *24*, 5979–5987. [CrossRef]
36. Kim, J.; Ahn, J.; Lee, M. Upregulation of MicroRNA-1246 Is Associated with BRAF Inhibitor Resistance in Melanoma Cells with Mutant BRAF. *Cancer Res. Treat.* **2017**, *49*, 947–959. [CrossRef]
37. Sun, Z.; Meng, C.; Wang, S.; Zhou, N.; Guan, M.; Bai, C.; Lu, S.; Han, Q.; Zhao, R.C. MicroRNA-1246 enhances migration and invasion through CADM1 in hepatocellular carcinoma. *BMC Cancer* **2014**, *14*, 616. [CrossRef]
38. Chijiiwa, Y.; Moriyama, T.; Ohuchida, K.; Nabae, T.; Ohtsuka, T.; Miyasaka, Y.; Fujita, H.; Maeyama, R.; Manabe, T.; Abe, A.; et al. Overexpression of microRNA-5100 decreases the aggressive phenotype of pancreatic cancer cells by targeting PODXL. *Int. J. Oncol.* **2016**, *48*, 1688–1700. [CrossRef]
39. Huang, H.; Jiang, Y.; Wang, Y.; Chen, T.; Yang, L.; He, H.; Lin, Z.; Liu, T.; Yang, T.; Kamp, D.W.; et al. miR-5100 promotes tumor growth in lung cancer by targeting Rab6. *Cancer Lett.* **2015**, *362*, 15–24. [CrossRef]
40. Long, D.; Lee, R.; Williams, P.; Chan, C.Y.; Ambros, V.; Ding, Y. Potent effect of target structure on microRNA function. *Nat. Struct. Mol. Biol.* **2007**, *14*, 287–294. [CrossRef]
41. McGill, G.G.; Horstmann, M.; Widlund, H.R.; Du, J.; Motyckova, G.; Nishimura, E.K.; Lin, Y.L.; Ramaswamy, S.; Avery, W.; Ding, H.F.; et al. Bcl2 regulation by the melanocyte master regulator Mitf modulates lineage survival and melanoma cell viability. *Cell* **2002**, *109*, 707–718. [CrossRef]
42. Haq, R.; Yokoyama, S.; Hawryluk, E.B.; Jönsson, G.B.; Frederick, D.T.; McHenry, K.; Porter, D.; Tran, T.N.; Love, K.T.; Langer, R.; et al. BCL2A1 is a lineage-specific anti-apoptotic melanoma oncogene that confers resistance to BRAF inhibition. *Proc. Natl. Acad. Sci. USA* **2013**, *110*, 4321–4326. [CrossRef] [PubMed]
43. Hartman, M.L.; Czyz, M. Pro-Survival Role of MITF in Melanoma. *J. Investig. Dermatol.* **2015**, *135*, 352–358. [CrossRef] [PubMed]
44. Changchien, C.R.; Wu, M.-C.; Tasi, W.-S.; Tang, R.; Chiang, J.-M.; Chen, J.-S.; Huang, S.-F.; Wang, J.-Y.; Yeh, C.-Y. Evaluation of prognosis for malignant rectal gastrointestinal stromal tumor by clinical parameters and immunohistochemical staining. *Dis. Colon Rectum* **2004**, *47*, 1922–1929. [CrossRef] [PubMed]
45. Hu, B.; Mitra, J.; van den Heuvel, S.; Enders, G.H. S and G_2 Phase Roles for Cdk2 Revealed by Inducible Expression of a Dominant-Negative Mutant in Human Cells. *Mol. Cell. Biol.* **2001**, *21*, 2755–2766. [CrossRef]
46. Bačević, K.; Lossaint, G.; Achour, T.N.; Georget, V.; Fisher, D.; Dulić, V. Cdk2 strengthens the intra-S checkpoint and counteracts cell cycle exit induced by DNA damage. *Sci. Rep.* **2017**, *7*, 13429. [CrossRef]
47. Goding, C.R.; Arnheiter, H. MITF—The first 25 years. *Genes Dev.* **2019**, *33*, 983–1007. [CrossRef]
48. Zarei, M.; Giannikou, K.; Du, H.; Liu, H.-J.; Duarte, M.; Johnson, S.; Nassar, A.H.; Widlund, H.R.; Henske, E.P.; Long, H.W.; et al. MITF is a driver oncogene and potential therapeutic target in kidney angiomyolipoma tumors through transcriptional regulation of CYR61. *Oncogene* **2021**, *40*, 112–126. [CrossRef]
49. Kim, N.; Kim, S.; Lee, M.W.; Jeon, H.J.; Ryu, H.; Kim, J.M.; Lee, H.J. MITF Promotes Cell Growth, Migration and Invasion in Clear Cell Renal Cell Carcinoma by Activating the RhoA/YAP Signal Pathway. *Cancers* **2021**, *13*, 2920. [CrossRef]

Article

Transcriptomic Changes Associated with *ERBB2* Overexpression in Colorectal Cancer Implicate a Potential Role of the Wnt Signaling Pathway in Tumorigenesis

Eman A. Abdul Razzaq [1,2], Khuloud Bajbouj [2,†], Amal Bouzid [1,†], Noura Alkhayyal [3], Rifat Hamoudi [1,2,4,*,‡] and Riyad Bendardaf [2,3,‡]

1. Sharjah Institute for Medical Research, University of Sharjah, Sharjah P.O. Box 27272, United Arab Emirates
2. Department of Clinical Sciences, College of Medicine, University of Sharjah, Sharjah P.O. Box 27272, United Arab Emirates
3. Oncology Unit, University Hospital Sharjah, Sharjah P.O. Box 72772, United Arab Emirates
4. Division of Surgery and Interventional Science, University College London, London WC1E 6BT, UK
* Correspondence: rhamoudi@sharjah.ac.ae
† These authors contributed equally to the work.
‡ These authors also contributed equally to the work.

Simple Summary: The present study identified cellular pathways and genes co-expressed with HER2 in colorectal cancer using whole transcriptomic analysis on colorectal cancer patients and cell lines. A comparison of the genes and pathways between patients and cell lines identified the Wnt signaling pathway and the homeobox gene NKX2-5 to be significant. This study sheds new light on the role of HER2 in colorectal cancer pathogenesis.

Abstract: Colorectal cancer (CRC) remains the third most common cause of cancer mortality worldwide. Precision medicine using OMICs guided by transcriptomic profiling has improved disease diagnosis and prognosis by identifying many CRC targets. One such target that has been actively pursued is an erbb2 receptor tyrosine kinase 2 (*ERBB2*) (Human Epidermal Growth Factor Receptor 2 (HER2)), which is overexpressed in around 3–5% of patients with CRC worldwide. Despite targeted therapies against HER2 showing significant improvement in disease outcomes in multiple clinical trials, to date, no HER2-based treatment has been clinically approved for CRC. In this study we performed whole transcriptome ribonucleic acid (RNA) sequencing on 11 HER2+ and 3 HER2− CRC patients with advanced stages II, III and IV of the disease. In addition, transcriptomic profiling was carried out on CRC cell lines (HCT116 and HT29) and normal colon cell lines (CCD841 and CCD33), ectopically overexpressing *ERBB2*. Our analysis revealed transcriptomic changes involving many genes in both CRC cell lines overexpressing *ERBB2* and in HER2+ patients, compared to normal colon cell lines and HER2− patients, respectively. Gene Set Enrichment Analysis indicated a role for HER2 in regulating CRC pathogenesis, with Wnt/β-catenin signaling being mediated via a HER2-dependent regulatory pathway impacting expression of the homeobox gene NK2 homeobox 5 (NKX2-5). Results from this study thus identified putative targets that are co-expressed with HER2 in CRC warranting further investigation into their role in CRC pathogenesis.

Keywords: *ERBB2*; HER2; colorectal cancer; RNA-seq; whole transcriptomic analysis; NKX2-5; Wnt signaling

1. Introduction

Despite improvements in early detection and treatment methods during the last two decades, colorectal cancer (CRC) remains the 3rd most common cause of cancer-related death worldwide with approximately 150,000 new CRC cases diagnosed in the United States annually [1]. Among them, approximately 20% of patients will have distant metastasis, and around 30% of patients with stage II and III disease will develop metastasis [2]. The five-year survival rate of CRC patients with distant metastasis is less than 15% [2]. The incidence of CRC in men and women under the age of 50 has steadily increased in the past two decades [1,2]. For patients with metastatic CRC (mCRC), chemotherapy remains the mainstay of treatment, but eventually, all patients develop resistance to therapy and experience treatment failure due to the intra-tumoral heterogeneity of CRC [3].

Precision medicine guided by tumor genomic profiling has transformed the cancer diagnosis, prognosis, and treatment paradigm over the past two decades. However, recent estimates suggest that fewer than 10% of cancer patients benefit from this approach [4,5]. The primary issues facing genomic profiling include firstly, actionable genomic alterations are not detected in a vast majority of cases [6], and even when they are, secondly, a significant proportion of patients fail to experience an antitumor response to the indicated targeted therapy [7].

One such target is the amplification of *ERBB2* (HER2), which occurs in approximately 3% of patients with metastatic CRC (mCRC) and 5% of patients with wild-type *NRAS* and *KRAS* tumors [8,9]. Several *ERBB2*-targeted therapies are either in different phases of clinical trials or approved for use in patients with *ERBB2*-positive breast and gastric and gastroesophageal tumors [8]. However, despite recommendations of the National Comprehensive Cancer Network guidelines and clinical evidence from phase II trials that anti-*ERBB2* therapies improve disease outcomes in *ERBB2*-positive mCRC patients, no *ERBB2*-directed approved therapies for patients with CRC are currently approved for clinical use [8–10].

The role of HER2 in carcinogenesis is most well-characterized in breast cancer [11,12]. HER2+ breast cancer is a historically aggressive subtype of breast cancer with a five-year survival rate of 30% [11,13,14]. The discovery that amplification or overexpression of *ERBB2* was associated with extremely poor survival in breast cancer led to efforts that resulted in the development of a monoclonal antibody (mAb) to HER2, trastuzumab [14,15]. However, whether *ERBB2* overexpression-mediated carcinogenesis follows similar mechanisms in breast and colon tissue is unknown.

Thus far, the majority of the studies related to *ERBB2* in cancers have focused on identifying the landscape of genomic amplification in *ERBB2* and defining therapeutic regimens to target these amplifications [8]. Our earlier study has shown that *ERBB2* mRNA and protein overexpression correlates with more aggressive colorectal cancer in the North African population [16]. However, the effect of the overexpression of *ERBB2* on the global transcriptomic profiles within CRC patients is not known. Therefore, the objective of the current study is to characterize whole transcriptomic changes associated with *ERBB2* overexpression in CRC cell lines and patient samples with a view to gain a deeper understanding of the role of *ERBB2* in CRC pathogenesis.

2. Materials and Methods

2.1. Patients and Tissue Specimens

Ethical approval for the study was provided by the Research and Ethics Committee (REC) of University Hospital Sharjah (UHS-HERC-055-25022019). All methods were performed in accordance with the relevant guidelines based on the Declaration of Helsinki and the Belmont Report. We obtained written informed consent from all study participants. This is a retrospective study of 14 patients with primary CRC. Patients with secondary cancers were excluded, whilst all primary CRC patients were included, regardless of age, gender, or tumor stage. The initial diagnosis was performed prior to and independently of our study to determine the Tumor, lymph Nodes, and Metastasis (TNM) score. Tissues were sectioned from formalin-fixed, paraffin-embedded (FFPE) biopsies for molecular and immunohistochemical analysis.

2.2. Immunohistochemistry

To begin with, 3 μm sections from the FFPE of 14 CRC patients' biopsies were immunohistochemically stained using the rabbit monoclonal antibody for HER2 (1:4000 dilution; ab214275, Abcam, Waltham, MA, USA) according to the manufacturer's instructions. An experienced pathologist (R.H.) scored the stained slides, following the consensus recommendations for HER2 scoring for CRC [17,18]. Briefly, scoring was performed on a 4-point scale—0, 1+, 2+, 3+ focusing on intensity and extent according to the Allred scoring system [19]. In this study, 0 and 1+ intensity were taken to be negative for HER2 expression and the study focused on the assessment of membranous HER2 expression.

2.3. Cell Culture

The CRC cell lines HT29 and HCT116, and two normal colon cell lines: CCD33 and CCD841, were obtained from Bio Medical Scientific Services (BIOMSS, Al Ain, United Arab Emirates), and cultured in Dulbecco's Modified Eagle's Medium (DMEM) supplemented with 10% fetal bovine serum (Sigma Aldrich, St. Louis, MO, USA), 1% Penicillin/Streptomycin (Sigma) and 20 mM L-Glutamine (Sigma) at 37 °C in 5% CO_2 incubator.

2.4. Transfection

The wild-type *ERBB2* expression construct was a gift from Mien-Chie Hung (Addgene plasmid #16257; https://www.addgene.org/16257, accessed on 15 December 2020) [20]. The cell lines were transfected with 5μg of pcDNA3-*ERBB2* plasmid construct using Lipofectamine 3000 reagent (ThermoFisher Scientific, Cambridge, MA, USA) according to the manufacturer's instructions. Cells transfected with the empty pcDNA3 vector served as the experimental control. The *ERBB2* expression level was checked 24 h post-transfection at the mRNA and protein levels using qRT-PCR and Western blotting, respectively.

2.5. RNA Isolation

RNA extraction was carried out from three sequential (3 μm) sections from the same FFPE block. A needle macrodissection was carried out to enrich the tumor's content. This was carried out by marking the tumor areas on the slides and carefully removing the unmarked non-tumor areas using a sterile needle, following which the marked areas were collected for molecular analysis. RNA was extracted using the RNA RecoverAll kit (ThermoFisher Scientific, Cambridge, MA, USA) according to the manufacturer's instructions. Genomic DNA removal was ensured by treating the RNA with Turbo DNase (ThermoFisher Scientific). For RNA extraction from cell lines, cells were pelleted at 12×10^3 g for 5 min and rinsed thrice with ice-cold 1X sterile PBS. RNA was extracted from the cell pellet as described above using the RNA RecoverAll kit, followed by genomic DNA removal using Turbo DNase. All RNA samples were stored at −800 °C until further use.

2.6. Quantitative Reverse Transcriptase-PCR (qRT-PCR)

qRT-PCR was performed using the Superscript First-strand Synthesis system (ThermoFisher Scientific). Real-time qPCR was performed in triplicates, using SYBR green (Solis BioDyne, Tartu, Estonia), on Quant Studio 3 (Applied Biosystem, Waltham, MA, USA). *ERBB2* and the reference genes (18S ribosomal RNA) were pre-amplified using the following primer sets [18]. *ERBB2*_sense: 5′-ACATGCTCCGCCACCTCTACCA-3′; *ERBB2* Antisense: 5′-GGATCTGCCTCACTTGGTTGTG-3′; 18SrRNA_sense: 5′-TGACTCAACACGGGAAACC-3′; 18SrRNA_antisense: 5′-TCGCTCCACCAACTAAGAAC-3′. A total of 40 PCR cycles were performed consisting of 15 s denaturation at 95 °C and a combined annealing and extension cycle of 10 min at 60 °C. The threshold cycle value (Ct) was normalized against the Ct value of internal control 18 s RNA.

2.7. Cell Lysis and Western Blot

Cell lysis and Western blot were performed as described previously [21]. Anti-HER2 (1:1000; Abcam) was used to probe the blots. All blots were subsequently stripped and re-probed for β-Actin (1:5000; Abcam, Waltham, MA, USA) to confirm equal loading.

2.8. Next-Generation RNA Sequencing

RNA sequencing was carried out on the indicated samples using a targeted AmpliSeq Transcriptome panel on Ion S5 XL System (ThermoFisher Scientific, Cambridge, MA, USA). In brief, ~30 ng of Turbo DNase treated RNA was used for cDNA synthesis using a SuperScript VILO cDNA Synthesis kit (ThermoFisher Scientific) followed by amplification using Ion AmpliSeq gene expression core panel primers. The enzymatic shearing was performed using FuPa reagent to obtain amplicons of ~200 bp and the sheared amplicons were ligated with the adapter and the unique barcodes. The prepared library was purified using Agencourt AMPure XP Beads (Beckman Coulter, Indianapolis, IN, USA) and the purified library was quantified using an Ion Library TaqMan™ Quantitation Kit (Applied Biosystems, Waltham, MA, USA). The libraries were further diluted to 100 pM and pooled equally with four individual samples per pool. The pooled libraries were amplified using emulsion PCR on Ion OneTouch™ 2 instrument (OT2) and the enrichment was performed on Ion OneTouch™ ES following the manufacturer's instructions. Thus, prepared template libraries were then sequenced with Ion S5 XL Semiconductor sequencer (ThermoFisher Scientific, Cambridge, MA, USA) using the Ion 540™ Chip.

2.9. Bioinformatics Analyses

RNA-seq data were analyzed using Ion Torrent Software Suite version 5.4 and the alignment was carried out using the Torrent Mapping Alignment Program (TMAP). TMAP is optimized for aligning the raw sequencing reads against the reference sequence derived from the hg19 (GRCh37) assembly, and the specificity and sensitivity were maintained by implementing a two-stage mapping approach by employing BWA-short, BWA-long, SSAHA [22], Super-maximal Exact Matching [23] and the Smith–Waterman algorithm [24] for optimal mapping. Raw read counts of the targeted genes were performed using Samtools (Samtools view–c–F 4–L bed_file bam_file) and the number of expressed transcripts was confirmed after Fragments Per Kilobase Million (FPKM) normalization. For technical variations, code-set content normalization was performed with the geometric median for all genes. Principal component analysis (PCA) was performed using the indicated samples with R statistical software. Differentially expressed gene (DEG) analysis was performed using R/Bioconductor package DESeq2 with raw read counts from the RNA sequencing data [25,26]. Genes with less than ten normalized read counts were excluded from further analysis. A fold change of 2 was set as the cutoff for differentially expressed gene identification. $p < 0.05$ was considered statistically significant. The DEGs were then subjected to Gene Set Enrichment Analysis (GSEA).

2.10. Analyses of Publicly Available Transcriptomic Data Sets for Breast Cancer

In order to compare the biological pathways and differentially expressed genes between the *ERBB2* over-represented in breast cancer (BC) and CRC, transcriptomic data sets of BC were searched and retrieved from the GEO (Gene Expression Omnibus) database (https://www.ncbi.nlm.nih.gov/geo/, accessed on 3 October 2022). The datasets were searched based on "Breast cancer" and "*ERBB2*" keywords. Then, datasets including BC patients with variable *ERBB2* expression based on the same platform Affymetrix Human Genome U133 Plus 2.0 Array were considered. Sixteen well-matched datasets were available, out of which fourteen were excluded for further analysis. Exclusion criteria were datasets performed *in vitro* cancer cell lines or in vivo study models using non-human species, repeated samples in super-series, and datasets exhibiting poor *ERBB2* expression values uncharacteristic of HER2+/HER2−. Two datasets that met the criteria were selected, including GSE29431 and GSE48391 (Supplementary Figure S1). The raw data and the probe annotation files were downloaded for further analysis.

2.11. Breast Cancer Microarray Data Analysis

A total of 65 BC patients were selected in our analysis including 48 samples with low *ERBB2* and 17 samples with high *ERBB2* expression. The Affymetrix microarray represents more than 54,000 probes where each gene is represented with different probes. The raw data were processed and normalized using in-house R script as previously described [27]. For normalization and adaptive filtering, Affymetrix Microarray Suite 5 (MAS5) and Gene Chip Robust Multiarray Averaging (GCRMA) packages in Bioconductor/R software were applied. Probes with a MAS5 value > 50 and coefficient of variation (CV) 10–100% in GCRMA among all samples of each dataset were identified to get only common variant probes. The filtered probes were then annotated and collapsed into the gene names list based on the maximum expression of probes for each gene. The unchanged probes, positive control probes, and unassigned probes were excluded from the downstream analysis. The mapped gene expression lists were subjected to Gene Set Enrichment Analysis (GSEA) to identify the activated and enriched biological pathways between high and low *ERBB2*-BC patients.

2.12. GSEA

GSEA was carried out separately for all resulting gene sets from the above different transcriptomic analyses including CRC patients, CRC cell lines, normal colon cell lines, and BC patients. First, the absolute GSEA was performed to identify the significantly enriched pathways among sets related to the C2: curated gene sets; C5: ontology gene sets including molecular function (MF) and biological process (BP); C6: oncogenic signature gene sets; and C7: immunologic signature gene sets. The results of the GSEA were ranked and selected according to the $p < 0.05$ as described previously [27,28]. Next, the selected significant pathways were further analyzed to identify the differentially enriched genes and the leading edge genes in each pathway. In order to further reduce the set of resulting genes, a systematic cross-reference of each gene enriched within statistically significant pathways was carried out. Finally, the genes with the highest frequency across the multiple significant pathways enriched between the HER2 positive and HER2 negative samples were identified.

2.13. Statistical Analysis

Functional data are presented as mean ± SD, except where otherwise stated. When two groups were compared, the student's *t*-test was used unless otherwise indicated. $p < 0.05$ was considered as statistically significant.

3. Results

3.1. Overexpression of ERBB2 Induces Distinct Transcriptional Profiles in the CRC Cell Lines HT29 and HCT116

We wanted to determine if ectopic overexpression of *ERBB2* in CRC cell lines induces genome-wide transcriptional changes. Hence, we screened different CRC cell lines to identify cell lines with low endogenous *ERBB2* expression. We initially determined the steady-state expression of *ERBB2* mRNA in the normal colon cell lines, CCD33 and CCD841, as well as the CRC adenoma cell lines HCT116 and HT29 using qRT-PCR. Compared to CCD33 and CCD841 cells, *ERBB2* expression was 6.28 ± 0.003 folds and 7.56 ± 1.54 folds less in HCT116 and 3.71 ± 0.01 4.43 ± 0.02 folds less in HT29, respectively (Figure 1A).

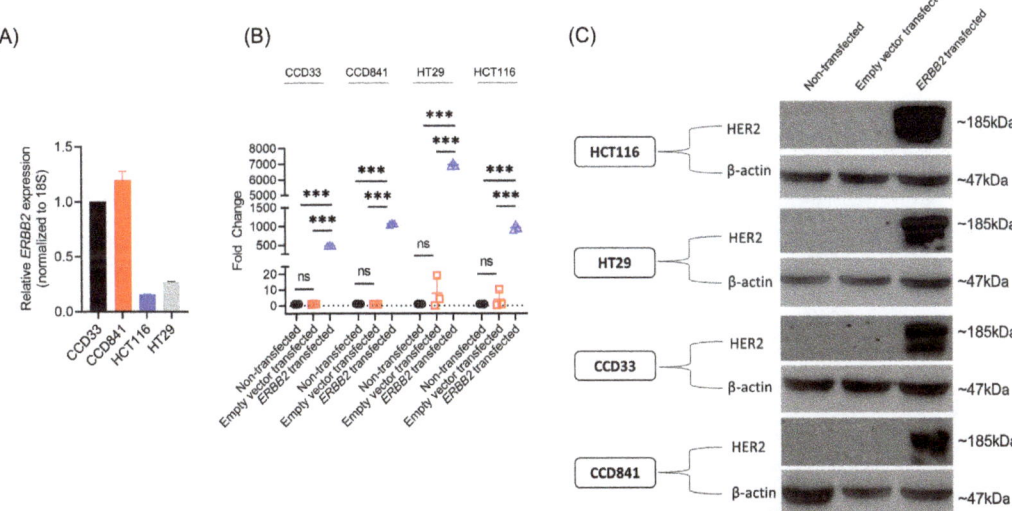

Figure 1. Validation of successful ectopic overexpression of *ERBB2* in the CRC (HCT116 and HT29) and normal colon (CCD33 and CCD841) cell lines. (**A**) Relative *ERBB2* expression in the normal colon cell lines CCD33 and CCD841 and the CRC cell lines HT29 and HCT116 as determined by qRT-PCR. Data were normalized to the expression of the internal control 18S rRNA gene and fold expressions were plotted relative to expression in the CCD33 cells. Data represent the mean ± SD of three independent experiments. (**B**) Relative *ERBB2* expression in non-transfected and either empty pcDNA3 vector or pcDNA3-*ERBB2* transfected HCT116, HT29, CCD33, and CCD841 cells as determined by qRT-PCR. Data were normalized to the expression of the internal control 18S rRNA gene and fold expressions were plotted relative to expression in the non-transfected cells. Data represent the mean ± SD of three independent experiments. *** $p < 0.001$; ns: not significant. (**C**) Same as B, but relative HER2 protein expression was determined in the different experimental conditions. Blots were re-probed with anti-β-Actin antibody to confirm equal loading across the lanes. The representative blots from three independent experiments are shown.

The CRC cell lines HT29 and HCT116 and normal colon cell lines CCD33 and CCD841 were transiently transfected with either an empty pcDNA3 vector or *ERBB2* expression construct. Successful transfection was confirmed by both qRT-PCR (Figure 1B), and Western blot analyses (Figure 1C).

RNA isolated from HCT116, HT29, CCD33, and CCD841 cells expressing an empty pcDNA3 vector (control) and those expressing ectopic *ERBB2* (*ERBB2*) were then subjected to RNA-seq in biological replicas. Multidimensional scaling using PCA was performed. Clusters distinguished by *ERBB2* expression levels in HCT116 and CCD841 cells were observed, confirming the reproducibility of the replicates and the unique transcriptomic profile associated with the ectopic expression of *ERBB2*; whereas samples were more staggered for the HT29 and CCD33 cells (Supplementary Figure S2A–D). The volcano plot of these data exhibited robust *ERBB2* whole transcriptomic changes in the CRC cell lines; HCT116 (1774 DEGs—730 upregulated and 1044 downregulated) and HT29 (1289 DEGs—430 upregulated and 859 downregulated) compared to the normal colon cell lines CCD33 (160 DEGs) and CCD841 (312 DEGs), (Supplementary Figure S3). In addition, the unsupervised hierarchical clustering analysis using the DEGs from each comparison exhibited clear subgroups of *ERBB2* transfected and control cell lines, for HCT116, HT29, CCD33, and CCD841 (Figure 2 and Figures S4–S6). The DEGs lists resulting from comparing *ERBB2* overexpression and control samples of HCT116, HT29, CCD33, and CCD841 are listed in Supplementary Table S1.

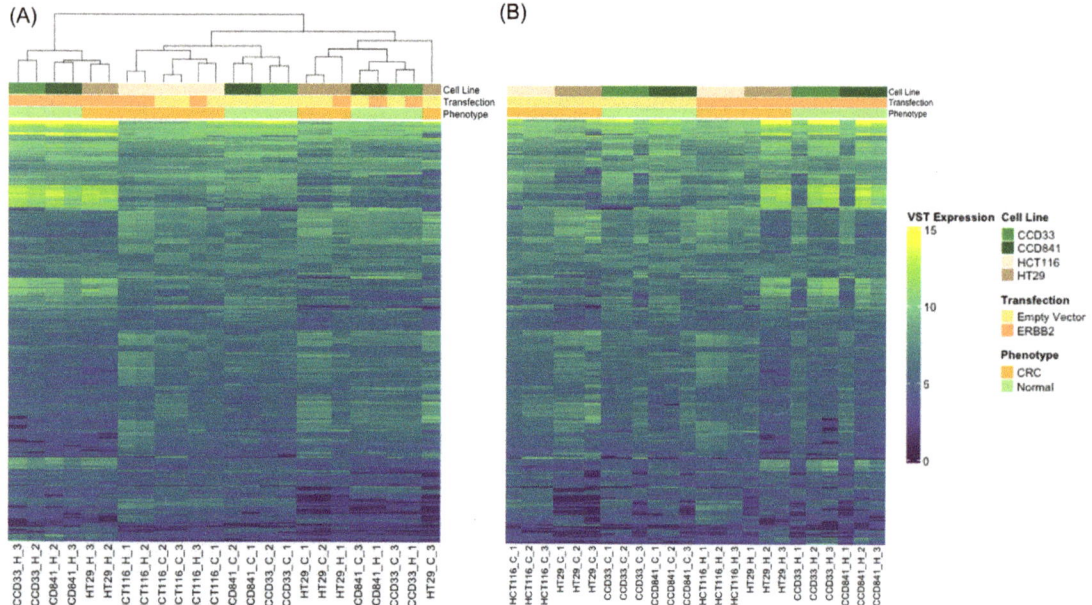

Figure 2. Heatmap of the differentially expressed genes in CRC (HCT116 and HT29) and normal colon (CCD33 and CCD841) cell lines transfected with empty vector or *ERBB2*, either clustered based on expression (**A**) or grouped based on transfection and phenotype (**B**).

3.2. Global Transcriptional Profiling in CRC Patients Based on HER2 Differential Expression

Given that cell lines exhibit a homogenous system, and a 2-D culture might not be a true replicate of an actual tumor, we next determined the genome-wide transcriptional patterns in CRC patients with varying HER2 protein expression. Of the 14 patients that were recruited, there were 8 females and 4 males with ages ranging from 37 to 86 years (mean ± SD, 63.15 ± 15.02 years). Histopathological examination identified most cases as adenocarcinomas.

Following the HER2 diagnostic criteria, 0 and 1+ staining scores were considered negative [29]. Among the CRC cases tested, 78.57% (11/14) of the cases were positive for HER2 (Score ≥ 2+; Figure 3A) whereas 21.43% (3/14) were negative for HER2 (≤1+; Figure 3B). All clinicopathological data are shown in Supplementary Table S2. No difference in classification and staging was observed with respect to gender, age, or HER2 expression.

RNA isolated from the 3 HER2− and 11 HER2+ biopsies were then subjected to whole RNA sequencing. Multidimensional scaling using PCA revealed clusters distinguished by *ERBB2* expression levels, confirming the unique transcriptomic profile associated with differential *ERBB2* expression (Figure 4A). The volcano plot of these data identified groups of differentially expressed genes, showing that 2701 were differentially expressed in HER2+ compared to HER2− CRC-patients of which 1344 were upregulated and 1357 were down-regulated (Figure 4B). Additionally, the unsupervised hierarchical clustering analysis based on the total DEGs showed that HER2+ and HER2− CRC-patients are, respectively, clustered as a single branch, further confirming the distinct transcriptional profiling between HER2+ and HER2− CRC-patients (Figure 4C, Supplementary Figure S7A). The differential HER2 protein expression in the 14 patient biopsies was further confirmed by the relative *ERBB2* gene expression in these samples (Supplementary Figure S7B). The DEGs between HER2+ and HER2− CRC patients are listed in Supplementary Table S3.

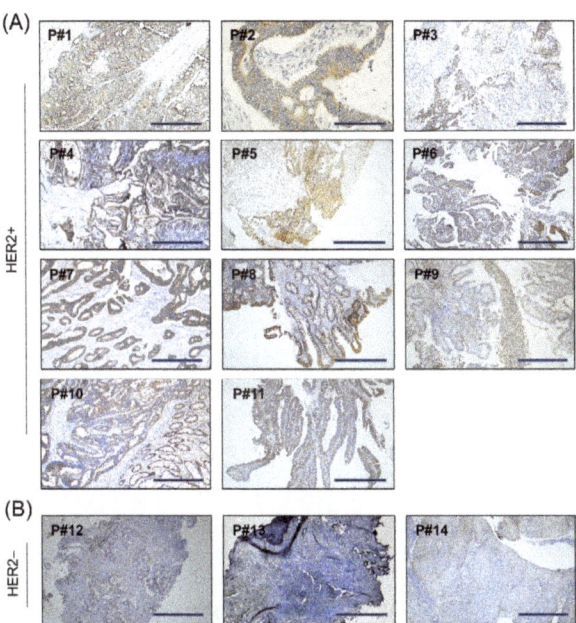

Figure 3. HER2 protein expression was detected by IHC staining in the 14 (P#1 to P#14) included patient samples. Images are representative IHC staining images of the 11 HER2+ (**A**) and 3 HER2− (**B**) patient samples. Scale bar, 100 µm.

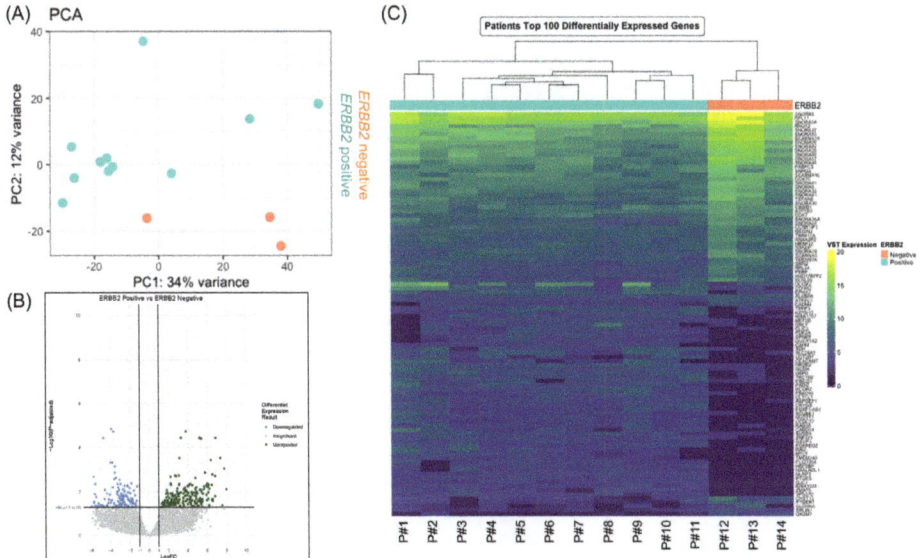

Figure 4. Genome-wide gene expression changes between *ERBB2*+ and *ERBB2*− CRC patients. (**A**) Principal component analysis (PCA) was performed to determine batch effects among the 14 patients' samples. Comparison of PC1 and PC2 variation sequestered the samples based on *ERBB2* expression. (**B**) Volcano plot of differentially expressed genes between *ERBB2*- and *ERBB2*+ patients' samples from input RNA-seq. Genes that are expressed significantly higher and lower based on log2 fold change in HER2+ samples are highlighted by green and blue dots, respectively. Unchanged transcripts are demarcated as grey circles ($p > 0.05$). (**C**) Heatmap of the top 100 differentially expressed genes.

3.3. GSEA of DEGs Revealed Distinctive ERBB2-Mediated Activation of Various Cellular Pathways Including Wnt Signaling and Regulation of Cellular Differentiation

To perform a functional interpretation of our transcriptomic analyses, the DEGs genes resulting from each comparison between HER2+ and HER2− CRC patients, HER2+ and control CRC cell lines, and HER2+ and normal colon cell lines were initially used as the input for the GSEA to identify the significantly enriched pathways ($p < 0.05$) among gene sets related to the following: C2, C5 (BP and MF), C6, and C7 collections. Given that the aim of the study was to define the putative role of HER2 in CRC pathogenesis, we chose the gene sets that contain functional pathways linked to cancer hallmarks and immune response. The results identified 98 significantly differentially activated pathways in HER2− compared to HER2− CRC patients (Supplementary Tables S4 and S5). The most significant pathways included the Wnt signaling pathway, T cell receptor signaling pathway, cell cycle, and cell differentiation pathways. Likewise, in HCT116 and HT29 CRC cell lines, the GSEA results showed, respectively, about 15 and 89 significant molecular functions and biological processes ontology gene sets (Supplementary Tables S4 and S5). Pathways related to cell signaling and leukocyte differentiation and migration were enriched in the HER2+ CRC cell lines. However, as expected, only a few significant activated cellular pathways were enriched in HER2+ compared to control normal colon cell lines (Supplementary Tables S4 and S5). Moreover, by overlapping the enriched pathways in CRC patients and cell lines, the results showed that the regulation of ion transport, cell–cell signaling, and cell proliferation were overexpressed between the two CRC systems.

An analysis of the leading-edge genes of the significant gene sets within the patients revealed that many were consistently represented by counting the number of times a gene occurs (gene frequency) across all the different pathways, suggesting that these genes strongly influenced the HER2-mediated expression pattern. The top 20 genes based on the gene frequency in the significantly activated cellular pathways between HER2+ and HER2− CRC patients showed key genes, including *NKX2-5, NKX6-1, WNT3A, WNT5A, NOG, SOX9, SOX18* (Figure 5A). The functional annotation of the top leading-edge genes showed highly significant enrichment of categories related to the regulation of cell differentiation, canonical Wnt signaling, cell development and maturation, and regulation of epithelial cell differentiation (Figure 5B).

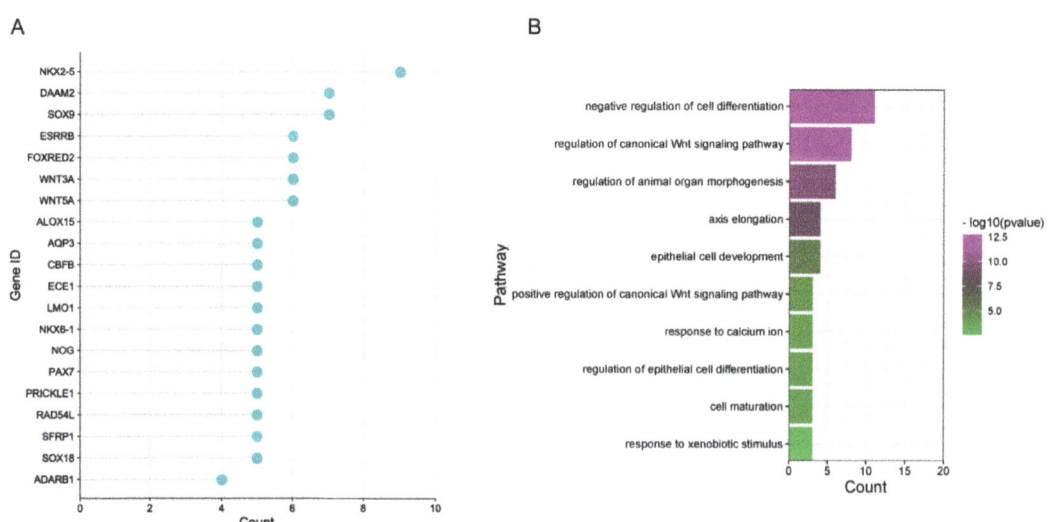

Figure 5. GSEA revealed significant enrichment of pathways related to cell differentiation and Wnt signaling. (**A**) The top 20 leading genes across the different enriched pathways based on frequency in HER2− vs. HER2+ CRC patients, and (**B**) the related pathways are shown.

3.4. Comparison of Cellular Pathways Revealed ERBB2-Mediated Enrichment of Pathways Related to Stem Cell Differentiation, Regulation of Wnt Signaling, and Immune Activation in Both Colorectal and Breast Cancers

Given the well-characterized role of *ERBB2* in breast cancer pathogenesis, we next analyzed breast cancer datasets available within the GEO database to determine if there are any similarities in the *ERBB2*-mediated enrichment of pathways in breast and colon cancer patients. Two independent datasets from different populations were selected: GSE29431 Caucasian/Spanish and GSE48391 Asian/Chinese (Supplementary Figure S1). These two datasets did not include the IHC scores for HER2. When comparing our transcriptomics data with the HER2 IHC, we found them to correlate if we take the average transcriptomics expression of the *ERBB2* mRNA. Therefore, we attempted to perform the stratification of HER2 from publicly available resources in a similar manner to the way we did with our patients' cohort from the mRNA transcriptome data, by taking the average HER2 expression from each group and stratifying them as HER2+ and HER2− accordingly. Therefore, followed by a quality control assessment based on HER2 expression in a particular patient compared to average HER2 expression in all patients within each cohort, 13 HER2− and 8 HER2+ were selected from the GSE29431 dataset and 35 HER2− and 9 HER2+ were selected from the GSE48391 dataset (Supplementary Table S6).

The GSEA was carried out to identify the significantly enriched pathways between HER2+ and HER2− breast cancer (BC) patients amongst gene sets related to the following: C2, C5 BP and MF, C6, and C7. Different significantly activated pathways and ontology gene sets were identified between HER2+ and HER2− BC patients (Supplementary Table S5). A comprehensive comparison between the significantly enriched pathways in HER2+/− BC and CRC patients was performed. The most significantly enriched pathways that are unique to BC patients include the VEGF signaling pathway, regulation of kinase activity MAPK pathway, and regulation of steroid metabolic process. On the other hand, twenty-six common activated pathways were observed between HER2+/HER2− breast cancer and CRC patients (Supplementary Table S7). This included pathways related to stem cell differentiation, regulation of Wnt signaling, and 17 related to immunological signature subsets including predominantly T cells, macrophages, and NK activation as depicted in Figure 6. In addition, the enrichment analysis identified unique immune-related pathways including macrophage activation and T-cell response, as shown in Supplementary Table S7. Analysis of the leading-edge genes underlying the enrichment of each gene set within the BC and CRC patient datasets (Figure 7A) revealed great resemblance in both lists of top genes (7 out of the top 20 genes) involved in the 26 common activated pathways of HER2+ vs. HER2− patients, and the top 20 frequent genes in activated cellular pathways between HER2+ and HER2− CRC patients (Figure 7B). The functional annotation of the top 20 genes between both colorectal and breast cancers revealed significant enrichment of categories related to the regulation of stem cell differentiation, protein catabolic process, regulation of peptidase and hydrolase activity, and regulation of Wnt signaling (Figure 7C). Moreover, by comparing the different enriched pathways between HER2+/− CRC patients, HER2+/− BC patients and cell lines, we noticed that, importantly, the response to calcium ions is a common enriched biological pathway between CRC patients, BC patients and CRC cell lines but absent in normal colon cell lines.

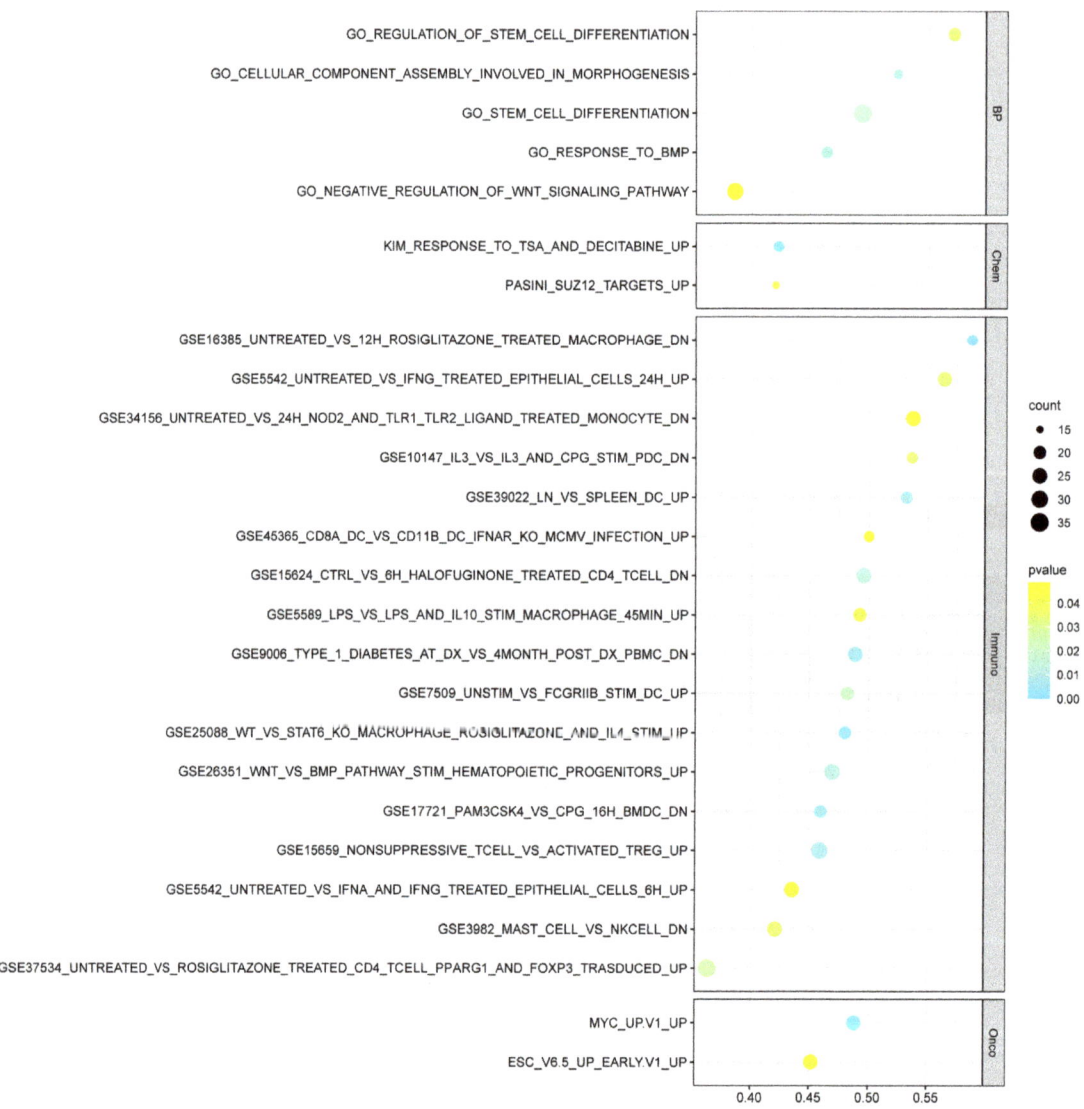

Figure 6. GSEA analysis of HER2+ and HER2− breast cancer and CRC patients revealed significant enrichment of common pathways. The figure shows the 26 significant pathways among CRC and BC in HER2+ vs. HER2− patients. *Immuno*, Immunologic Signature; *Onco*, Oncogenic Signature; *BP GO*, Biological Process; *Chem*, Chemical and Genetic Perturbations.

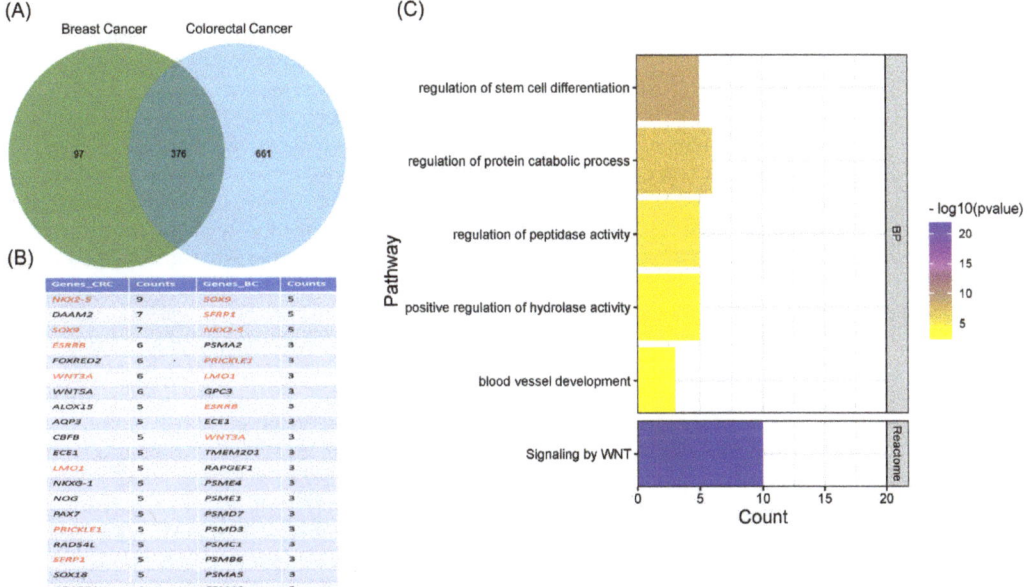

Figure 7. Top genes involved in the activated pathways of CRC and BC HER2+ vs. HER2− patients revealed a significant resemblance. (**A**) Venn diagram representing the overlap of genes overrepresented in enriched pathways in CRC and breast cancer patients. (**B**) Table showing the top 20 enriched genes in breast cancer and CRC patients. Genes shown in *red* were common in both CRC and breast cancer patients. (**C**) Related pathways of the top 20 enriched genes are shown. Representation is based on frequency in HER2− vs. HER2+ CRC patients.

4. Discussion

Our earlier work has shown that HER2 overexpression is correlated with more aggressive disease in CRC patients [16], indicating that stratification of patients according to HER2 status might be beneficial in the early detection and subsequent therapeutic management of patients with metastatic CRC. The results from the current study reveal that HER2 overexpression is associated with distinct global transcriptomic profiling, the characterization of which might lead to the identification of putative diagnostic and prognostic biomarkers.

In our study, there were fewer up and downregulated genes in HT29 compared to HCT116, indicating that the transcriptomics perturbation post-HER2-overexpression was less in HT29 in comparison to HCT116. The expression levels of APC, GSK3B, and CTNNB1 were not significant, which indicates that perhaps there may be a link between *ERBB2* expression and TP53 mutational status (rather than APC), since HT29 harbors TP53 mutation.

About half of the top 100 differentially expressed genes between HER2+ and HER2− CRC patients exhibited a more prominent difference in expression and were overexpressed in the HER2+ samples. A vast majority of them were non-coding RNA (small nucleolar RNAs, snoRNAs) that are normally involved in the biogenesis of other RNAs. That they were among the top differentially expressed genes indicates that expression of the snoRNAs may be regulated by HER2 and that they might be involved in the CRC pathogenesis in HER2+ patients. Indeed, it has been reported that numerous snoRNAs, including tumor-promoting and tumor-suppressing snoRNAs, are not only dysregulated in tumors but also show associations with clinical prognosis [30]. In addition, aberrant expression of snoRNAs has been reported in cell transformation, tumorigenesis, and metastasis, indicating that snoRNAs may be considered as biomarkers and/or therapeutic targets of cancer [31]. Even CRC associations between snoRNAs and CRC development have been reported [32–34].

For example, SNORA21 promotes CRC cell proliferation by regulating cancer-associated pathways such as Hippo and Wnt signaling pathways, and overexpression of SNORA21 has been reported to be associated with distant metastasis in CRC [35].

Colorectal cancer demonstrates hyperactivation of the Wnt pathway, which is involved in tumorigenesis, stemness, and metastatic progression [36,37]. One of the enriched pathways in our analysis was the regulation of stem cell differentiation. HER2-overexpression in gastric cancer cells results in increased stemness and invasiveness [38]. Furthermore, this increased HER2-mediated stemness is regulated by Wnt/β-catenin signaling [39]. Our analysis also revealed significant enrichment of pathways related to positive and negative regulation of Wnt signaling and the enrichment of the Wnt signaling pathway genes *WNT3A* and *WNT5A*. These results would thus indicate that HER2 overexpression in CRC cells might result in poorly differentiated tumors that are more invasive.

One of the leading-edge genes that were differentially expressed between HER2+ and HER2− patients and cell lines was a homeobox gene of the NKL subclass, cardiac transcription factor *NKX2-5*. *NKX2-5* is one of the earliest known transcription factors required for cardiac cell specification and proliferation [40–42].

NKX2-5 is expressed in several types of tumors [43–46], but its precise role in tumorigenesis is unknown. Another family member of the NKL, *NKX2-1*, has been reported to mediate p53-induced tumor suppression [47–50], Indeed, in the context of CRC, *NKX2-5* functions as a conditional tumor suppressor gene via activating the p53-mediated p21WAF1/CIP1 expression [51]. It has been predicted via bioinformatic analysis and confirmed by chromatin immunoprecipitation analysis that in hepatocellular carcinoma cells the promoter of *ERBB2* binds to the transcription factor *NKX2-5*, resulting in a negative regulatory effect [52]. Interestingly, promoter hypermethylation of *NKX2-6* has been identified as a candidate biomarker associated with differential methylation in HER2+ breast cancer and breast carcinogenesis [53]. Whether a similar mechanism is operant for the HER2-mediated downregulation of *NKX2-5* remains to be determined.

During cardiogenesis, *NKX2-5* potentiates Wnt signaling by regulating the expression of the R-spondin3 [54]. In the current study, we also saw the enrichment of pathways related to positive and negative regulation of Wnt signaling. Analysis of the RNAseq data showed significantly more *NKX2-5* expression in HER2− patients compared to HER2+ patients (Supplementary Table S3). Whether HER2 overexpression drives negative regulation of *NKX2-5*-mediated Wnt signaling, ultimately resulting in well-differentiated less invasive tumors in CRC, remains to be determined.

It is important to note that the transcriptomic analysis in CRC cell lines was conducted post-transient transfection in CRC cell lines. It is highly plausible that short-term transfection (to an unphysiological expression level) in the CRC cell lines will induce transcriptomic changes that will be different from those observed in cell lines that have evolved under HER2-overexpression selective pressure. Hence, comparing our data to gene expression changes observed in HER2-amplified or HER2-mutant cell lines would indeed be informative in this context. However, this shortcoming is potentially offset to a large extent by three facts—(a) our analysis also involved HER2− and HER2+ patient samples, which are a better model compared to any of the homogeneous cell line models; (b) our usage of two different CRC cell lines and two normal colon cell lines; and (c) our observation that similar pathways were being enriched when the patient samples and the HER2-overexpressing cell lines were compared.

Despite the current study having inherent weaknesses in low sample size and lack of validation in a wider population, the results do highlight the importance of transcriptional profiling of HER2+ and HER2− CRC patients in identifying potential biomarkers that play a role in CRC pathogenesis. Additional studies are warranted in different population cohorts as the incidence of HER2+ CRC patients varies widely based on geographical location. It would be intricate and intriguing to investigate whether and how HER2 regulates *NKX2-5* in wild-type and p53 mutant CRC and their subsequent effects on Wnt signaling and CRC invasives.

5. Conclusions

The study identified genes and pathways related to HER2 expression in colorectal cancer in both patients and colorectal cancer cell lines. Comparison of whole transcriptome RNA sequencing analysis in patients with HER2+ and HER2− identified unique immune-related pathways including macrophage activation, and T-cell response as well as cancer-related pathways including Wnt signaling and MYC-related pathways. The Wnt signaling pathway was also significant in the two different colorectal cell lines (HCT116 and HT29) when compared with normal colon cell lines (CCD841 and CCD33), ectopically overexpressing *ERBB2*. Differential gene expression analysis identified many Wnt signaling pathway genes including *WNT5A*, *WNT9A*, *WNT3A*, *WNT16*, and *WNT10A*. Gene Set Enrichment Analysis indicated a role for HER2 in regulating CRC pathogenesis, with Wnt/β-catenin signaling being mediated via a HER2-dependent regulatory pathway impacting expression of the homeobox gene *NKX2-5*. Results from this study thus identified putative targets that are co-expressed with HER2 in colorectal cancer, which are different from HER2 in breast cancer, warranting further investigation into their role in colorectal cancer pathogenesis.

Supplementary Materials: The following supporting information can be downloaded at https://www.mdpi.com/article/10.3390/cancers15010130/s1, Table S1: Differentially expressed genes from the whole transcriptomic analysis of the CRC cell lines; Table S2: Clinicopathological characteristics of the patients. Table S3: Differentially expressed genes from the whole transcriptomic analysis of the patient samples; Table S4: Summary of the total activated cellular pathways identified by GSEA in HER2+ vs. HER2− colorectal cancer (CRC) patients, CRC cell lines HCT116 and HT29, normal colon cell lines CCD33 and CCD841, and breast cancer patients ($p < 0.05$); Table S5: Summary of the total activated cellular pathways identified by GSEA in ERBB2+ vs. ERBB2− colorectal cancer (CRC) patients, CRC cell lines HCT116 and HT29, normal colon cell lines CCD33 and CCD841, and breast cancer (BC) patients ($p < 0.05$); Table S6: Details of breast cancer datasets analyzed to determine correlation of ERBB2-mediated enrichment of pathways in breast and colon cancer patients. N, total number of breast cancer patients; Table S7: The 26 common activated pathways among CRC and BC HER2+ vs. HER2− patients obtained from Gene Set Enrichment Analysis; Figure S1: Schematic representation of the process used to select the HER2− and HER2+ breast cancer patient datasets; Figure S2: Principal component analysis (PCA) in different cell lines; Figure S3: Volcano plot of differentially expressed genes between control (empty vector) and *ERBB2* overexpressing cells; Figure S4: Heat map of the differentially expressed genes between control (empty vector) and *ERBB2* overexpressing normal colon cell lines; Figure S5: Heat map of the differentially expressed genes between control (empty vector) and *ERBB2* overexpressing CRC cell line HCT116; Figure S6: Heat map of the differentially expressed genes between control (empty vector) and *ERBB2* overexpressing CRC cell line HT29; Figure S7: Heat map of the differentially expressed genes between *ERBB2* negative and positive CRC patients and relative *ERBB2* expressions in the patient samples.

Author Contributions: Conceptualization, E.A.A.R., R.H. and R.B.; methodology, E.A.A.R., K.B., R.H. and R.B.; validation, E.A.A.R., K.B. and A.B.; formal analysis, E.A.A.R., A.B. and R.H.; investigation, E.A.A.R., K.B., N.A., R.H. and R.B.; resources, R.H. and R.B.; data curation, E.A.A.R., N.A. and A.B.; writing—original draft preparation, E.A.A.R., R.H. and R.B.; writing—review and editing, E.A.A.R., K.B., A.B., R.H. and R.B.; supervision, R.H. and R.B.; project administration, R.H. and R.B.; funding acquisition, R.B. All authors have read and agreed to the published version of the manuscript.

Funding: R.B. and R.H. are funded by Al-Jalila Foundation (grant code: AJF2018112). R.H. is funded by the University of Sharjah (grant code: 22010902103).

Institutional Review Board Statement: Ethical approval for the study was provided by the Research and Ethics Committee (REC) of University Hospital Sharjah (UHS-HERC-055-25022019). All methods were performed in accordance with the relevant guidelines based on the Declaration of Helsinki and the Belmont Report.

Informed Consent Statement: Written informed consent was obtained from all subjects involved in the in vivo validation of the study.

Data Availability Statement: The RNA sequencing data generated in this work have been deposited in figshare and can be obtained from the following link: https://doi.org/10.6084/m9.figshare.21578763, accessed on 17 November 2022. The breast cancer datasets used in the manuscript are deposited in Gene Expression Omnibus (https://www.ncbi.nlm.nih.gov/geo, accessed on 3 October 2022) and are accessible through GEO Series accession numbers GSE29431 and GSE48391. All other supporting data of this study are either included in the manuscript or available on request from the corresponding author.

Acknowledgments: The authors would like to thank the Al Jalila Foundation and University Hospital Sharjah for their support in this study.

Conflicts of Interest: The authors declare no conflict of interest.

References

1. Torre, L.A.; Siegel, R.L.; Ward, E.M.; Jemal, A. Global Cancer Incidence and Mortality Rates and Trends—An Update. *Cancer Epidemiol. Biomarkers. Prev.* **2016**, *25*, 16–27. [CrossRef]
2. Siegel, R.L.; Miller, K.D.; Fuchs, H.E.; Jemal, A. Cancer statistics, 2022. *CA Cancer J. Clin.* **2022**, *72*, 7–33. [CrossRef]
3. Meyerhardt, J.A.; Mayer, R.J. Systemic therapy for colorectal cancer. *N. Engl. J. Med.* **2005**, *352*, 476–487. [CrossRef] [PubMed]
4. Marquart, J.; Chen, E.Y.; Prasad, V. Estimation of the Percentage of US Patients With Cancer Who Benefit From Genome-Driven Oncology. *JAMA Oncol.* **2018**, *4*, 1093–1098. [CrossRef] [PubMed]
5. Prasad, V. Perspective: The precision-oncology illusion. *Nature* **2016**, *537*, S63. [CrossRef]
6. Meric-Bernstam, F.; Brusco, L.; Shaw, K.; Horombe, C.; Kopetz, S.; Davies, M.A.; Routbort, M.; Piha-Paul, S.A.; Janku, F.; Ueno, N.; et al. Feasibility of Large-Scale Genomic Testing to Facilitate Enrollment Onto Genomically Matched Clinical Trials. *J. Clin. Oncol.* **2015**, *33*, 2753–2762. [CrossRef] [PubMed]
7. Schwaederle, M.; Zhao, M.; Lee, J.J.; Lazar, V.; Leyland-Jones, B.; Schilsky, R.L.; Mendelsohn, J.; Kurzrock, R. Association of Biomarker-Based Treatment Strategies With Response Rates and Progression-Free Survival in Refractory Malignant Neoplasms: A Meta-analysis. *JAMA Oncol.* **2016**, *2*, 1452–1459. [CrossRef] [PubMed]
8. Nowak, J.A. HER2 in Colorectal Carcinoma: Are We There yet? *Surg. Pathol. Clin.* **2020**, *13*, 485–502. [CrossRef]
9. Strickler, J.H.; Yoshino, T.; Graham, R.P.; Siena, S.; Bekaii-Saab, T. Diagnosis and Treatment of *ERBB2*-Positive Metastatic Colorectal Cancer: A Review. *JAMA Oncol.* **2022**, *8*, 760–769. [CrossRef]
10. May, M.; Raufi, A.G.; Sadeghi, S.; Chen, K.; Iuga, A.; Sun, Y.; Ahmed, F.; Bates, S.; Manji, G.A. Prolonged Response to HER2-Directed Therapy in Three Patients with HER2-Amplified Metastatic Carcinoma of the Biliary System: Case Study and Review of the Literature. *Oncologist* **2021**, *26*, 640–646. [CrossRef]
11. Harari, D.; Yarden, Y. Molecular mechanisms underlying *ERBB2*/HER2 action in breast cancer. *Oncogene* **2000**, *19*, 6102–6114. [CrossRef] [PubMed]
12. Schechter, A.L.; Stern, D.F.; Vaidyanathan, L.; Decker, S.J.; Drebin, J.A.; Greene, M.I.; Weinberg, R.A. The neu oncogene: An erb-B-related gene encoding a 185,000-Mr tumour antigen. *Nature* **1984**, *312*, 513–516. [CrossRef]
13. Slamon, D.J.; Clark, G.M.; Wong, S.G.; Levin, W.J.; Ullrich, A.; McGuire, W.L. Human breast cancer: Correlation of relapse and survival with amplification of the HER-2/neu oncogene. *Science* **1987**, *235*, 177–182. [CrossRef]
14. Swain, S.M.; Shastry, M.; Hamilton, E. Targeting HER2-positive breast cancer: Advances and future directions. *Nat. Rev. Drug. Discov.* **2022**, 1–26. [CrossRef]
15. Ng, C.K.; Martelotto, L.G.; Gauthier, A.; Wen, H.C.; Piscuoglio, S.; Lim, R.S.; Cowell, C.F.; Wilkerson, P.M.; Wai, P.; Rodrigues, D.N.; et al. Intra-tumor genetic heterogeneity and alternative driver genetic alterations in breast cancers with heterogeneous HER2 gene amplification. *Genome Biol.* **2015**, *16*, 107. [CrossRef]
16. Abdul Razzaq, E.A.; Venkatachalam, T.; Bajbouj, K.; Rahmani, M.; Mahdami, A.; Rawat, S.; Mansuri, N.; Alhashemi, H.; Hamoudi, R.A.; Bendardaf, R. HER2 overexpression is a putative diagnostic and prognostic biomarker for late-stage colorectal cancer in North African patients. *Libyan J. Med.* **2021**, *16*, 1955462. [CrossRef] [PubMed]
17. Owen, D.R.; Wong, H.L.; Bonakdar, M.; Jones, M.; Hughes, C.S.; Morin, G.B.; Jones, S.J.M.; Renouf, D.J.; Lim, H.; Laskin, J.; et al. Molecular characterization of *ERBB2*-amplified colorectal cancer identifies potential mechanisms of resistance to targeted therapies: A report of two instructive cases. *Cold Spring Harb. Mol. Case Stud.* **2018**, *4*, a002535. [CrossRef]
18. Pye, H.; Butt, M.A.; Funnell, L.; Reinert, H.W.; Puccio, I.; Rehman Khan, S.U.; Saouros, S.; Marklew, J.S.; Stamati, I.; Qurashi, M.; et al. Using antibody directed phototherapy to target oesophageal adenocarcinoma with heterogeneous HER2 expression. *Oncotarget* **2018**, *9*, 22945–22959. [CrossRef] [PubMed]
19. Choudhury, K.R.; Yagle, K.J.; Swanson, P.E.; Krohn, K.A.; Rajendran, J.G. A robust automated measure of average antibody staining in immunohistochemistry images. *J. Histochem. Cytochem.* **2010**, *58*, 95–107. [CrossRef]
20. Li, Y.M.; Pan, Y.; Wei, Y.; Cheng, X.; Zhou, B.P.; Tan, M.; Zhou, X.; Xia, W.; Hortobagyi, G.N.; Yu, D.; et al. Upregulation of CXCR4 is essential for HER2-mediated tumor metastasis. *Cancer Cell* **2004**, *6*, 459–469. [CrossRef]
21. Chaudhury, A.; Hussey, G.S.; Ray, P.S.; Jin, G.; Fox, P.L.; Howe, P.H. TGF-beta-mediated phosphorylation of hnRNP E1 induces EMT via transcript-selective translational induction of Dab2 and ILEI. *Nat. Cell Biol.* **2010**, *12*, 286–293. [CrossRef]

22. Ning, Z.; Cox, A.J.; Mullikin, J.C. SSAHA: A fast search method for large DNA databases. *Genome Res.* **2001**, *11*, 1725–1729. [CrossRef]
23. Li, H. Exploring single-sample SNP and INDEL calling with whole-genome de novo assembly. *Bioinformatics* **2012**, *28*, 1838–1844. [CrossRef] [PubMed]
24. Smith, T.F.; Waterman, M.S.; Fitch, W.M. Comparative biosequence metrics. *J. Mol. Evol.* **1981**, *18*, 38–46. [CrossRef] [PubMed]
25. Love, M.I.; Huber, W.; Anders, S. Moderated estimation of fold change and dispersion for RNA-seq data with DESeq2. *Genome Biol.* **2014**, *15*, 550. [CrossRef] [PubMed]
26. Ritchie, M.E.; Phipson, B.; Wu, D.; Hu, Y.; Law, C.W.; Shi, W.; Smyth, G.K. limma powers differential expression analyses for RNA-sequencing and microarray studies. *Nucleic Acids Res.* **2015**, *43*, e47. [CrossRef]
27. Hamoudi, R.A.; Appert, A.; Ye, H.; Ruskone-Fourmestraux, A.; Streubel, B.; Chott, A.; Raderer, M.; Gong, L.; Wlodarska, I.; De Wolf-Peeters, C.; et al. Differential expression of NF-kappaB target genes in MALT lymphoma with and without chromosome translocation: Insights into molecular mechanism. *Leukemia* **2010**, *24*, 1487–1497. [CrossRef]
28. Subramanian, A.; Tamayo, P.; Mootha, V.K.; Mukherjee, S.; Ebert, B.L.; Gillette, M.A.; Paulovich, A.; Pomeroy, S.L.; Golub, T.R.; Lander, E.S.; et al. Gene set enrichment analysis: A knowledge-based approach for interpreting genome-wide expression profiles. *Proc. Natl. Acad. Sci. USA* **2005**, *102*, 15545–15550. [CrossRef]
29. Liu, F.; Ren, C.; Jin, Y.; Xi, S.; He, C.; Wang, F.; Wang, Z.; Xu, R.H.; Wang, F. Assessment of two different HER2 scoring systems and clinical relevance for colorectal cancer. *Virchows. Arch* **2020**, *476*, 391–398. [CrossRef]
30. Huang, Z.H.; Du, Y.P.; Wen, J.T.; Lu, B.F.; Zhao, Y. snoRNAs: Functions and mechanisms in biological processes, and roles in tumor pathophysiology. *Cell Death Discov.* **2022**, *8*, 259. [CrossRef]
31. Liang, J.; Wen, J.; Huang, Z.; Chen, X.P.; Zhang, B.X.; Chu, L. Small Nucleolar RNAs: Insight Into Their Function in Cancer. *Front. Oncol.* **2019**, *9*, 587. [CrossRef] [PubMed]
32. Fang, X.; Yang, D.; Luo, H.; Wu, S.; Dong, W.; Xiao, J.; Yuan, S.; Ni, A.; Zhang, K.J.; Liu, X.Y.; et al. SNORD126 promotes HCC and CRC cell growth by activating the PI3K-AKT pathway through FGFR2. *J. Mol. Cell Biol.* **2017**, *9*, 243–255. [CrossRef] [PubMed]
33. Liu, Y.; Zhao, C.; Wang, G.; Chen, J.; Ju, S.; Huang, J.; Wang, X. SNORD1C maintains stemness and 5-FU resistance by activation of Wnt signaling pathway in colorectal cancer. *Cell Death Discov.* **2022**, *8*, 200. [CrossRef] [PubMed]
34. Wu, H.; Qin, W.; Lu, S.; Wang, X.; Zhang, J.; Sun, T.; Hu, X.; Li, Y.; Chen, Q.; Wang, Y.; et al. Long noncoding RNA ZFAS1 promoting small nucleolar RNA-mediated 2′-O-methylation via NOP58 recruitment in colorectal cancer. *Mol. Cancer* **2020**, *19*, 95. [CrossRef]
35. Yoshida, K.; Toden, S.; Weng, W.; Shigeyasu, K.; Miyoshi, J.; Turner, J.; Nagasaka, T.; Ma, Y.; Takayama, T.; Fujiwara, T.; et al. SNORA21—An Oncogenic Small Nucleolar RNA, with a Prognostic Biomarker Potential in Human Colorectal Cancer. *EBioMedicine* **2017**, *22*, 68–77. [CrossRef]
36. Schatoff, E.M.; Leach, B.I.; Dow, L.E. Wnt Signaling and Colorectal Cancer. *Curr. Colorectal. Cancer Rep.* **2017**, *13*, 101–110. [CrossRef]
37. Zhao, H.; Ming, T.; Tang, S.; Ren, S.; Yang, H.; Liu, M.; Tao, Q.; Xu, H. Wnt signaling in colorectal cancer: Pathogenic role and therapeutic target. *Mol. Cancer* **2022**, *21*, 144. [CrossRef]
38. de Sousa, E.M.; Vermeulen, L.; Richel, D.; Medema, J.P. Targeting Wnt signaling in colon cancer stem cells. *Clin. Cancer Res.* **2011**, *17*, 647–653. [CrossRef]
39. Jung, D.H.; Bae, Y.J.; Kim, J.H.; Shin, Y.K.; Jeung, H.C. HER2 Regulates Cancer Stem Cell Activities via the Wnt Signaling Pathway in Gastric Cancer Cells. *Oncology* **2019**, *97*, 311–318. [CrossRef]
40. Chung, I.M.; Rajakumar, G. Genetics of Congenital Heart Defects: The NKX2-5 Gene, a Key Player. *Genes* **2016**, *7*, 6. [CrossRef]
41. Lyons, I.; Parsons, L.M.; Hartley, L.; Li, R.; Andrews, J.E.; Robb, L.; Harvey, R.P. Myogenic and morphogenetic defects in the heart tubes of murine embryos lacking the homeo box gene Nkx2-5. *Genes Dev.* **1995**, *9*, 1654–1666. [CrossRef] [PubMed]
42. Tanaka, M.; Chen, Z.; Bartunkova, S.; Yamasaki, N.; Izumo, S. The cardiac homeobox gene Csx/Nkx2.5 lies genetically upstream of multiple genes essential for heart development. *Development* **1999**, *126*, 1269–1280. [CrossRef]
43. Hwang, C.; Jang, S.; Choi, D.K.; Kim, S.; Lee, J.H.; Lee, Y.; Kim, C.D.; Lee, J.H. The role of nkx2.5 in keratinocyte differentiation. *Ann. Dermatol.* **2009**, *21*, 376–381. [CrossRef]
44. Nagel, S.; Kaufmann, M.; Drexler, H.G.; MacLeod, R.A. The cardiac homeobox gene NKX2-5 is deregulated by juxtaposition with BCL11B in pediatric T-ALL cell lines via a novel t(5;14)(q35.1;q32.2). *Cancer Res.* **2003**, *63*, 5329–5334.
45. Penha, R.C.C.; Buexm, L.A.; Rodrigues, F.R.; de Castro, T.P.; Santos, M.C.S.; Fortunato, R.S.; Carvalho, D.P.; Cardoso-Weide, L.C.; Ferreira, A.C.F. NKX2.5 is expressed in papillary thyroid carcinomas and regulates differentiation in thyroid cells. *BMC Cancer* **2018**, *18*, 498. [CrossRef]
46. Shibata, K.; Kajiyama, H.; Yamamoto, E.; Terauchi, M.; Ino, K.; Nawa, A.; Kikkawa, F. Establishment and characterization of an ovarian yolk sac tumor cell line reveals possible involvement of Nkx2.5 in tumor development. *Oncology* **2008**, *74*, 104–111. [CrossRef]
47. Chen, M.J.; Chen, P.M.; Wang, L.; Shen, C.J.; Chen, C.Y.; Lee, H. Cisplatin sensitivity mediated by NKX2-1 in lung adenocarcinoma is dependent on p53 mutational status via modulating TNFSF10 expression. *Am. J. Cancer Res.* **2020**, *10*, 1229–1237. [PubMed]
48. Chen, P.M.; Wu, T.C.; Cheng, Y.W.; Chen, C.Y.; Lee, H. NKX2-1-mediated p53 expression modulates lung adenocarcinoma progression via modulating IKKbeta/NF-kappaB activation. *Oncotarget* **2015**, *6*, 14274–14289. [CrossRef]

49. Tsai, L.H.; Chen, P.M.; Cheng, Y.W.; Chen, C.Y.; Sheu, G.T.; Wu, T.C.; Lee, H. LKB1 loss by alteration of the NKX2-1/p53 pathway promotes tumor malignancy and predicts poor survival and relapse in lung adenocarcinomas. *Oncogene* **2014**, *33*, 3851–3860. [CrossRef]
50. Winslow, M.M.; Dayton, T.L.; Verhaak, R.G.; Kim-Kiselak, C.; Snyder, E.L.; Feldser, D.M.; Hubbard, D.D.; DuPage, M.J.; Whittaker, C.A.; Hoersch, S.; et al. Suppression of lung adenocarcinoma progression by Nkx2-1. *Nature* **2011**, *473*, 101–104. [CrossRef]
51. Li, H.; Wang, J.; Huang, K.; Zhang, T.; Gao, L.; Yang, S.; Yi, W.; Niu, Y.; Liu, H.; Wang, Z.; et al. Nkx2.5 Functions as a Conditional Tumor Suppressor Gene in Colorectal Cancer Cells via Acting as a Transcriptional Coactivator in p53-Mediated p21 Expression. *Front. Oncol.* **2021**, *11*, 648045. [CrossRef]
52. Chen, G.; Jiang, J.; Wang, X.; Feng, K.; Ma, K. lncENST Suppress the Warburg Effect Regulating the Tumor Progress by the Nkx2-5/*ERBB2* Axis in Hepatocellular Carcinoma. *Comput. Math. Methods Med.* **2021**, *2021*, 6959557. [CrossRef]
53. Lindqvist, B.M.; Wingren, S.; Motlagh, P.B.; Nilsson, T.K. Whole genome DNA methylation signature of HER2-positive breast cancer. *Epigenetics* **2014**, *9*, 1149–1162. [CrossRef]
54. Cambier, L.; Plate, M.; Sucov, H.M.; Pashmforoush, M. Nkx2-5 regulates cardiac growth through modulation of Wnt signaling by R-spondin3. *Development* **2014**, *141*, 2959–2971. [CrossRef]

Disclaimer/Publisher's Note: The statements, opinions and data contained in all publications are solely those of the individual author(s) and contributor(s) and not of MDPI and/or the editor(s). MDPI and/or the editor(s) disclaim responsibility for any injury to people or property resulting from any ideas, methods, instructions or products referred to in the content.

Review

EZH2: An Accomplice of Gastric Cancer

Wuhan Yu [1,2,†], Ning Liu [1,3,†], Xiaogang Song [1,4], Lang Chen [1,2], Mancai Wang [1,2], Guohui Xiao [1,2], Tengfei Li [1,2], Zheyuan Wang [1,2,*] and Youcheng Zhang [1,2,*]

1 The Second Hospital of Lanzhou University, Lanzhou 730030, China
2 The Second General Surgery Department, The Second Hospital of Lanzhou University, Lanzhou 730030, China
3 The Neurology Department, The Second Hospital of Lanzhou University, Lanzhou 730030, China
4 The Cardiology Department, The Second Hospital of Lanzhou University, Lanzhou 730030, China
* Correspondence: wangzhy1985@hotmail.com (Z.W.); zhangyouchengphd@163.com (Y.Z.)
† These authors contributed equally to this work.

Simple Summary: Enhancer of zeste homolog 2 (EZH2) modifies the trimethylation of Lys-27 of histone 3, affecting downstream target genes' expression. It was reported that EZH2 is highly expressed in gastric cancer and may be a potential prognostic molecule and promising therapeutic target. We aim to present the value of EZH2 research in gastric cancer by focusing on the crucial events of EZH2 involvement in gastric cancer progression. Therefore, in this review, we present the two main functions of EZH2: histone methylation modification and DNA methylation by EZH2; the molecular mechanism of the action of EZH2 in regulating target genes; a detailed description of the mechanism of EZH2 in gastric cancer-related events. Finally, progress in the development of EZH2 inhibitors is summarized. This review article provides researchers studying the epigenetics of gastric cancer with research ideas to find new targets for studying gastric cancer pathogenesis.

Abstract: Gastric cancer is the fifth most common cancer and the third leading cause of cancer deaths worldwide. Understanding the factors influencing the therapeutic effects in gastric cancer patients and the molecular mechanism behind gastric cancer is still facing challenges. In addition to genetic alterations and environmental factors, it has been demonstrated that epigenetic mechanisms can also induce the occurrence and progression of gastric cancer. Enhancer of zeste homolog 2 (EZH2) is the catalytic subunit of the polycomb repressor complex 2 (PRC2), which trimethylates histone 3 at Lys-27 and regulates the expression of downstream target genes through epigenetic mechanisms. It has been found that EZH2 is overexpressed in the stomach, which promotes the progression of gastric cancer through multiple pathways. In addition, targeted inhibition of EZH2 expression can effectively delay the progression of gastric cancer and improve its resistance to chemotherapeutic agents. Given the many effects of EZH2 in gastric cancer, there are no studies to comprehensively describe this mechanism. Therefore, in this review, we first introduce EZH2 and clarify the mechanisms of abnormal expression of EZH2 in cancer. Secondly, we summarize the role of EZH2 in gastric cancer, which includes the association of the EZH2 gene with genetic susceptibility to GC, the correlation of the EZH2 gene with gastric carcinogenesis and invasive metastasis, the resistance to chemotherapeutic drugs of gastric cancer mediated by EZH2 and the high expression of EZH2 leading to poor prognosis of gastric cancer patients. Finally, we also clarify some of the current statuses of drug development regarding targeted inhibition of EZH2/PRC2 activity.

Keywords: gastric cancer; enhancer of zeste homolog 2; H3K27; epigenetics

Citation: Yu, W.; Liu, N.; Song, X.; Chen, L.; Wang, M.; Xiao, G.; Li, T.; Wang, Z.; Zhang, Y. EZH2: An Accomplice of Gastric Cancer. *Cancers* 2023, *15*, 425. https://doi.org/10.3390/cancers15020425

Academic Editors: Marie-Pierre Buisine and Sang Kil Lee

Received: 17 November 2022
Revised: 30 December 2022
Accepted: 5 January 2023
Published: 9 January 2023

Copyright: © 2023 by the authors. Licensee MDPI, Basel, Switzerland. This article is an open access article distributed under the terms and conditions of the Creative Commons Attribution (CC BY) license (https://creativecommons.org/licenses/by/4.0/).

1. Introduction

Gastric cancer (GC) is the fifth most diagnosed malignancy worldwide, with more than 1 million new cases annually [1]. There is a lack of methods to diagnose GC early, so many patients are diagnosed at a later stage, which leads to a high mortality rate for

GC patients [1]. It was reported that more than 784,000 GC patients died worldwide in 2018, making it the third most common cause of death among oncologic diseases [2]. The complicated pathogenesis, late diagnosis and lack of effective treatment for GC lead to the poor prognosis of patients. To fundamentally prevent and treat GC, it is highly significant to understand the pathogenesis of GC [3]. In addition to genetic changes and environmental factors, it has been proven that epigenetic inheritance guides the occurrence and development of cancer and is a hallmark of gastric malignancies [4]. It was known that the polycomb group (PcG) was one of the most important epigenetic regulators, which influences the expression of many genes involved in the development of the body [5]. As a core member of the PcG family, EZH2 plays a vital role in cell proliferation, differentiation and tumor formation through H3K27me3-mediated downstream gene silencing. The expression of H3K27me3 in GC tissues is significantly increased, and it is the most common type of histone methylation modification in GC studies, which is closely related to the pathogenesis of GC and the prognosis of patients [4,6]. In conclusion, EZH2 plays a role in the pathogenesis of GC through H3K27me3. For these reasons, EZH2 can be considered an exciting target for developing targeted therapies for GC. Therefore, this article focuses on the relationship between EZH2 and the occurrence, metastasis, and drug resistance of GC and further explores the mechanisms of development of GC from the epigenetic factors.

2. Overview of EZH2

PcG proteins are a group of transcriptional repressors that regulate target genes through chromatin modification and can induce tumor development. These proteins chemically and functionally represent the core proteins of the polycomb repressive complexes (PRCs) [5]. PRCs are enzyme complexes that modify lysine residues on histones [7,8]. There are two major PRCs in mammals: Polycomb repressive complex 1 (PCR1) and polycomb repressive complex 2 (PCR2). PRC1 consists of ring finger protein 1 (RING 1) (RING1A or RING1B) and PcG ring finger protein (PCGF1-6) that monoubiquitinates lysine 119 on histone H2A (H2AK119ub1) [9]. PRC2 complexes are histone methyltransferases (HM-Tases) that are dependent on S-adenosyl-L-methionine (SAM) and contain four major core subunits: EZH2/1, suppressor of zeste 12 (SUZ12), embryonic ectoderm development 1-4 (EED1-4) and RBAP46/48 [9,10]. It catalyzes the mono-methylation, di-methylation and tri-methylation of lysine 27 on histone H3 (H3K27me1, H3K27me2 and H3K27me3) [11]. PRC2 is further divided into two different subclasses: PRC2.1 and PRC2.2. In addition to the four major core structures, the former includes Jumonji AT-rich interactive domain 2 (JARID2) and adipocyte enhancer binding protein 2 (AEBP2), the latter of which includes PCL1-3 and c17orf96/c10orf12 [11] (Figure 1). At present, the accessory proteins have been shown to regulate PRC2 activity and play a role in cells by localizing PRC2 to chromatin. However, the exact function of these proteins is unknown.

EZH2 is a critical functional member of the PRCs family, which is located in chromosome 7q35 and consists of 20 exons containing 746 amino acid residues [12] (Figure 2C). It has five structural domains, including the EED-interacting structural domain (EID), structural domain I, structural domain II (SANT2L), a cysteine-rich structural domain (CXC structural domain), and a three-chested structural domain (SET structural domain) [12,13] (Figure 2D). Its most critical function is to inhibit gene expression by promoting histone methylation and DNA methylation in the nucleus.

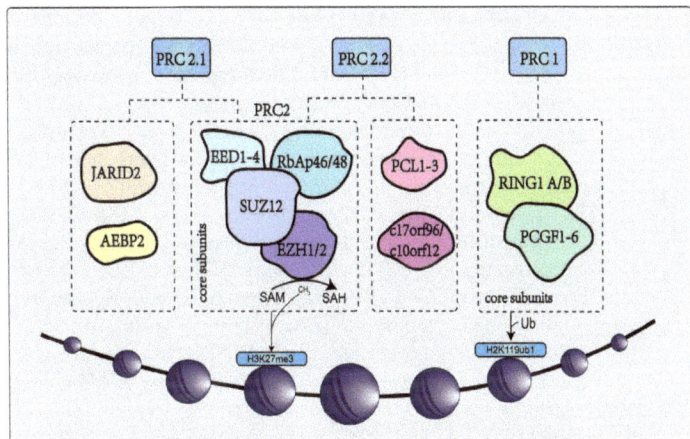

Figure 1. The core structure of PRCs, including PRC1 and PRC2. PRC1 complexes are E3 ubiquitin ligases that monoubiquitinate lysine 119 of histone H2A (H2AK119ub1), consisting mainly of two core subunits. PRC2 consists of four major core subunits and binds to different non-core subunits divided into PRC2.1 and PRC2.2. PRC2 catalyzes the monomethylation, dimethylation, and trimethylation of lysine 27 on histone H3 (H3K27me1, H3K27me2, and H3K27me3). SAM provides the methyl group for the reaction catalyzed by histone methyltransferase.

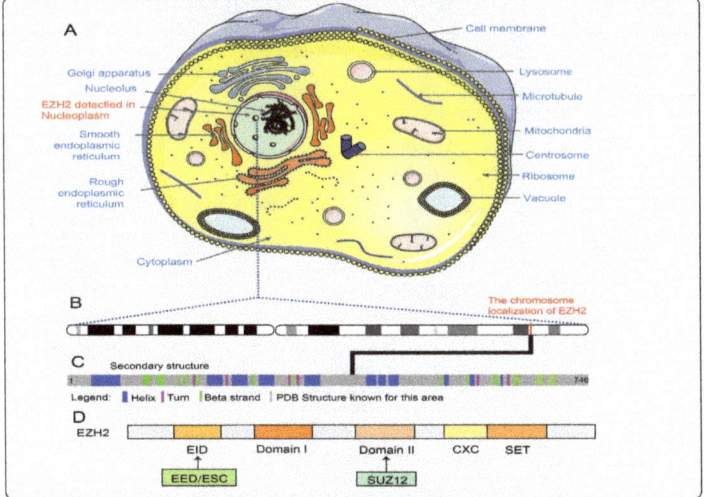

Figure 2. The gene and protein structure of the EZH2. (**A**) The location of the EZH2 protein in the cell. (**B**) The chromosome localization of EZH2. (**C**) The protein secondary structure of EZH2. (**D**) Schematic representation of the organization of the five functional domains in EZH2 is depicted. The EID structural domain is the binding site for the EED subunit in the RC2 complex. The domain II structural domain is the linkage site for the SUZ12 subunit in the PRC2 complex. The SET structural domain is the site that exerts methyl transfer activity and is also the binding site for SAM. The CXC structural domain also contributes to methyl transfer activity, whereas the function of the domain I structural domain is not known.

3. Histone Modification of EZH2

Histones are an essential part of the nucleosome, the basic structure of chromosomes. The modification of histones can change the loose or agglutination state of chromatin which has a regulatory effect on gene expression similar to that of the DNA genetic code [14]. It was clarified that H3K27me3 is considered a key epigenetic event that can make the chromosome structure denser to inhibit the expression of target genes [15]. EZH2, a histone methyltransferase in the human genome, catalyzes the lysine trimethylation of histone 3 at position 27 (H3K27me3), which leads to the silencing of its target genes involved in cell proliferation, cell differentiation, and cancer development [16,17]. However, it is noteworthy that EZH2 alone cannot exert its methyltransferase biological activity. EZH2 must be combined with at least two non-catalytic partners, SUZ12 and EED, to obtain a strong histone methyltransferase activity [18–20]. In addition, EZH2 typically forms PRC2 complexes to perform histone-modifying processes, which is the classical mode of action of EZH2 and will be explicitly described below. EZH2 is overactive in cancer cells through functionally acquired mutations and overexpression. In a study on prostate cancer, elevated levels of EZH2 and H3K27me3 were associated with poor prognosis in metastatic prostate cancer. However, the deletion of EZH2 inhibited the growth of prostate cancer cells [21]. The overexpression of EZH2 and elevation of H3K27me3 in solid cancers, including breast, gastric, endometrial, ovarian, melanoma, bladder, kidney, colorectal, and lung cancers, as well as hematological malignancies such as T-cell and B-cell lymphomas [22,23]. In conclusion, the histone modification of EZH2 is closely related to tumorigenesis. The overexpression of EZH2 leads to an increase in H3K27me3, which inhibits tumor suppressor genes including p16 and E-cadherin, and drives cellular differentiation [24] (Table 1).

Table 1. The genes that are regulated by EZH2 through histone modifications.

Genes	Mechanism of Action of EZH2	The Role of Genes	Reference
METTL3	EZH2 overexpression leads to increases in H3K27me3, up-regulating the expression of METTL3	Drug resistance	[25]
P16	EZH2 overexpression leads to increases in H3K27me3, inhibiting the expression of P16	Inhibition of tumor growth; Drive cellular differentiation	[26]
E-cadherin	EZH2 overexpression leads to increases in H3K27me3, inhibiting E-cadherin	Inhibition of tumor growth	[26]
HIF-1α	EZH2 stabilizes the expression of HIF-1α	Promotion of tumor growth and metabolism favoring glycolysis	[5]
INK4B-ARF-INK4A	EZH2 suppresses the expression of INK4B–ARF–INK4A	Induce cell cycle progression and inhibit cell senescence	[27]

Note: METTL3: methyltransferase-like 3; HIF-1α: hypoxia inducible factor-1.

4. DNA Methylation of EZH2

DNA methylation is another significant mechanism of epigenetic regulation. DNA hypermethylation at promoter site CpG islands is thought to promote tumorigenesis through transcriptional silencing of tumor suppressor genes [28]. There are some DNA methyltransferases (DNMTs), including DNMT1, DNMT3A, and DNMT3B, which assist in the pattern of DNA methylation. Some studies reported that these genes play a crucial role in various human cancers [26]. Vire et al. found that EZH2 interacts directly with DNMTs and recruits DNMT to the promoter regions of target genes in cancer cells, affecting DNA methylation status, suggesting that EZH2 is involved in DNA methylation [17]. (Table 2). For these reasons, EZH2 is not only essential in histone modification, but also plays a vital role in DNA methylation as EZH2 is thought to be a recruitment platform for DNMT. In fact, histone methylation can help to guide DNA methylation patterns. In addition, DNA methylation might serve as a template for some histone modifications after DNA replication [29]. In general, this suggests that histone modifications and DNA methylation synergistically regulate epigenetic states and promote tumor development.

Table 2. The genes that are regulated by EZH2 through DNA methylation.

Genes	Mechanism of Action of EZH2	The Role of Genes	Reference
BMPR1B	EZH2 represses the expression of BMPR1B	Inhibition of growth of tumors	[30]
VASH1	EZH2 represses VASH1	Promotion of the angiogenesis of tumors	[31]
SRBC	EZH2 plays a substantial role in silencing SRBC	Inhibition of tumor growth; involved in tumor resistance against chemotherapeutic agents	[32]
RASSF5	EZH2 inhibits RASSF5	Suppression of cell growth	[33]
ITGB2	EZH2 inhibits ITGB2	Contribute to natural killer cell development and function	[34]

Note: *BMPR1B*: bone morphogenetic protein receptor type-1B; *VASH1*: vasohibin1; *SRBC*: serum deprivation response factor-related gene product that binds to the c-kinase; *RASSF5*: ras association domain family member 5; *ITGB2*: integrin beta 2.

5. Mechanism of Abnormal Expression of EZH2 in Cancer

Epigenetic abnormalities are key factors in the development and progression of cancer [35]. It has been widely recognized that EZH2 can be regulated in different kinds of human cancers at both the transcriptional and translational levels [36]. Next, we will introduce the mechanism of EZH2 in cancer in different ways (Figure 3).

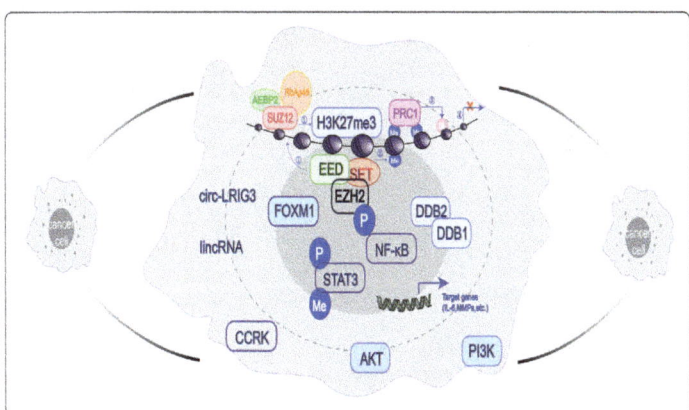

Figure 3. Mechanism of abnormal expression of EZH2 in cancer. The figure mainly shows the regulation of EZH2 at the level of transcription and translation. DDB: damage specific DNA binding protein; CCRK: cell cycle-related kinas; FOXM1: forkhead box protein M1.

6. Regulation of EZH2 in Transcription Levels

6.1. Transcription Inhibition Dependent on PcG Family

EZH2 acts primarily as a histone methyltransferase through its SET domain which can inhibit or co-activate transcription in a PRC2-dependent or independent manner. PcG protein, an important epigenetic regulator, is considered a transcriptional repressor and a key regulator of cell fate in cancer development. They include two complexes, PRC1 and PRC2 [37]. It has been reported that EZH2 can be used as a catalytic subunit of PRC2, the complex that participates in the transcriptional inhibition of multiple target genes, including the inactivation of more than 200 tumor suppressor genes [12]. In the process of gene silencing, EZH2, SUZ12, EED, retinoblastoma suppressor associated protein 46/48 (RbAp46/48), and AEBP2 first polymerize to form a PRC2 complex in the nucleus.

Subsequently, the catalytic domain SETB (su(var)3-9, enhancer of zeste, trithorax) in EZH2 will catalyze trimethylation of lysine at position 27 of histone H3 (H3K27me3) that binds to the target gene. In addition, the subunit homologue of PCR1, that has the same function as PRC2, recognizes and binds to the methylation site of histone H3. Finally, RING 1, the catalytic subunit of PRC1, further catalyzes lysine monobiquitination at position 119 of histone H2A and inhibits the transcriptional elongation reaction that is dependent on RNA polymerase, thus inhibiting the transcription of target genes [26]. In recent years, with the completion of whole genome sequencing and chromatin immunoprecipitation sequencing, scientists have found that some genes can still be recognized by PRC2 and inhibit transcription even those that lack H2AK119ub1 modification sites [38]. In addition, PRC1 can also independently polymerize to the target gene and mediate H2AK119ub1 generation without the involvement of PRC2 and the methylation of H3 histone [39]. Meanwhile, it was found that the EZH2 target gene labeled with H3K27me3 by PRC2 in normal cells has a corresponding relationship with the abnormal hypermethylation gene in cancer cells. Therefore, it is speculated that under the condition of the existence of carcinogenic inducements, the abnormal increase of EZH2 expression will change the target gene of H3K27me3 marker mediated by PRC2 into a higher methylation state during normal development, thus changing the activity of the target gene and causing the transformation of normal cells into cancer cells [40].

6.2. Transcription Inhibition Independent of the PcG Family

There is another non-classical EZH2 regulatory mechanism in cells which is a PRC2-independent way to directly combine with other factors to form transcription complexes to activate the transcription of downstream target genes. It was reported that cycle-related kinases activate the EZH2/NF-κB signaling pathway. Subsequently, EZH2 interacts with NF-κB to form the EZH2–p-p65-Ser536 complex, which binds to the IL-6 promoter and leads to the accumulation of immunosuppressive myeloid-derived suppressor cells (MDSCs), providing a tumor immunosuppressive microenvironment and inducing drug resistance [41]. Moreover, EZH2, along with the potential epigenetic regulator lysine-specific demethylase 2B (KDM2B), dampens colorectal cancer migration, invasion, and maintenance of cancer stem cells via the PI3K/AKT pathway [42]. Kim also found that Akt-mediated serine phosphorylation of EZH2 at position 21 occurs in glioma stem cells, allowing EZH2 to interact with the transcriptional activation factor STAT3 to induce lysine methylation at position 180 [43]. It follows that EZH2 directly interacts physically with several transcription factors in tumor cells and exerts histone methyltransferase activity independently of the PcG family.

It has been reported that RNA may also be involved in the independent regulation of EZH2 transcription. There is negative regulation of the expression of EZH2 by microRNA (miRNA). In tumor cells, down-regulation of miRNA was significantly correlated with EZH2 over-expression because of direct inhibition of the transcription and translation process of EZH2 by miRNA [44]. Moreover, EZH2 can bind to the STAT3 signaling pathway via a novel circular RNA named circ-LRIG3, resulting in methylation and phosphorylation, and promoting the development and progression of hepatocellular carcinoma [45]. In addition, the functions of EZH2 that switches from histone methyltransferase to non-histone methyltransferase rely on long non-coding RNA (lncRNA) p21. Sun reported that STAT3/lncRNA HOTAIR interacts with pEZH2-Ser21 to regulate the growth of head and neck squamous cell carcinoma, thereby improving the anti-tumor efficacy of cisplatin and cetuximab therapy [46,47]. It can be seen that EZH2 plays different roles in regulating the transcription of target genes by interacting with different transcription factors.

7. Regulation of EZH2 Translation and Post-Translational Modifications

It has been reported that the two main mechanisms responsible for protein diversity, mRNA and PTMs, lead to the number of proteins far exceeding the estimated DNA coding capabilities [48]. At present, increasing evidence supports that EZH2 can also be regulated by post-translational modifications (PTMs) in the development of cancer. Several studies have confirmed the importance of PTMs in the regulation of tumors by EZH2, particularly those involved in phosphorylation, acetylation, ubiquitination, sumoylation, and O-GlcNAcylation [37]. The scientists found that phosphorylation of T416 mediated by cyclin-dependent kinase 2 promoted EZH2 recruitment at target gene promoters, and the methylation of K307 mediated by SMYD2 enhanced the stability of EZH2 [49,50]. Glycosylation of O-GlcNAc occurs at multiple sites in the EZH2 molecule, such as S73, S84, S87, T313, and S729, which can regulate the level of free EZH2 [51]. In addition, ubiquitination is also vital for EZH2 regulation. For example, the degradation of K63 ubiquitination mediated by tumor necrosis factor receptor-associated factor 6 can reduce EZH2 levels [52]. Collectively, studying the regulation of EZH2 by different types of PTMs and their regulation in carcinogenesis, as well as elucidating its intrinsic molecular mechanism, will open up a potential and promising approach for tumor therapy.

8. Role of EZH2 in GC

The occurrence of GC is a multi-stage evolutionary process in which gene research is crucial for the diagnosis and treatment of GCs (Table 3). EZH2 alters the cellular memory system and regulates transcription by targeting silencing mechanisms, which is closely related to the occurrence and development of GCs. Next, we mainly introduce the expression and prognostic effects of EZH2 in GC, and discuss the possible mechanism of EZH2 guiding the progression of GC from aspects of metastasis metabolism and drug resistance.

Table 3. The genes related to EZH2 and gastric carcinogenesis.

Genes	Mechanism of Action of EZH2	The Role of Genes	Reference
E-cadherin	EZH2 causes the silencing of the E-cadherin gene	E-cadherin is involved in epithelial–mesenchymal transition, causing GC metastasis	[53]
PTEN	EZH2 downregulates PTEN expression	PTEN causes GC metastasis	[54]
P21	p21 increases when EZH2 is knocked down	p21 inhibits proliferation and invasion of GC cells	[55]
P16	p16 increases when EZH2 is knocked down	p16 promotes GC cellular senescence	[56]
INK4/ARF	EZH2 silencing results in the activation of INK4/ARF	INK4/ARF causes cell cycle arrest and induces senescence in GC cells	[57]

PTEN: Phosphatase and tensin homolog deleted on chromosome ten.

9. Association of the EZH2 Gene with Genetic Susceptibility to GC

In recent years, the relationship between EZH2 single-nucleotide polymorphisms (SNPs) and tumor genetic susceptibility has attracted the attention of many scholars. Variants of SNPs in EZH2 can affect the function of EZH2 and its downstream targets by altering EZH2 transcription and H3K27 trimethylation. Thus SNPs associated with poorer prognosis would be predictors of higher EZH2 expression [58]. Joan found that the frequency of the rs2302427 (D185H) allele of the EZH2 gene was 3.7% and 5.2% in the case and control groups, respectively, and that the EZH2 heterozygous genotype significantly reduced the risk of prostate cancer (OR = 0.63, P = 0.0085) [59]. Crea, F. et al. showed that the rs3757441 genotype of the EZH2 gene was associated with progression-free survival time and overall survival time in colon cancer. More importantly, some reports focused on the involvement of EZH2 SNPs in GC susceptibility. Zhou et al. investigated the relationship between SNP variants in the EZH2 gene and genetic susceptibility to GC using a case-control study [60]. Their study confirmed that EZH2 variants were significantly associated with GC risk, which provides a new perspective on the susceptibility factors of EZH2 gene variants in gastric carcinogenesis. Their study genotyped EZH2 in 311 cases of GC and 425 cases of the Chinese Han population and found 5 SNPs (rs12670401, rs6464926, rs2072407, rs734005, rs734004) in the EZH2 gene that were significantly associated with the risk of GC development. Among them, the rs12670401 genotype CC and rs6464926 TT can increase the risk of GC. rs12670401 is located at the intron region of the EID that binds with EED, and rs6464926 is located in the D2 intron region of the SUZ12-binding site. Therefore, these two SNP loci polymorphisms may indirectly affect the binding efficiency of EZH2 to EED and SUZ12, thus affecting the formation of PRC2 and further preventing it from playing the role of a histone methyltransferase. The other three loci, rs734004 genotype CG, rs734005 genotype TC, and rs2072407 genotype TC, all reduced the risk of gastric carcinogenesis. In another study, and Lee genotyped 23 tag SNPs of EZH2 in 2349 Korean participants. The SNP genotypes of 1100 patients with GC and 1249 healthy controls were compared to conduct a statistical test for their GC risk correlation and epistasis. The results showed that EZH2 SNPs were associated with susceptibility to GC, the depth of primary tumor invasion, and lymph node metastasis [61], which was consistent with the findings of Sun et al. [62]. In conclusion, gene polymorphisms in EZH2 play a crucial role in the occurrence and development of gastric cancer. However, the current research on the relationship between the gene polymorphisms in EZH2 and gastric cancer is not comprehensive enough, and the research area of relevant experiments is also limited and needs further research.

10. Correlation of EZH2 Gene with Gastric Carcinogenesis and Invasive Metastasis

The correlation between the EZH2 gene and cell invasion and metastasis of cancer is a research hotspot. Some studies have shown that the mutation or overexpression of the EZH2 gene is often directly related to the progression of malignant tumors. In recent years, the increased expression of EZH2 in tissues of GC is significantly associated with gastric carcinogenesis, progression, invasion, and metastasis.

One of the key events is that EZH2 promotes epithelial–mesenchymal transition (EMT) events. EMT is a phenomenon that leads epithelial cells to gradually acquire a mesenchymal phenotype. During this process, the expression of specific epithelial proteins called "epithelial markers" is deficient, such as E-cadherin (encoded by the CDH1 gene), claudin, or occludin. Moreover, there is also increased expression of proteins called "interstitial markers", such as vimentin or N-cadherin [63]. EMT is a crucial factor in promoting cancer cell progression [64]. It has been reported that low EMT expression blocks GC metastasis through epigenetic modifications. Fujii showed that overexpression of EZH2 caused silencing of the E-cadherin gene in MKN1 cells, while knockdown of EZH2 reversed E-cadherin deletion and downregulated the invasive ability of GC cells [65]. In addition, it has been found that non-coding RNAs (ncRNA) are upstream molecules that regulate EZH2 and can affect downstream signaling pathways that mediate EMT events. Liu showed that HOTAIR recruits and binds PRC2 to inhibit miR34a by epigenetic inheritance, which controls the targets C-Met (HGF/C-Met/Snail pathway) and Snail, contributing to the EMT process in GC cells and accelerating tumor metastasis [66]. TP73-AS1 plays an oncogenic role in Epstein-Barr virus-associated GC (EBVaGC). Using RIP analysis, the authors found that EZH2 directly targets TP73-AS1, suggesting that TP73-AS2 may interact with EZH2 to silence WIFI in an epigenetic manner and trigger EMT events. Furthermore, TP73-AS1 knockdown suppressed EZH2 binding and H3K27me3 levels in the WIF1 promoter, and WIF1 transcription was enhanced [67]. Interestingly, overexpression of EZH2 could reverse the up-regulation of WIF1 mRNA and protein levels induced by TP73-AS1 knockdown, thereby mediating EBVaGC progression. Carvalho et al. [68] revealed the relationship between the miR-101-EZH2 pathway and EMT. The data show that miR-101-downregulated GC cases displayed concomitant EZH2 overexpression (at the RNA and protein levels), which in turn correlates with E-cadherin deletion/abnormal expression. In vitro experiments showed that after transient depletion of EZH2 in KatoII cells using RNAi (Kato-siEZH2), EZH2 transcript levels were reduced, resulting in increased CDH1 mRNA levels. E-cadherin was restored to the plasma membrane of EZH2-deficient cells compared to non-silenced siRNA cells. This strongly suggests that the histone methyltransferase EZH2 mediates the dysfunction of E-cadherin in GC. In addition, EZH2 directly binds to key tumor oncogenes and triggers signaling pathways for EMT events. EZH2 binds to the PTEN locus and downregulates PTEN expression, which activates the Akt pathway, stabilizes vimentin, downregulates E-cadherin, and protects Sox2 and Oct4 from degradation. Thus, this ultimately leads to the acquisition of EMT and pluripotent phenotype in GC cells [54].

EMT-activated transcription factors (EMT-ATF), such as the Snail, Twist, and E-box binding zinc finger protein (ZEB) families, are significant regulators of EMT. The ncRNAs are dysregulated in GC, which can cause EMT events by regulating EMT-ATF. For example, lncRNA CCAT2 downregulates E-calmodulin expression and upregulates ZEB2 expression, which promotes EMT in GC cells. In addition, CCAT2 interacts with EZH2 to regulate the expression of E-cadherin and large tumor suppressor homolog 2 (LATS2) [69,70]. In addition, EZH2 interacts with EMT-ATF to form a multimolecular complex that contributes to the silencing of E-cadherin. Moreover, EZH2 is required to help to recruit Snail-Ring1A/B to E-cadherin promoter sites [71]. The most striking feature of cells that undergo EMT is enhanced cell motility, thus promoting tumor cell invasion and metastasis. In addition, EMT mediates immune evasion and drug resistance in tumor cells. This shows that EMT provides various benefits for tumor growth [72,73].

Another critical event is that EZH2 suppresses tumor growth suppressor genes. Several growth-suppressing genes, including p21, p16, and p27, and pro-death genes, including F-box protein 32 protein (FBOX32), are downstream targets of EZH2 and participate in proliferation, cell cycle arrest, senescence, and apoptosis of tumor cells, and ultimately determines the cell fate [74,75]. Currently, the most studied classical pathway involved in cell cycle alteration is the p53/p21 signaling pathway [76]. As a member of the Cip/Kip family of cyclin kinase inhibitors (CKIs), p21 is a major effector of various tumor inhibition pathways with anti-proliferative activity. It mainly binds and inhibits cyclin-dependent kinases (CDKs) to regulate their biological activity, leading to growth arrest at specific stages of the cell cycle. p21, a key downstream regulator of EZH2, is significantly increased in GC cells with the knockdown of EZH2, resulting in the inhibition of proliferation and invasion of GC cells. In contrast, in GC cells without knockdown of EZH2, EZH2 binds directly to the p21 promoter region. It mediates H3K27me3 modifications that mediate transcriptional repression of P21. Thus, this suggests that EZH2 acts as an oncogene in GC cells by regulating p21 [55]. CXXC finger protein 4 (CXXC4) is a newly discovered GC suppressor and was identified as a new target of EZH2. EZH2 promotes the activation of Wnt signaling by downregulating the expression of CXXC4. The induced aberrant activation of typical Wnt/β-catenin signaling is one of the drivers of progression in many cancers, including GC [77]. Moreover, other tumor suppressor genes that have been studied in GC are targeted by EZH2 for inhibition. For example, EZH2 targeting inhibits CDH1, and the invasion of GC cells is enhanced [65,78]. However, the administration of exogenous CDH1 prevented invasion. In addition, RUNX3 controls the proliferation of gastric epithelial cells. After the knockdown of EZH2, RUNX3 expression was upregulated, and GC cell proliferation was inhibited [78]. These studies suggest that EZH2 can promote the development of gastric carcinogenesis by downregulating the expression of downstream tumor suppressor genes. However, their specific mechanisms of action are unclear and need to be uncovered by further studies.

There is another way of inhibiting cell proliferation by inducing cellular senescence. During senescence, EZH2 is explicitly downregulated in senescent cells, and the deletion of EZH2 has an impact on histone methylation patterns. Deletion of the INK4/ARF gene on chromosome 9p21 is one of the most common cytogenetic events in human cancers [79]. INK4/ARF encodes $p15^{INK4b}$, $p14^{ARF}$, and $p16^{INK4a}$, which are known as common key reprogramming regulators and are inducers of cellular progression toward senescence [80]. Specifically, EZH2 silencing resulted in the loss of H3K27me3 and activation of INK4/ARF genes to some extent, which led to upregulated expression of $p15^{INK4b}$, $p14^{ARF}$, and $p16^{INK4a}$, causing cell cycle arrest and inducing senescence in GC cells [57]. These results clarified that GC cells could escape senescence by recruiting EZH2 to the INK4/ARF locus. Similarly, Bai found that EZH2 suppressed the senescent state in the human GC cells SGC-7901. There is a recovery of phenotypic features of cellular senescence when EZH2 is depleted in cells. Moreover, p21 and p16 were activated to some extent upon EZH2 depletion [55]. EZH2 knockdown causing cellular senescence is an exciting topic. Ito et al. [56] further delved into the mechanism of cellular senescence induced by EZH2 disruption in two broad phases: first, depletion of EZH2 in proliferating cells rapidly initiates a DNA damage response, but significant changes in H3K27me3 do not accompany this phase. Second, concomitant with the eventual deletion of H3K27me3 at a later stage, it induces p16 (CDKN2A) gene expression and effectively activates the senescence-associated secretory phenotype genes (SASP). Here, the gradual depletion of the H3K27me3 marker can be seen as a molecular "timer" that provides a window for cellular repair of DNA damage. Cellular senescence plays an essential physiological role in tumor suppression. Blocking tumor cell cycle progression is one of the important ways to fight against tumors.

In general, EZH2 plays an active role in EMT events and inhibition of cellular senescence. Moreover, these two significant events facilitate cell invasion and stable growth [81]. EZH2 is an active participant in the occurrence of these two major events that cause the progression of GC. Therefore, given the important role of EZH2 in GC, targeted inhibition of EZH2 expression is a promising measure for treating GC.

11. EZH2 Mediates Resistance to Chemotherapeutic Drugs of GC

Chemotherapy is an important measure for treating advanced GC [82]. Due to the emergence of insensitivity and multi-drug resistance (MDR), chemotherapy is not effective in the treatment of GC patients. First-line chemotherapy drugs commonly used clinically for GC, such as oxaliplatin and capecitabine combined chemotherapy, failed in 95% of non-operative patients with GC. Unfortunately, second-line chemotherapeutic agents, such as mitomycin C, irinotecan, adriamycin, methotrexate, or etoposide, also fail to provide better efficacy for patients with GC [83]. Therefore, the drug resistance of GC chemotherapy drugs significantly shortens the survival of GC patients. Thus, there is a great need to understand the mechanisms of GC chemotherapy resistance. The possible mechanisms of drug resistance in GC include reduced drug uptake by GC cells or increased drug efflux; a reduced proportion of active agents in tumor cells due to a reduction in pro-drug activation or an enhancement in drug inactivation; expression and functional changes of molecular targets of anticancer drugs; change in the ability of cancer cells to repair DNA damage induced by anticancer drugs; and expression/function of pro-apoptotic factors or up-regulation of anti-apoptotic genes [83].

There are few studies on the EZH2 gene and chemotherapy sensitivity of GC. Some scholars have found that EZH2 is significantly up-regulated in drug-resistant GC cell lines and is involved in regulating the sensitivity of GC to chemotherapy drugs, which is the most widely studied among platinum-based drugs. Zhou investigated the effect of EHZ2 on cisplatin resistance in AGS/DDP cells. He found that EZH2 expression levels were significantly higher in AGS/DDP cells than in the parental cells [84]. In addition, the silencing of EZH2 using siRNA increased the intracellular concentration of cisplatin in AGS/DDP cell lines, which significantly reversed the resistance to cisplatin in AGS/DDP cell lines [84]. In another study, Wang et al. used two GC drug-resistant cell lines, namely vincristine (VCR)-resistant cell lines (SGC7901/VCR) and adriamycin (ADR)-resistant cell lines (SGC7901/ADR) and compared them with the parental cell line SGC7901; miR-126 expression was shown to be decreased in these two resistant cell lines. In contrast, the upregulation of miR-126 expression increased the sensitivity of SGC7901/VCR and SGC7901/ADR cells to VCR and ADR. Mechanistically, the enhancer of EZH2 was identified as a direct target of miR-126, and the silencing of EZH2 reflected the role of miR-126 in drug resistance. However, restoring the EZH2 gene blocks the inhibitory effect of miR-126 on GC [85]. This suggests that the EZH2 gene indeed plays a crucial role in chemotherapeutic agents for GC. In addition, about 20% of lncRNAs can bind and silence the EZH2 gene to increase drug resistance in GC cells. For example, the lncRNA UCA1 up-regulates the level of EZH2 in GC. The over-expressed EZH2 activates the PI3K/AKT pathway [86], which affects the expression of multiple drug resistance-associated and anti-apoptotic proteins and plays a very important role in chemoresistance [86]. Moreover, the knockdown of the EZH2 gene decreased GC cell proliferation, increasing cisplatin-induced apoptosis and caspase-3 expression, inhibiting UCA1-induced upregulation of PI3K/AKT [87]. The scientists found that the lncRNA PCAT-1 is highly expressed in DDP-resistant tissues and cells in GC, which promotes DDP resistance in GC cells by recruiting EZH2 to epigenetically suppress PTEN expression and regulate the miR-128/ZEB1 axis [88,89]. In conclusion, EZH2 is regulated by multiple lncRNAs and is involved in the resistance of GC cells to platinum-based chemotherapy drugs. A final event shared among the mechanisms of action of many antitumor drugs is the activation of apoptosis [83]. It was reported that the apoptotic factors or anti-apoptotic pathways were inhibited in GC cell lines after the knockdown of EZH2. Thus, silencing of EHZ2 could effectively reverse chemotherapeutic drug resistance in GC cells. Based on the above results, targeted silencing of EZH2 can effectively reverse the resistance of GC cells to chemotherapy drugs.

12. High Expression of EZH2 Leads to Poor Prognosis for GC Patients

There are many factors affecting the prognosis of GC patients, such as the size of the mass, the depth of tumor infiltration, the pathological type of the tumor, the degree of differentiation of tumor cells, the degree of choroidal invasion, and the metastasis of lymph nodes. It has been reported that EZH2 expression is closely related to the above prognostic factors of GC [6,54,90]. Specifically, EZH2 expression was increased in GC tissues, and the higher the expression level, the higher the malignant degree of the tumor and the worse the prognosis. In a study of patients with GC, the tissue samples of 105 patients with primary GC were included for immunohistochemical detection. Among them, 72 patients had a positive expression of EZH2 protein in GC tissues that was higher than that in para-carcinoma tissue [54]. In addition, the overexpression of EZH2 mRNA and protein detected using qRT-PCR and immunohistochemistry was closely associated with tumor size, lymphatic invasion, and TNM stage. These evidences strongly demonstrated that EZH2 is closely related to the prognosis of GC patients. The Kaplan–Meier method was used to analyze the expression of EZH2 and H3K27me3 proteins and their correlation with the prognosis of GC patients. The results showed that these two highly expressed proteins were commonly found in patients with advanced GC and those with lymph node metastases. Moreover, the overall survival rate of patients with high expression was significantly lower than that of those with low expression. These results are similar to those of other authors [90,91]. Meanwhile, in another more convincing meta-analysis, 872 GC patients were included in this study. The results showed that the expression level of EZH2 protein in GC was higher than that in normal gastric tissue, and positively correlated with tumor-node-metastasis (TNM) stage and lymph node metastasis. The overall survival rate of patients with positive EZH2 expression was shorter than that of patients with negative EZH2 expression [92]. Pan found that high expression of EZH2 in GC tissues was regulated by IL-6/STAT3 signaling. STAT3 acted as a transcription factor to enhance the transcriptional activity of EZH2 by binding to the relevant promoter region ($-214\sim-206$), and there is a positive functional loop between STAT3 and EZH2 [91]. Regarding prognosis, STAT3 was positively correlated with EZH2 expression in GC cells and tissues, and activation of EZH2 and STAT3 was significantly associated with the TNM stage and low patient survival. Furthermore, the combination of siSTAT3 and the EZH2-specific inhibitor 3-deazaneplanocin A (DZNep) increased the apoptosis rate of GC cells, suggesting that the combination of siSTAT3 and EZH2 inhibitors may contribute to the potential epigenetic treatment of GC patients. In general, these evidences strongly confirmed the view that high expression of EZH2 may be involved in gastric carcinogenesis.

Therefore, the EZH2 protein may be a valuable biomarker for the diagnosis and prognosis of GC. It is meaningful that targeted inhibition of EZH2 expression could contribute to the potential epigenetic therapy against GC patients.

13. Current Status of EZH2 Inhibitor Research

Given that abnormal expression of EZH2 plays an essential role in tumor cell proliferation, invasion, metastasis, and drug resistance [93], targeted inhibition of EZH2/PRC2 is considered an attractive target for cancer therapy. Here, we review some typical EZH2 inhibitors and their current application status.

DZNep, an inhibitor of S-adenosyl-l-homocysteine (SAH) hydrolase, induces the accumulation of SAH in cells, directly inhibiting histone methyltransferase activity and then indirectly degrading the PRC2 complex. It was one of the first small molecules to evaluate the inhibitory effect of EZH2 [94]. Some studies have shown that DZNep exhibits good anti-tumor properties by inhibiting EZH2 in breast cancer [95], lung cancer [96], prostate cancer [97], colon cancer [98], and other cancer cells. However, DZNep lacks specificity in human tissues because it affects all SAM-dependent processes by blocking the methionine cycle and the regeneration of SAM [99].

In this context, several highly selective EZH2 inhibitors were developed (EPZ005687, GSK126, EPZ6438), which almost always have a 2-pyridone group in their structure [100]. This is mainly because the 2-pyridone group is essential for enzyme inhibition, occupying the site of the common substrate (SAM or the by-product SAH) in the binding pocket of the enzyme [100]. The first highly selective inhibitor was EPZ005687, that competes with SAM and does not compete with peptide substrates, and thereby does not disrupt protein–protein interactions between PRC2 subunits. Most importantly, EPZ005687 was 500 times more selective to the PRC2/EZH2 complex than the other 15 methyltransferases and approximately 50 times more selective to PRC2/EZH2 than PRC2/EZH1 [101]. In lymphomas, EPZ005687 potently reduced H3K27me3 levels in EZH2 mutation-containing cells [94].

Another highly selective inhibitor is GSK126, which has a similar core structure to EPZ005687, but GSK126 is more than 1000 times more selective for EZH2 than the other 20 human methyltransferases [101]. GSK126 effectively suppresses the expression of H3K27me3 in cells with EZH2 mutations. McCabe et al. showed that GSK126 effectively inhibited the proliferation of EZH2 mutant diffuse large B cell lymphoma (DLBCL) cell lines and markedly inhibits the growth of EZH2 mutant DLBCL xenografts in mice. In addition, tumor growth was essentially completely inhibited (91–100% inhibition) in subcutaneous xenografts utilizing the more aggressive KARPAS-422 and Pfeiffer cells when high doses of GSK126 were administered (300 mg/kg twice a week) [102]. In 2014, GSK126 entered phase I clinical trials in patients with various lymphomas and solid tumors [101]. In 2019, the results of the phase 1 clinical study of GSK126 (NCT02082977) were published [103]. However, the trial was forced to stop due to the deficiency of clinical activity of the drug. This is mainly because of the limitation of twice-weekly administration and the pharmacokinetic characteristics of the drug itself. Nevertheless, GSK126 remains a promising agent. There is a mistaken perception that GSK126 may inhibit tumor immunity due to GSK126 treatment essentially inhibiting tumor growth in immunodeficient hosts but not in immunocompetent hosts. In another study, however, GSK126 exerted anti-tumor activity in immunologically active hosts combined with drugs (gemcitabine and 5-fluorouracil) that deplete myeloid-derived suppressor cells [104]. These results provide a new strategy for the use of GSK126 in the clinic.

In 2013, UNC1999 was reported to be the first oral bio-availability inhibitor with high in vitro potency. UNC1999 showed high efficacy in vitro against wild-type and mutant EZH2 and EZH1, effectively reducing the level of H3K27me3 in cells and selectively killing diffuse large B-cell lymphoma cell lines containing EZH2^{Y641N} mutation with low cytotoxicity [105]. Currently, UNC1999 is used to study mixed lineage leukemia [106].

As a result of this effort, Tazemetostat, an oral, first-in-class inhibitor of EZH2, was introduced to strongly inhibit wild-type and mutant EZH2 enzyme activity with its improved potency, pharmacokinetic parameters, and oral bioavailability [107]. Tazemetostat is a direct inhibitor of EZH2, which can competitively bind to the SET domain of EZH2 protein directly with SAM. Tazemetostat has mainly been used in lymphoma and epithelioid sarcoma (ES) studies [108]. The results of a phase I clinical study conducted in France showed that tazemetostat exhibited significant antitumor activity in patients with non-Hodgkin lymphoma (NHL) and advanced solid tumors [109]. The recommended dose of tazemetostat was 800 mg, administered twice a day, and is more effective in NHL. Ribrag et al. [110] suggested that tazemetostat showed apparent clinical efficacy in DLBCL, follicular lymphoma (FL), and marginal zone lymphoma (MZL) patients, showing good anti-tumor activity, with the best effect in patients with stable FL. In a study of tazemetostat for the treatment of epithelioid sarcoma (ES), phase II clinical data showed an overall efficacy rate of 15% and a disease control rate of 26% with tazemetostat administration (2.4 to 18.4 months). Further analysis of clinical study data presented by Epizyme at the 2019 American Society of Clinical Oncology showed that tazemetostat improved life expectancy in ES patients and those first-treatment ES patients taking tazemetostat had better outcomes than patients with relapsed or refractory ES [111]. These results suggest that tazemetostat shows significant anti-ES activity and may be a new treatment option for ES patients. Ultimately, tazemetostat received accelerated approval in January 2020 in the USA for treating adults and adolescents aged ≥ 16 years with locally advanced or metastatic ES not eligible for complete resection [111].

The polyprotein nature of the PRC2 complex and EZH2 function is significantly dependent on those of other core subunits, such as EED. Therefore, scientists developed a compound that binds to EED. On the one hand, the possibility of weakening PRC2 complex function by interfering with the close protein-protein interaction (PPI) between EZH2 and EED has facilitated the development of different chemical types as inhibitors of EZH2−EED. These chemical agents exert a methyltransferase inhibitory activity on PRC2 by impeding the scaffolding role of EED on the SET domain of EZH2 [112]. High-throughput screening demonstrated the "druggability" of the H3K27me3 recognition cavity of EED as a means of heterologous inhibition of EZH2 catalysis. Eventually, the Novartis company developed an EED-binding agent called MAK683 [113], which is currently in phase I/II study (NCT02900651) for the treatment of patients with DLBCL, nasopharyngeal carcinoma, or other advanced solid tumors of malignancy [114]. Some of the information on these inhibitors is summarized in Table 4.

Table 4. Pre-clinical and clinical trial status of drugs related to EZH2.

Drug	Role	Phase	Reference(s)
DZNep	SAH hydrolase inhibitor	pre-clinical	[94,99]
EPZ005687	Inhibitor of EZH2 T641 and A677 mutants	pre-clinical	[101]
GSK126 (GSK2816126)	SAM-competitive inhibitors of EZH2	Phase I	[102]
Tazemetostat (EPZ-6438, E7438)	SAM-competitive inhibitors of EZH2	Phase I/II	[107]
UNC1999	SAM-competitive inhibitors of EZH2 and EZH1	pre-clinical	[105]
MAK683 (EED226)	Selective EED inhibitor	Phase I/II	[113]

In conclusion, some inhibitors targeting EZH2 have achieved some success, but most anti-tumor studies of EZH2 inhibitors are still in the preliminary stages. With the continuous deepening of the research and the gradual extension to the clinic, the molecular mechanism of the anti-tumor effect of EZH2 small molecule inhibitors will be further clarified. In the future, it is necessary to develop efficient, highly selective, and low-toxicity EZH2 inhibitors, which are an important target for cancer therapy.

14. Conclusions

EZH2 is highly expressed in GC, which has been proven to be associated with poor prognosis in GC. These epigenetic disorders are frequently mutated by multiple factors in GC and other cancers, resulting in the uncontrolled expression of many downstream cancer-associated genes. Therefore, it is meaningful that targeting these epigenetic regulators may have positive implications for treating some tumors. Many drugs targeting EZH2/PRC2 are being developed and evaluated in clinical trials. However, most are still in preclinical studies or phase 1/2 clinical trials, with only tazemetostat approved for the treatment of epithelioid sarcoma (ES) and preliminary evidence of efficacy in follicular lymphoma (FL). Although the role of EZH2 in GC has achieved positive results, most of the research has studied the pre-clinical stage of targeted treatment of EZH2, which has not yet broken through to the clinical stage. This is mainly because many of the modification enzymes of EZH2 and the exact sites of PTMs are unknown. In addition, it is not known whether rare types of PTMs exist in EZH2 such as succinylation, malonylation, crotonylation, propionylation, and butyrylation. More importantly, there are no clinical trials targeting EZH2 PTMs for cancer treatment. At present, targeted therapies for EZH2 are mostly focused on the hematologic and lymphatic systems, such as B-cell lymphoma and non-Hodgkin's lymphoma. We may be inspired by studies that have achieved some results, such as combining EZH2 inhibitors with immunotherapy, chemotherapy, targeted therapy, endocrine therapy, and other therapies that may achieve complementary or synergistic anti-tumor effects [115]. In conclusion, as a novel target for GC treatment, EZH2 has become a research hotspot, and its functions and effects have been continuously revealed. In the future, it is necessary to further study its mechanism of action and develop therapeutic drugs based on this target.

Author Contributions: W.Y. and N.L. wrote the manuscript; L.C. and M.W. illustrated the picture in the manuscript; G.X., T.L. and X.S. reviewed the manuscript. Z.W. provided funding acquisition. Y.Z. supervised the manuscript. All authors have read and agreed to the published version of the manuscript.

Funding: This work was funded by the Gansu Province Youth Science and Technology Fund program (20JR10RA759); Chinese Foundation for Hepatitis Prevention and Control—TianQing liver disease research fund subject (TQGB20190165); Science and Technology Projects in Cheng-guan District of Lanzhou City (2014-4-4); Lanzhou City Science and Technology Development Guidance Plan (2019-ZD-71); and Gansu Provincial Administration of Traditional Chinese Medicine Scientific Research Project (GZKZ-2020-7).

Conflicts of Interest: The authors declare that the research was conducted in the absence of any commercial or financial relationships that could be construed as a potential conflict of interest.

References

1. Smyth, E.C.; Nilsson, M.; Grabsch, H.I.; van Grieken, N.C.; Lordick, F. Gastric cancer. *Lancet* **2020**, *396*, 635–648. [CrossRef] [PubMed]
2. Cai, Z.; Liu, Q. Understanding the Global Cancer Statistics 2018: Implications for cancer control. *Sci. China Life Sci.* **2021**, *64*, 1017–1020. [CrossRef] [PubMed]
3. Canale, M.; Casadei-Gardini, A.; Ulivi, P.; Arechederra, M.; Berasain, C.; Lollini, P.L.; Fernández-Barrena, M.G.; Avila, M.A. Epigenetic Mechanisms in Gastric Cancer: Potential New Therapeutic Opportunities. *Int. J. Mol. Sci.* **2020**, *21*, 5500. [CrossRef]
4. Ramezankhani, R.; Solhi, R.; Es, H.A.; Vosough, M.; Hassan, M. Novel molecular targets in gastric adenocarcinoma. *Pharmacol. Ther.* **2021**, *220*, 107714. [CrossRef] [PubMed]
5. Papale, M.; Ferretti, E.; Battaglia, G.; Bellavia, D.; Mai, A.; Tafani, M. EZH2, HIF-1, and Their Inhibitors: An Overview on Pediatric Cancers. *Front. Pediatr.* **2018**, *6*, 328. [CrossRef] [PubMed]
6. He, L.J.; Cai, M.Y.; Xu, G.L.; Li, J.J.; Weng, Z.J.; Xu, D.Z.; Luo, G.Y.; Zhu, S.L.; Xie, D. Prognostic significance of overexpression of EZH2 and H3k27me3 proteins in gastric cancer. *Asian Pac. J. Cancer Prev. APJCP* **2012**, *13*, 3173–3178. [CrossRef] [PubMed]
7. Wang, J.; Wang, G.G. No Easy Way Out for EZH2: Its Pleiotropic, Noncanonical Effects on Gene Regulation and Cellular Function. *Int. J. Mol. Sci.* **2020**, *21*, 9501. [CrossRef]

8. Stairiker, C.J.; Thomas, G.D.; Salek-Ardakani, S. EZH2 as a Regulator of CD8+ T Cell Fate and Function. *Front. Immunol.* **2020**, *11*, 593203. [CrossRef]
9. Chittock, E.C.; Latwiel, S.; Miller, T.C.; Müller, C.W. Molecular architecture of polycomb repressive complexes. *Biochem. Soc. Trans.* **2017**, *45*, 193–205. [CrossRef]
10. Kuzmichev, A.; Nishioka, K.; Erdjument-Bromage, H.; Tempst, P.; Reinberg, D. Histone methyltransferase activity associated with a human multiprotein complex containing the Enhancer of Zeste protein. *Genes Dev.* **2002**, *16*, 2893–2905. [CrossRef]
11. Margueron, R.; Li, G.; Sarma, K.; Blais, A.; Zavadil, J.; Woodcock, C.L.; Dynlacht, B.D.; Reinberg, D. Ezh1 and Ezh2 maintain repressive chromatin through different mechanisms. *Mol. Cell* **2008**, *32*, 503–518. [CrossRef] [PubMed]
12. Simon, J.A.; Lange, C.A. Roles of the EZH2 histone methyltransferase in cancer epigenetics. *Mutat. Res.* **2008**, *647*, 21–29. [CrossRef] [PubMed]
13. Duan, R.; Du, W.; Guo, W. EZH2: A novel target for cancer treatment. *J. Hematol. Oncol.* **2020**, *13*, 104. [CrossRef] [PubMed]
14. Millán-Zambrano, G.; Burton, A.; Bannister, A.J.; Schneider, R. Histone post-translational modifications—Cause and consequence of genome function. *Nat. Rev. Genet.* **2022**, *23*, 563–580. [CrossRef]
15. Chen, Y.; Ren, B.; Yang, J.; Wang, H.; Yang, G.; Xu, R.; You, L.; Zhao, Y. The role of histone methylation in the development of digestive cancers: A potential direction for cancer management. *Signal Transduct. Target. Ther.* **2020**, *5*, 143. [CrossRef]
16. Zhang, T.; Cooper, S.; Brockdorff, N. The interplay of histone modifications—Writers that read. *EMBO Rep.* **2015**, *16*, 1467–1481. [CrossRef]
17. Viré, E.; Brenner, C.; Deplus, R.; Blanchon, L.; Fraga, M.; Didelot, C.; Morey, L.; Van Eynde, A.; Bernard, D.; Vanderwinden, J.M.; et al. The Polycomb group protein EZH2 directly controls DNA methylation. *Nature* **2006**, *439*, 871–874. [CrossRef]
18. Cyrus, S.; Burkardt, D.; Weaver, D.D.; Gibson, W.T. PRC2-complex related dysfunction in overgrowth syndromes: A review of EZH2, EED, and SUZ12 and their syndromic phenotypes. *Am. J. Med. Genetics. Part C Semin. Med. Genet.* **2019**, *181*, 519–531. [CrossRef]
19. Hsu, J.H.; Rasmusson, T.; Robinson, J.; Pachl, F.; Read, J.; Kawatkar, S.; O'Donovan, D.H.; Bagal, S.; Code, E.; Rawlins, P.; et al. EED-Targeted PROTACs Degrade EED, EZH2, and SUZ12 in the PRC2 Complex. *Cell Chem. Biol.* **2020**, *27*, 41–46.e17. [CrossRef]
20. Zhu, K.; Du, D.; Yang, R.; Tao, H.; Zhang, H. Identification and Assessments of Novel and Potent Small-Molecule Inhibitors of EED-EZH2 Interaction of Polycomb Repressive Complex 2 by Computational Methods and Biological Evaluations. *Chem. Pharm. Bull.* **2020**, *68*, 58–63. [CrossRef]
21. Varambally, S.; Dhanasekaran, S.M.; Zhou, M.; Barrette, T.R.; Kumar-Sinha, C.; Sanda, M.G.; Ghosh, D.; Pienta, K.J.; Sewalt, R.G.; Otte, A.P.; et al. The polycomb group protein EZH2 is involved in progression of prostate cancer. *Nature* **2002**, *419*, 624–629. [CrossRef] [PubMed]
22. Kim, K.H.; Roberts, C.W. Targeting EZH2 in cancer. *Nat. Med.* **2016**, *22*, 128–134. [CrossRef] [PubMed]
23. Hanaki, S.; Shimada, M. Targeting EZH2 as cancer therapy. *J. Biochem.* **2021**, *170*, 1–4. [CrossRef] [PubMed]
24. Bracken, A.P.; Kleine-Kohlbrecher, D.; Dietrich, N.; Pasini, D.; Gargiulo, G.; Beekman, C.; Theilgaard-Mönch, K.; Minucci, S.; Porse, B.T.; Marine, J.C.; et al. The Polycomb group proteins bind throughout the INK4A-ARF locus and are disassociated in senescent cells. *Genes Dev.* **2007**, *21*, 525–530. [CrossRef] [PubMed]
25. Li, F.; Chen, S.; Yu, J.; Gao, Z.; Sun, Z.; Yi, Y.; Long, T.; Zhang, C.; Li, Y.; Pan, Y.; et al. Interplay of m(6) A and histone modifications contributes to temozolomide resistance in glioblastoma. *Clin. Transl. Med.* **2021**, *11*, e553. [CrossRef] [PubMed]
26. Eich, M.L.; Athar, M.; Ferguson, J.E., 3rd; Varambally, S. EZH2-Targeted Therapies in Cancer: Hype or a Reality. *Cancer Res.* **2020**, *80*, 5449–5458. [CrossRef]
27. Han Li, C.; Chen, Y. Targeting EZH2 for cancer therapy: Progress and perspective. *Curr. Protein Pept. Sci.* **2015**, *16*, 559–570. [CrossRef]
28. Timp, W.; Feinberg, A.P. Cancer as a dysregulated epigenome allowing cellular growth advantage at the expense of the host. *Nature Rev. Cancer* **2013**, *13*, 497–510. [CrossRef]
29. Cedar, H.; Bergman, Y. Linking DNA methylation and histone modification: Patterns and paradigms. *Nat. Rev. Genet.* **2009**, *10*, 295–304. [CrossRef]
30. Lee, J.; Son, M.J.; Woolard, K.; Donin, N.M.; Li, A.; Cheng, C.H.; Kotliarova, S.; Kotliarov, Y.; Walling, J.; Ahn, S.; et al. Epigenetic-mediated dysfunction of the bone morphogenetic protein pathway inhibits differentiation of glioblastoma-initiating cells. *Cancer Cell* **2008**, *13*, 69–80. [CrossRef]
31. Lu, C.; Han, H.D.; Mangala, L.S.; Ali-Fehmi, R.; Newton, C.S.; Ozbun, L.; Armaiz-Pena, G.N.; Hu, W.; Stone, R.L.; Munkarah, A.; et al. Regulation of tumor angiogenesis by EZH2. *Cancer Cell* **2010**, *18*, 185–197. [CrossRef] [PubMed]
32. Rezaei, S.; Hosseinpourfeizi, M.A.; Moaddab, Y.; Safaralizadeh, R. Contribution of DNA methylation and EZH2 in SRBC down-regulation in gastric cancer. *Mol. Biol. Rep.* **2020**, *47*, 5721–5727. [CrossRef] [PubMed]
33. Li, S.; Teng, J.; Li, H.; Chen, F.; Zheng, J. The Emerging Roles of RASSF5 in Human Malignancy. *Anti-Cancer Agents Med. Chem.* **2018**, *18*, 314–322. [CrossRef] [PubMed]
34. Tiffen, J.; Gallagher, S.J.; Filipp, F.; Gunatilake, D.; Emran, A.A.; Cullinane, C.; Dutton-Register, K.; Aoude, L.; Hayward, N.; Chatterjee, A.; et al. EZH2 Cooperates with DNA Methylation to Downregulate Key Tumor Suppressors and IFN Gene Signatures in Melanoma. *J. Investig. Dermatol.* **2020**, *140*, 2442–2454.e2445. [CrossRef]
35. Asano, T. Drug Resistance in Cancer Therapy and the Role of Epigenetics. *J. Nippon. Med. Sch. Nippon. Ika Daigaku Zasshi* **2020**, *87*, 244–251. [CrossRef]

36. Batool, A.; Jin, C.; Liu, Y.X. Role of EZH2 in cell lineage determination and relative signaling pathways. *Front. Biosci. (Landmark Ed.)* **2019**, *24*, 947–960. [CrossRef]
37. Lu, H.; Li, G.; Zhou, C.; Jin, W.; Qian, X.; Wang, Z.; Pan, H.; Jin, H.; Wang, X. Regulation and role of post-translational modifications of enhancer of zeste homologue 2 in cancer development. *Am. J. Cancer Res.* **2016**, *6*, 2737–2754.
38. Ku, M.; Koche, R.P.; Rheinbay, E.; Mendenhall, E.M.; Endoh, M.; Mikkelsen, T.S.; Presser, A.; Nusbaum, C.; Xie, X.; Chi, A.S.; et al. Genomewide analysis of PRC1 and PRC2 occupancy identifies two classes of bivalent domains. *PLoS Genet.* **2008**, *4*, e1000242. [CrossRef]
39. Leeb, M.; Pasini, D.; Novatchkova, M.; Jaritz, M.; Helin, K.; Wutz, A. Polycomb complexes act redundantly to repress genomic repeats and genes. *Genes Dev.* **2010**, *24*, 265–276. [CrossRef]
40. Chen, X.; Pan, X.; Zhang, W.; Guo, H.; Cheng, S.; He, Q.; Yang, B.; Ding, L. Epigenetic strategies synergize with PD-L1/PD-1 targeted cancer immunotherapies to enhance antitumor responses. *Acta Pharm. Sin. B* **2020**, *10*, 723–733. [CrossRef]
41. Sun, S.; Yu, F.; Xu, D.; Zheng, H.; Li, M. EZH2, a prominent orchestrator of genetic and epigenetic regulation of solid tumor microenvironment and immunotherapy. *Biochim. Biophys. Acta Rev. Cancer* **2022**, *1877*, 188700. [CrossRef] [PubMed]
42. Sanches, J.G.P.; Song, B.; Zhang, Q.; Cui, X.; Yabasin, I.B.; Ntim, M.; Li, X.; He, J.; Zhang, Y.; Mao, J.; et al. The Role of KDM2B and EZH2 in Regulating the Stemness in Colorectal Cancer Through the PI3K/AKT Pathway. *Front. Oncol.* **2021**, *11*, 637298. [CrossRef] [PubMed]
43. Kim, E.; Kim, M.; Woo, D.H.; Shin, Y.; Shin, J.; Chang, N.; Oh, Y.T.; Kim, H.; Rheey, J.; Nakano, I.; et al. Phosphorylation of EZH2 activates STAT3 signaling via STAT3 methylation and promotes tumorigenicity of glioblastoma stem-like cells. *Cancer Cell* **2013**, *23*, 839–852. [CrossRef]
44. Yan, K.S.; Lin, C.Y.; Liao, T.W.; Peng, C.M.; Lee, S.C.; Liu, Y.J.; Chan, W.P.; Chou, R.H. EZH2 in Cancer Progression and Potential Application in Cancer Therapy: A Friend or Foe? *Int. J. Mol. Sci.* **2017**, *18*, 1172. [CrossRef] [PubMed]
45. Tumes, D.J.; Onodera, A.; Suzuki, A.; Shinoda, K.; Endo, Y.; Iwamura, C.; Hosokawa, H.; Koseki, H.; Tokoyoda, K.; Suzuki, Y.; et al. The polycomb protein Ezh2 regulates differentiation and plasticity of CD4(+) T helper type 1 and type 2 cells. *Immunity* **2013**, *39*, 819–832. [CrossRef] [PubMed]
46. Sun, S.; Wu, Y.; Guo, W.; Yu, F.; Kong, L.; Ren, Y.; Wang, Y.; Yao, X.; Jing, C.; Zhang, C.; et al. STAT3/HOTAIR Signaling Axis Regulates HNSCC Growth in an EZH2-dependent Manner. *Clin. Cancer Res. Off. J. Am. Assoc. Cancer Res.* **2018**, *24*, 2665–2677. [CrossRef]
47. Hao, A.; Wang, Y.; Stovall, D.B.; Wang, Y.; Sui, G. Emerging Roles of LncRNAs in the EZH2-regulated Oncogenic Network. *Int. J. Biol. Sci.* **2021**, *17*, 3268–3280. [CrossRef]
48. Vu, L.D.; Gevaert, K.; De Smet, I. Protein Language: Post-Translational Modifications Talking to Each Other. *Trends Plant Sci.* **2018**, *23*, 1068–1080. [CrossRef]
49. Nie, L.; Wei, Y.; Zhang, F.; Hsu, Y.H.; Chan, L.C.; Xia, W.; Ke, B.; Zhu, C.; Deng, R.; Tang, J.; et al. CDK2-mediated site-specific phosphorylation of EZH2 drives and maintains triple-negative breast cancer. *Nat. Commun.* **2019**, *10*, 5114. [CrossRef]
50. Zeng, Y.; Qiu, R.; Yang, Y.; Gao, T.; Zheng, Y.; Huang, W.; Gao, J.; Zhang, K.; Liu, R.; Wang, S.; et al. Regulation of EZH2 by SMYD2-Mediated Lysine Methylation Is Implicated in Tumorigenesis. *Cell Rep.* **2019**, *29*, 1482–1498.e1484. [CrossRef]
51. Lo, P.W.; Shie, J.J.; Chen, C.H.; Wu, C.Y.; Hsu, T.L.; Wong, C.H. O-GlcNAcylation regulates the stability and enzymatic activity of the histone methyltransferase EZH2. *Proc. Natl. Acad. Sci. USA* **2018**, *115*, 7302–7307. [CrossRef] [PubMed]
52. Lu, W.; Liu, S.; Li, B.; Xie, Y.; Izban, M.G.; Ballard, B.R.; Sathyanarayana, S.A.; Adunyah, S.E.; Matusik, R.J.; Chen, Z. SKP2 loss destabilizes EZH2 by promoting TRAF6-mediated ubiquitination to suppress prostate cancer. *Oncogene* **2017**, *36*, 1364–1373. [CrossRef] [PubMed]
53. Qiao, Y.; Jiang, X.; Lee, S.T.; Karuturi, R.K.; Hooi, S.C.; Yu, Q. FOXQ1 regulates epithelial-mesenchymal transition in human cancers. *Cancer Res.* **2011**, *71*, 3076–3086. [CrossRef]
54. Gan, L.; Xu, M.; Hua, R.; Tan, C.; Zhang, J.; Gong, Y.; Wu, Z.; Weng, W.; Sheng, W.; Guo, W. The polycomb group protein EZH2 induces epithelial-mesenchymal transition and pluripotent phenotype of gastric cancer cells by binding to PTEN promoter. *J. Hematol. Oncol.* **2018**, *11*, 9. [CrossRef] [PubMed]
55. Bai, J.; Chen, J.; Ma, M.; Cai, M.; Xu, F.; Wang, G.; Tao, K.; Shuai, X. Inhibiting enhancer of zeste homolog 2 promotes cellular senescence in gastric cancer cells SGC-7901 by activation of p21 and p16. *DNA Cell Biol.* **2014**, *33*, 337–344. [CrossRef]
56. Ito, T.; Teo, Y.V.; Evans, S.A.; Neretti, N.; Sedivy, J.M. Regulation of Cellular Senescence by Polycomb Chromatin Modifiers through Distinct DNA Damage- and Histone Methylation-Dependent Pathways. *Cell Rep.* **2018**, *22*, 3480–3492. [CrossRef]
57. Jie, B.; Weilong, C.; Ming, C.; Fei, X.; Xinghua, L.; Junhua, C.; Guobin, W.; Kaixiong, T.; Xiaoming, S. Enhancer of zeste homolog 2 depletion induces cellular senescence via histone demethylation along the INK4/ARF locus. *Int. J. Biochem. Cell Biol.* **2015**, *65*, 104–112. [CrossRef]
58. Ling, Z.; You, Z.; Hu, L.; Zhang, L.; Wang, Y.; Zhang, M.; Zhang, G.; Chen, S.; Xu, B.; Chen, M. Effects of four single nucleotide polymorphisms of EZH2 on cancer risk: A systematic review and meta-analysis. *OncoTargets Ther.* **2018**, *11*, 851–865. [CrossRef]
59. Breyer, J.P.; McReynolds, K.M.; Yaspan, B.L.; Bradley, K.M.; Dupont, W.D.; Smith, J.R. Genetic variants and prostate cancer risk: Candidate replication and exploration of viral restriction genes. *Cancer Epidemiol. Biomark. Prev.* **2009**, *18*, 2137–2144. [CrossRef]
60. Zhou, Y.; Du, W.D.; Wu, Q.; Liu, Y.; Chen, G.; Ruan, J.; Xu, S.; Yang, F.; Zhou, F.S.; Tang, X.F.; et al. EZH2 genetic variants affect risk of gastric cancer in the Chinese Han population. *Mol. Carcinog.* **2014**, *53*, 589–597. [CrossRef]

61. Lee, S.W.; Park, D.Y.; Kim, M.Y.; Kang, C. Synergistic triad epistasis of epigenetic H3K27me modifier genes, EZH2, KDM6A, and KDM6B, in gastric cancer susceptibility. *Gastric Cancer* **2019**, *22*, 640–644. [CrossRef] [PubMed]
62. Sun, B.; Lin, Y.; Wang, X.; Lan, F.; Yu, Y.; Huang, Q. Single Nucleotide Polymorphism of the Enhancer of Zeste Homolog 2 Gene rs2072408 is Associated with Lymph Node Metastasis and Depth of Primary Tumor Invasion in Gastric Cancer. *Clin. Lab.* **2016**, *62*, 2099–2105. [CrossRef] [PubMed]
63. Zhang, J.; Hu, Z.; Horta, C.A.; Yang, J. Regulation of epithelial-mesenchymal transition by tumor microenvironmental signals and its implication in cancer therapeutics. *Semin. Cancer Biol.* **2022**, *88*, 46–66. [CrossRef]
64. Marrelli, D.; Marano, L.; Ambrosio, M.R.; Carbone, L.; Spagnoli, L.; Petrioli, R.; Ongaro, A.; Piccioni, S.; Fusario, D.; Roviello, F. Immunohistochemical Markers of the Epithelial-to-Mesenchymal Transition (EMT) Are Related to Extensive Lymph Nodal Spread, Peritoneal Dissemination, and Poor Prognosis in the Microsatellite-Stable Diffuse Histotype of Gastric Cancer. *Cancers* **2022**, *14*, 6023. [CrossRef] [PubMed]
65. Fujii, S.; Ochiai, A. Enhancer of zeste homolog 2 downregulates E-cadherin by mediating histone H3 methylation in gastric cancer cells. *Cancer Sci.* **2008**, *99*, 738–746. [CrossRef] [PubMed]
66. Liu, Y.W.; Sun, M.; Xia, R.; Zhang, E.B.; Liu, X.H.; Zhang, Z.H.; Xu, T.P.; De, W.; Liu, B.R.; Wang, Z.X. LincHOTAIR epigenetically silences miR34a by binding to PRC2 to promote the epithelial-to-mesenchymal transition in human gastric cancer. *Cell Death Dis.* **2015**, *6*, e1802. [CrossRef]
67. He, Z.C.; Yang, F.; Guo, L.L.; Wei, Z.; Dong, X. LncRNA TP73-AS1 promotes the development of Epstein-Barr virus associated gastric cancer by recruiting PRC2 complex to regulate WIF1 methylation. *Cell. Signal.* **2021**, 110094. [CrossRef] [PubMed]
68. Wang, H.J.; Ruan, H.J.; He, X.J.; Ma, Y.Y.; Jiang, X.T.; Xia, Y.J.; Ye, Z.Y.; Tao, H.Q. MicroRNA-101 is down-regulated in gastric cancer and involved in cell migration and invasion. *Eur. J. Cancer* **2010**, *46*, 2295–2303. [CrossRef]
69. Tiwari, N.; Tiwari, V.K.; Waldmeier, L.; Balwierz, P.J.; Arnold, P.; Pachkov, M.; Meyer-Schaller, N.; Schübeler, D.; van Nimwegen, E.; Christofori, G. Sox4 is a master regulator of epithelial-mesenchymal transition by controlling Ezh2 expression and epigenetic reprogramming. *Cancer Cell* **2013**, *23*, 768–783. [CrossRef]
70. Liu, X.; Wang, C.; Chen, Z.; Jin, Y.; Wang, Y.; Kolokythas, A.; Dai, Y.; Zhou, X. MicroRNA-138 suppresses epithelial-mesenchymal transition in squamous cell carcinoma cell lines. *Biochem. J.* **2011**, *440*, 23–31. [CrossRef]
71. Sun, S.; Yu, F.; Zhang, L.; Zhou, X. EZH2, an on-off valve in signal network of tumor cells. *Cell. Signal.* **2016**, *28*, 481–487. [CrossRef] [PubMed]
72. Kudo-Saito, C.; Shirako, H.; Takeuchi, T.; Kawakami, Y. Cancer metastasis is accelerated through immunosuppression during Snail-induced EMT of cancer cells. *Cancer Cell* **2009**, *15*, 195–206. [CrossRef] [PubMed]
73. Ashrafizadeh, M.; Zarrabi, A.; Hushmandi, K.; Kalantari, M.; Mohammadinejad, R.; Javaheri, T.; Sethi, G. Association of the Epithelial-Mesenchymal Transition (EMT) with Cisplatin Resistance. *Int. J. Mol. Sci.* **2020**, *21*, 4002. [CrossRef] [PubMed]
74. Zhang, Y.; Tong, T. FOXA1 antagonizes EZH2-mediated CDKN2A repression in carcinogenesis. *Biochem. Biophys. Res. Commun.* **2014**, *453*, 172–178. [CrossRef] [PubMed]
75. Fiskus, W.; Wang, Y.; Sreekumar, A.; Buckley, K.M.; Shi, H.; Jillella, A.; Ustun, C.; Rao, R.; Fernandez, P.; Chen, J.; et al. Combined epigenetic therapy with the histone methyltransferase EZH2 inhibitor 3-deazaneplanocin A and the histone deacetylase inhibitor panobinostat against human AML cells. *Blood* **2009**, *114*, 2733–2743. [CrossRef]
76. Engeland, K. Cell cycle regulation: p53-p21-RB signaling. *Cell Death Differ.* **2022**, *29*, 946–960. [CrossRef]
77. Lu, H.; Sun, J.; Wang, F.; Feng, L.; Ma, Y.; Shen, Q.; Jiang, Z.; Sun, X.; Wang, X.; Jin, H. Enhancer of zeste homolog 2 activates wnt signaling through downregulating CXXC finger protein 4. *Cell Death Dis.* **2013**, *4*, e776. [CrossRef]
78. Fujii, S.; Ito, K.; Ito, Y.; Ochiai, A. Enhancer of zeste homologue 2 (EZH2) down-regulates RUNX3 by increasing histone H3 methylation. *J. Biol. Chem.* **2008**, *283*, 17324–17332. [CrossRef]
79. O'Driscoll, M. INK4a/ARF-dependent senescence upon persistent replication stress. *Cell Cycle* **2013**, *12*, 1997–1998. [CrossRef]
80. Hirosue, A.; Ishihara, K.; Tokunaga, K.; Watanabe, T.; Saitoh, N.; Nakamoto, M.; Chandra, T.; Narita, M.; Shinohara, M.; Nakao, M. Quantitative assessment of higher-order chromatin structure of the INK4/ARF locus in human senescent cells. *Aging Cell* **2012**, *11*, 553–556. [CrossRef]
81. Qi, L.N.; Xiang, B.D.; Wu, F.X.; Ye, J.Z.; Zhong, J.H.; Wang, Y.Y.; Chen, Y.Y.; Chen, Z.S.; Ma, L.; Chen, J.; et al. Circulating Tumor Cells Undergoing EMT Provide a Metric for Diagnosis and Prognosis of Patients with Hepatocellular Carcinoma. *Cancer Res.* **2018**, *78*, 4731–4744. [CrossRef] [PubMed]
82. Ye, S.; Wang, L.; Zuo, Z.; Bei, Y.; Liu, K. The role of surgery and radiation in advanced gastric cancer: A population-based study of Surveillance, Epidemiology, and End Results database. *PLoS ONE* **2019**, *14*, e0213596. [CrossRef] [PubMed]
83. Marin, J.J.; Al-Abdulla, R.; Lozano, E.; Briz, O.; Bujanda, L.; Banales, J.M.; Macias, R.I. Mechanisms of Resistance to Chemotherapy in Gastric Cancer. *Anti-Cancer Agents Med. Chem.* **2016**, *16*, 318–334. [CrossRef]
84. Zhou, W.; Wang, J.; Man, W.Y.; Zhang, Q.W.; Xu, W.G. siRNA silencing EZH2 reverses cisplatin-resistance of human non-small cell lung and gastric cancer cells. *Asian Pac. J. Cancer Prev. APJCP* **2015**, *16*, 2425–2430. [CrossRef]
85. Wang, P.; Li, Z.; Liu, H.; Zhou, D.; Fu, A.; Zhang, E. MicroRNA-126 increases chemosensitivity in drug-resistant gastric cancer cells by targeting EZH2. *Biochem. Biophys. Res. Commun.* **2016**, *479*, 91–96. [CrossRef]
86. Wang, Z.Q.; Cai, Q.; Hu, L.; He, C.Y.; Li, J.F.; Quan, Z.W.; Liu, B.Y.; Li, C.; Zhu, Z.G. Long noncoding RNA UCA1 induced by SP1 promotes cell proliferation via recruiting EZH2 and activating AKT pathway in gastric cancer. *Cell Death Dis.* **2017**, *8*, e2839. [CrossRef] [PubMed]

87. Dai, Q.; Zhang, T.; Pan, J.; Li, C. LncRNA UCA1 promotes cisplatin resistance in gastric cancer via recruiting EZH2 and activating PI3K/AKT pathway. *J. Cancer* **2020**, *11*, 3882–3892. [CrossRef]
88. Li, H.; Ma, X.; Yang, D.; Suo, Z.; Dai, R.; Liu, C. PCAT-1 contributes to cisplatin resistance in gastric cancer through epigenetically silencing PTEN via recruiting EZH2. *J. Cell. Biochem.* **2020**, *121*, 1353–1361. [CrossRef]
89. Guo, Y.; Yue, P.; Wang, Y.; Chen, G.; Li, Y. PCAT-1 contributes to cisplatin resistance in gastric cancer through miR-128/ZEB1 axis. *Biomed. Pharmacother. Biomed. Pharmacother.* **2019**, *118*, 109255. [CrossRef]
90. Lee, H.; Yoon, S.O.; Jeong, W.Y.; Kim, H.K.; Kim, A.; Kim, B.H. Immunohistochemical analysis of polycomb group protein expression in advanced gastric cancer. *Hum. Pathol.* **2012**, *43*, 1704–1710. [CrossRef]
91. Pan, Y.M.; Wang, C.G.; Zhu, M.; Xing, R.; Cui, J.T.; Li, W.M.; Yu, D.D.; Wang, S.B.; Zhu, W.; Ye, Y.J.; et al. STAT3 signaling drives EZH2 transcriptional activation and mediates poor prognosis in gastric cancer. *Mol. Cancer* **2016**, *15*, 79. [CrossRef] [PubMed]
92. Guo, L.; Yang, T.F.; Liang, S.C.; Guo, J.X.; Wang, Q. Role of EZH2 protein expression in gastric carcinogenesis among Asians: A meta-analysis. *Tumour Biol. J. Int. Soc. Oncodev. Biol. Med.* **2014**, *35*, 6649–6656. [CrossRef] [PubMed]
93. Chi, P.; Allis, C.D.; Wang, G.G. Covalent histone modifications–miswritten, misinterpreted and mis-erased in human cancers. *Nat. Rev. Cancer* **2010**, *10*, 457–469. [CrossRef] [PubMed]
94. Knutson, S.K.; Wigle, T.J.; Warholic, N.M.; Sneeringer, C.J.; Allain, C.J.; Klaus, C.R.; Sacks, J.D.; Raimondi, A.; Majer, C.R.; Song, J.; et al. A selective inhibitor of EZH2 blocks H3K27 methylation and kills mutant lymphoma cells. *Nat. Chem. Biol.* **2012**, *8*, 890–896. [CrossRef] [PubMed]
95. Hayden, A.; Johnson, P.W.; Packham, G.; Crabb, S.J. S-adenosylhomocysteine hydrolase inhibition by 3-deazaneplanocin A analogues induces anti-cancer effects in breast cancer cell lines and synergy with both histone deacetylase and HER2 inhibition. *Breast Cancer Res. Treat.* **2011**, *127*, 109–119. [CrossRef]
96. Kemp, C.D.; Rao, M.; Xi, S.; Inchauste, S.; Mani, H.; Fetsch, P.; Filie, A.; Zhang, M.; Hong, J.A.; Walker, R.L.; et al. Polycomb repressor complex-2 is a novel target for mesothelioma therapy. *Clin. Cancer Res. Off. J. Am. Assoc. Cancer Res.* **2012**, *18*, 77–90. [CrossRef]
97. Crea, F.; Hurt, E.M.; Mathews, L.A.; Cabarcas, S.M.; Sun, L.; Marquez, V.E.; Danesi, R.; Farrar, W.L. Pharmacologic disruption of Polycomb Repressive Complex 2 inhibits tumorigenicity and tumor progression in prostate cancer. *Mol. Cancer* **2011**, *10*, 40. [CrossRef]
98. Sha, M.; Mao, G.; Wang, G.; Chen, Y.; Wu, X.; Wang, Z. DZNep inhibits the proliferation of colon cancer HCT116 cells by inducing senescence and apoptosis. *Acta Pharm. Sin. B* **2015**, *5*, 188–193. [CrossRef]
99. Miranda, T.B.; Cortez, C.C.; Yoo, C.B.; Liang, G.; Abe, M.; Kelly, T.K.; Marquez, V.E.; Jones, P.A. DZNep is a global histone methylation inhibitor that reactivates developmental genes not silenced by DNA methylation. *Mol. Cancer Ther.* **2009**, *8*, 1579–1588. [CrossRef]
100. Dockerill, M.; Gregson, C.; DH, O.D. Targeting PRC2 for the treatment of cancer: An updated patent review (2016–2020). *Expert Opin. Ther. Pat.* **2021**, *31*, 119–135. [CrossRef]
101. Fioravanti, R.; Stazi, G.; Zwergel, C.; Valente, S.; Mai, A. Six Years (2012–2018) of Researches on Catalytic EZH2 Inhibitors: The Boom of the 2-Pyridone Compounds. *Chem. Rec.* **2018**, *18*, 1818–1832. [CrossRef] [PubMed]
102. McCabe, M.T.; Ott, H.M.; Ganji, G.; Korenchuk, S.; Thompson, C.; Van Aller, G.S.; Liu, Y.; Graves, A.P.; Della Pietra, A., 3rd; Diaz, E.; et al. EZH2 inhibition as a therapeutic strategy for lymphoma with EZH2-activating mutations. *Nature* **2012**, *492*, 108–112. [CrossRef]
103. Yap, T.A.; Winter, J.N.; Giulino-Roth, L.; Longley, J.; Lopez, J.; Michot, J.M.; Leonard, J.P.; Ribrag, V.; McCabe, M.T.; Creasy, C.L.; et al. Phase I Study of the Novel Enhancer of Zeste Homolog 2 (EZH2) Inhibitor GSK2816126 in Patients with Advanced Hematologic and Solid Tumors. *Clin. Cancer Res. Off. J. Am. Assoc. Cancer Res.* **2019**, *25*, 7331–7339. [CrossRef] [PubMed]
104. Huang, S.; Wang, Z.; Zhou, J.; Huang, J.; Zhou, L.; Luo, J.; Wan, Y.Y.; Long, H.; Zhu, B. EZH2 Inhibitor GSK126 Suppresses Antitumor Immunity by Driving Production of Myeloid-Derived Suppressor Cells. *Cancer Res.* **2019**, *79*, 2009–2020. [CrossRef]
105. Konze, K.D.; Ma, A.; Li, F.; Barsyte-Lovejoy, D.; Parton, T.; Macnevin, C.J.; Liu, F.; Gao, C.; Huang, X.P.; Kuznetsova, E.; et al. An orally bioavailable chemical probe of the Lysine Methyltransferases EZH2 and EZH1. *ACS Chem. Biol.* **2013**, *8*, 1324–1334. [CrossRef] [PubMed]
106. Xu, B.; On, D.M.; Ma, A.; Parton, T.; Konze, K.D.; Pattenden, S.G.; Allison, D.F.; Cai, L.; Rockowitz, S.; Liu, S.; et al. Selective inhibition of EZH2 and EZH1 enzymatic activity by a small molecule suppresses MLL-rearranged leukemia. *Blood* **2015**, *125*, 346–357. [CrossRef] [PubMed]
107. Knutson, S.K.; Warholic, N.M.; Wigle, T.J.; Klaus, C.R.; Allain, C.J.; Raimondi, A.; Porter Scott, M.; Chesworth, R.; Moyer, M.P.; Copeland, R.A.; et al. Durable tumor regression in genetically altered malignant rhabdoid tumors by inhibition of methyltransferase EZH2. *Proc. Natl. Acad. Sci. USA* **2013**, *110*, 7922–7927. [CrossRef] [PubMed]
108. Izutsu, K.; Ando, K.; Nishikori, M.; Shibayama, H.; Teshima, T.; Kuroda, J.; Kato, K.; Imaizumi, Y.; Nosaka, K.; Sakai, R.; et al. Phase II study of tazemetostat for relapsed or refractory B-cell non-Hodgkin lymphoma with EZH2 mutation in Japan. *Cancer Sci.* **2021**, *112*, 3627–3635. [CrossRef] [PubMed]
109. Italiano, A.; Soria, J.C.; Toulmonde, M.; Michot, J.M.; Lucchesi, C.; Varga, A.; Coindre, J.M.; Blakemore, S.J.; Clawson, A.; Suttle, B.; et al. Tazemetostat, an EZH2 inhibitor, in relapsed or refractory B-cell non-Hodgkin lymphoma and advanced solid tumours: A first-in-human, open-label, phase 1 study. *Lancet Oncol.* **2018**, *19*, 649–659. [CrossRef]

110. Ribrag, V.; Soria, J.C.; Michot, J.M.; Schmitt, A.; Postel-Vinay, S.; Bijou, F.; Thomson, B.; Keilhack, H.; Blakemore, S.J.; Reyderman, L.; et al. Phase 1 Study of Tazemetostat (EPZ-6438), an Inhibitor of Enhancer of Zeste-Homolog 2 (EZH2): Preliminary Safety and Activity in Relapsed or Refractory Non Hodgkin Lymphoma (NHL) Patients. *Blood* **2015**, *126*, 3. [CrossRef]
111. Hoy, S.M. Tazemetostat: First Approval. *Drugs* **2020**, *80*, 513–521. [CrossRef] [PubMed]
112. Tomassi, S.; Romanelli, A.; Zwergel, C.; Valente, S.; Mai, A. Polycomb Repressive Complex 2 Modulation through the Development of EZH2-EED Interaction Inhibitors and EED Binders. *J. Med. Chem.* **2021**, *64*, 11774–11797. [CrossRef] [PubMed]
113. Qi, W.; Zhao, K.; Gu, J.; Huang, Y.; Wang, Y.; Zhang, H.; Zhang, M.; Zhang, J.; Yu, Z.; Li, L.; et al. An allosteric PRC2 inhibitor targeting the H3K27me3 binding pocket of EED. *Nat. Chem. Biol.* **2017**, *13*, 381–388. [CrossRef] [PubMed]
114. Huang, Y.; Sendzik, M.; Zhang, J.; Gao, Z.; Sun, Y.; Wang, L.; Gu, J.; Zhao, K.; Yu, Z.; Zhang, L.; et al. Discovery of the Clinical Candidate MAK683: An EED-Directed, Allosteric, and Selective PRC2 Inhibitor for the Treatment of Advanced Malignancies. *J. Med. Chem.* **2022**, *65*, 5317–5333. [CrossRef] [PubMed]
115. Li, C.; Wang, Y.; Gong, Y.; Zhang, T.; Huang, J.; Tan, Z.; Xue, L. Finding an easy way to harmonize: A review of advances in clinical research and combination strategies of EZH2 inhibitors. *Clin. Epigenet.* **2021**, *13*, 62. [CrossRef] [PubMed]

Disclaimer/Publisher's Note: The statements, opinions and data contained in all publications are solely those of the individual author(s) and contributor(s) and not of MDPI and/or the editor(s). MDPI and/or the editor(s) disclaim responsibility for any injury to people or property resulting from any ideas, methods, instructions or products referred to in the content.

Review

An Update of G-Protein-Coupled Receptor Signaling and Its Deregulation in Gastric Carcinogenesis

Huan Yan [1,2,3,†], Jing-Ling Zhang [1,2,3,†], Kam-Tong Leung [4], Kwok-Wai Lo [1], Jun Yu [2,5], Ka-Fai To [1,2] and Wei Kang [1,2,3,*]

1. Department of Anatomical and Cellular Pathology, State Key Laboratory of Translational Oncology, Sir Y.K. Pao Cancer Center, Prince of Wales Hospital, The Chinese University of Hong Kong, Hong Kong 999077, China
2. State Key Laboratory of Digestive Disease, Institute of Digestive Disease, The Chinese University of Hong Kong, Hong Kong 999077, China
3. CUHK-Shenzhen Research Institute, The Chinese University of Hong Kong, Shenzhen 518000, China
4. Department of Pediatrics, The Chinese University of Hong Kong, Hong Kong 999077, China
5. Department of Medicine and Therapeutics, The Chinese University of Hong Kong, Hong Kong 999077, China
* Correspondence: weikang@cuhk.edu.hk; Tel.: +852-35051505; Fax: +852-26497286
† These authors contributed equally to this work.

Simple Summary: Gastric cancer (GC) ranks as one of the most life-threatening malignancies worldwide, and over one billion new cases and 783,000 deaths were reported last year. The incidence of GC is exceptionally high in Asian countries. Multiple oncogenic signaling pathways are aberrantly activated and implicated in gastric carcinogenesis, leading to the malignant phenotype acquisition. G-protein-coupled receptor (GPCR) signaling is one of them, and the aberrant activation of GPCRs and G proteins promotes GC progression. The activated GPCRs/G proteins might serve as useful biomarkers for early diagnosis, prognostic prediction, and even clinically therapeutic targets. This review summarized the recent research progress of GPCRs and highlighted their mechanisms in tumorigenesis, especially in GC initiation and progression.

Abstract: G-protein-coupled receptors (GPCRs) belong to a cell surface receptor superfamily responding to a wide range of external signals. The binding of extracellular ligands to GPCRs activates a heterotrimeric G protein and triggers the production of numerous secondary messengers, which transduce the extracellular signals into cellular responses. GPCR signaling is crucial and imperative for maintaining normal tissue homeostasis. High-throughput sequencing analyses revealed the occurrence of the genetic aberrations of GPCRs and G proteins in multiple malignancies. The altered GPCRs/G proteins serve as valuable biomarkers for early diagnosis, prognostic prediction, and pharmacological targets. Furthermore, the dysregulation of GPCR signaling contributes to tumor initiation and development. In this review, we have summarized the research progress of GPCRs and highlighted their mechanisms in gastric cancer (GC). The aberrant activation of GPCRs promotes GC cell proliferation and metastasis, remodels the tumor microenvironment, and boosts immune escape. Through deep investigation, novel therapeutic strategies for targeting GPCR activation have been developed, and the final aim is to eliminate GPCR-driven gastric carcinogenesis.

Keywords: G-protein-coupled receptor; G protein; gastric cancer; targeted therapy

1. Introduction

Gastric cancer (GC) is a substantial global health burden, accounting for the fifth most commonly diagnosed cancer and the third leading cause of fatal malignancies worldwide. Incidence rates are markedly increased in Eastern Asia, especially in Mongolia, Japan, and Korea, which are strongly associated with various predisposing and etiological factors, according to several migrant studies [1,2]. Most GC-related deaths occur due to late diagnosis, lymph node metastasis, and refractory after surgery. Thus, numerous efforts have been made to develop useful prognosis markers for early detection and therapeutic targets to improve clinical outcomes. Heterogeneity represents one of the biggest challenges in GC treatment owing to the histological categories and diverse molecular drivers. The well-established histological classification divides gastric carcinomas into diffuse and intestinal types [3]. The Cancer Genome Atlas (TCGA) network also reaffirmed our understanding of molecular categories by analyzing the dysregulated pathways identified in multiomics data. This study developed a robust molecular classification scheme comprising Epstein-Barr virus (EBV), microsatellite instability (MSI), chromosomal instability (CIN), and genomically stable (GS) tumors [4]. In the past two decades, trastuzumab and chemotherapy were used as the first-line treatment, and the combination of ramucirumab and paclitaxel was used in second-line treatment [5]. However, the clinical applicability remains quite limited. There is an urgent need to uncover more targetable pathways to develop more accurate diagnosis makers against nonspecific symptoms in early-stage GC and optimize existing therapy for precision medicine.

Since G protein-coupled receptors (GPCRs) were reported in cellular transformation in 1986, emerging evidence shows that these membrane-embedded receptors regulate many biological processes and are crucial targets against several human malignancies [6]. The involvement of GPCRs in GC is emerging due to the identification of genomic aberrations that lurk at different stages and subtypes of GC and promote tumor initiation and progression [7]. This review recapitulated the current knowledge related to the aberrated regulation of the GPCR pathway in GC, including the common tactic hijacked by tumor cells for their growth, metastasis, and immune evasion. Moreover, we will discuss the advances in the current treatment strategies and summarize the ongoing clinical trials that attempt to translate biological findings into clinical applications.

2. Basic Knowledge of GPCRs

GPCRs comprise over 800 members accounting for about 4% of human genes. They have various structures and signal transduction. Based on their specific characteristics, GPCR members are further classified into different subgroups and participate in various physiological processes, whereas the aberrant expression and abnormal activation of GPCRs are associated with tumor progression.

2.1. Structure and Classification of GPCRs

GPCRs have seven transmembrane α-helices (TM1-7) that connect the N-terminal extracellular domain (ECD), three extracellular and intracellular loops: ECL1-3, ICL1-3, and the C-terminus (Figure 1). They are classified into six groups based on their structural and functional similarities, whereas only four groups (A, B, C, and F) are found in vertebrates. The Rhodopsin-like class A, which has 719 members, represents humans' most common but diverse group. Half of the class A members serve as sensor receptors primarily in smell and vision. In contrast, diffusible ligands, such as peptides, lipids, hormones, and nucleotides, can trigger the other receptors. Class B includes secretin and adhesion receptors, which have a similar sequence in 7TM but different sequences in ECD. The secretin subgroup contains receptors for polypeptide gut hormones, such as the GLP-1 receptor, glucagon receptor, and parathyroid hormone receptor. Research has focused on the adhesion receptors by defining the mechanisms of ligand binding sites and the GPCR autoproteolysis-inducing domain (GAIN)-mediated receptor activation. The metabotropic glutamate family (class C) is characterized by a large ECD, consisting of γ-aminobutyric acid B receptors (GABA$_B$), metabotropic glutamate receptors (mGluRs), and a calcium-sensing receptor (CasR). The frizzled/taste family (class F) includes frizzled and smoothened proteins that can be activated by the lipo-glycoproteins of the Wnt and Hedgehog families [8–10].

reassociation and the signaling termination by accelerating intrinsic GTPase activity. Notably, agonist-activated GPCRs are also phosphorylated by GRKs and interact with β-arrestin, resulting in signaling desensitization and GPCR endocytosis. The endocytic β-arrestin-GPCR complex can be modulated by multiple factors and undergo degradation or recycling. (**B**) GPCR-associated crosstalk on the membrane and GPCR-EGFR crosstalk contain EGFR ligand-dependent transactivation and EGFR ligand-independent transactivation. The following pathways are the Wnt and Shh pathways. (**C**) The main pathways targeted by the multiple effectors in (a) consist of the following signaling pathways: Hippo pathway, MAPK pathway, Shh pathway, and Wnt pathway. Abbreviations: AC, adenylyl cyclase; AKT, protein kinase B; CREB, cAMP response element-binding protein; EGF, epidermal growth factor; EGFR, EGF receptor; ERK, extracellular signal-regulated kinase; GEF, guanine exchange factor; GLI, glioma-associated oncogenes; GPCR, G protein-coupled receptor; GRK, G protein-coupled receptor kinase; JNK, c-jun N-terminal kinase; LATS, large tumor suppressor kinase; MAPK, mitogen-activated protein kinase; mTOR, mammalian target of rapamycin; PDEs, phosphodiesterases; PI3K, phosphatidylinositol-3-kinase; PKA, Protein Kinase A; PLCβ, Phospholipase C β; ROCK, Rho-associated protein kinase; Shh, sonic hedgehog protein; SMO, Smoothened protein; SuFu, suppressor of fused; TAZ, transcriptional coactivator with PDZ-binding motif; TCF/LEF, T-cell factor/lymphoid enhancer factor; TEAD, transcriptional enhanced associate domain; YAP, yes-associated protein.

β-arrestins undergo conformational changes when recognizing the GRK-phosphorylated GPCRs. Then, they enhance the process of desensitization, internalization, and clathrin-mediated endocytosis of the activated GPCRs. As scaffold proteins, β-arrestins facilitate GPCR-stimulated signal transduction. As one of the most prominent and earliest examples, the GPCR-mediated extracellular signal-regulated kinase 1/2 (ERK1/2) activation is a β-arrestin-dependent and G protein-independent signaling [16]. The genetic ablation or inactivation of several G proteins induces a zero functional state for the G protein and abolishes the β-arrestin-mediated signaling in response to GPCR activation [??] However, it was reported that β-arrestins are not required for ERK1/2 phosphorylation despite their crucial roles in receptor internalization [23]. Indeed, the cumulative impact of GPCR-induced ERK1/2 activation is tightly controlled by β-arrestins and G proteins [24]. Moreover, GPCRs scaffold several signaling proteins for Wnt [25], the hedgehog (Hh) [26], and Notch [27] pathways (Figure 2B,C).

2.3. Diversification of GPCR Machinery

GPCRs are sophisticated dynamic machines rather than static on-and-off switches. When they are engaged with different ligands, receptors, and regulatory partners, they may exhibit specific conformations and undergo subcellular distributions. Exploring the dynamic nature of GPCRs is vital to elucidate the mechanisms underlying allosteric modulation, biased agonism, oligomerization, and sustained and compartmentalized signaling. These mechanisms convey novel insights into drug discovery.

Allosteric ligand binding sites in GPCRs are potential new targets for modulating GPCR functions and improving drug selectivity. These modulators augment (positive allosteric modulators [PAMs]) or reduce (negative allosteric modulators [NAMs]) the affinity and efficacy of endogenous agonists [28]. The discovery of allosteric modulators has sparked interest in central nervous system (CNS) diseases, though with limited success [29]. MK-7622 is a PAM, selectively binding with the M_1 muscarinic receptor in the CNS, which has been stopped because it fails to improve recognition and increases adverse effects like diarrhea [30,31].

Another ligand-receptor dynamic is biased agonism, a mechanism in which the active conformational states of the receptors are stabilized by some ligands, resulting in distinct cellular signaling profiles [32]. There are three different modes of biased signaling, including the same receptor bound with other ligands adopting distinct conformations (ligand bias), varying stoichiometric ratios of signaling effectors in distinguished cells (system bias), and GPCR stimulation within divergent intracellular compartments (location bias) [17].

However, substantial evidence on this is limited. The endogenous ligands, CCL9 and CCL21, have been considered equipotent for activating CCR7-G protein coupling and calcium mobilization. However, both ligands cause the distinct conformation of CCR7. Only CCL9 can promote robust receptor desensitization after coupling the β-arrestins and efficiently accelerating ERK1/2 phosphorylation, which CCL21 cannot achieve [33]. Besides, small molecules targeting TRV130 and PZM21 have been utilized to improve analgesia with fewer side effects because of the biased receptor μ-OR activity, potent Gαi signaling profile, and limited β-arrestin recruitment [34,35]. These two examples can partly explain how ligands trigger the biased mechanism of GPCRs. Revealing the structural features of GPCRs under multiple activation states and different cellular backgrounds may be required to understand the biased signaling.

Receptor oligomerization conveys much more diversities in the function and physiological roles of GPCRs. However, unlike the oligomer tyrosine kinase receptors and ion channels, the formation of GPCR multimers remains controversial [36,37]. It has been found that dimerization was necessary for some GPCRs, such as the class C members. The first tangible evidence for GPCR dimerization was that the gonadotrophin-releasing hormone (GnRH) antagonist-conjugated bivalent antibody played an essential role in biphasic receptor formation [38,39]. The emergence of heteromers was associated with the preferential pattern of receptors in different tissues and cell types [40]. The balance between the monomers and heteromers of GPCRs may contribute to diseases [41,42]. Unraveling the pattern of GPCR heteromers will provide pharmacotherapeutic targets to benefit disease management.

Compartmentalized signaling may partly explain why the GPCRs can activate a typical profile of secondary messengers and kinases. In addition to locating the membrane, the GPCRs might be desensitized and undergo β-arrestin-mediated endocytosis and intracellular signaling [43–45]. Notably, some of the mechanisms are studied in the digestive systems. Recently, PAR2 endosomal may underlie the sustained hyperexcitability of nociceptors in patients with irritable bowel syndrome (IBS). The IBS supernatants and trypsin could persistently activate PAR2 in the colonic mucosa in a clathrin-mediated, endocytosis-dependent fashion [46,47]. The inhibitors of clathrin-mediated endocytosis and targeted PAR2 antagonists suppressed PAR2 endosomal signal [48].

2.4. Dysregulated GPCR Signaling in Tumors

Based on the recent pan-cancer analysis, GPCR signaling was among the 55 pathways most significantly mutated. Mutations and the aberrant expression of GPCRs and G proteins contribute to various diseases, including neurodegenerative, reproductive, immunological, and metabolic disorders, as well as cancers and infectious diseases [49,50]. The dysregulated GPCR signaling may exert a significant tumorigenic effect, as those alterations frequently co-occur in well-characterized oncogenes, such as tyrosine and serine-threonine kinase Ras-family members [51]. In-depth omics analysis approaches, like MutSig2CV and GISTIC (Genomic Identification of Significant Targets in Cancers), have comprehensively investigated the mutations and copy number variations (CNVs) of GPCRs and G proteins in 33 TCGA patient groups. Remarkably, mutated GPCRs and G proteins have been significantly identified in GI malignancies, even though these tumors' mutation rates are not typically high [52]. Therefore, the relevance between these mutations and biological outcomes is vastly underestimated [53,54].

2.5. GPCR Mutation and Abundant Expression

GPCRs are mutated in over 20% of all sequenced samples [55,56]. Unlike the mutated hotspots in G proteins, GPCRs exhibit diverse mutations across different cancer types. The three-dimensional structures of GPCRs and their interaction elements were evaluated to acquire a mutational landscape of GPCRs in cancers [51]. The bulk of the alterations occurs in the conserved 7TM via the visualization of the representative GPCR 3D structure, such as the ionic lock switch E/DRY arginine motif, G protein-binding sites, and the tyrosine toggle switch motif NPxxY, and ligand-binding site. GPCR mutations impair GPCR signaling by altering the basal activity, ligand binding affinity, G-protein interaction, and cell-surface expression. As with thyroid-stimulating hormone receptors like *HCRT2*, *P2RY12*, *LPAR4*, and *GPR174*, frequent mutations in the DRY motif may result in constitutive activation due to conformational changes in TM3, TM5, and TM6 [51]. Mutations of the connection between the NPxxY motif on TM7 and a conserved tyrosine in TM5 could stabilize the inactive-state conformations of the α_{1B}- and β_2-adrenoceptors, which may account for lower agonist potency in transducing the downstream IP1 and cAMP signaling, respectively [57]. Understanding the mutated structural features will shed new light on GPCR malfunctions and devise possible therapeutic strategies [58].

The large and ever-grossing body of sequencing by pan-cancer analysis suggests that the frequently mutated GPCR families are adhesion-related GPCRs, such as the glutamate metabotropic receptors (*GRM1-8*, class C) [59], lysophosphatidic acid (LPA) receptors (*LPAR1-6*), sphingosine-1-phosphate (S1P) receptors (*S1PR1-5*), and muscarinic receptors (*CHRM1-5*, class A). However, most adhesion receptors are orphan receptors with unknown ligands and physiological functions [60,61]. GPCR genetic alternations were found in melanoma by exon capture and massively parallel sequencing. *GPR98* and *GRM3* were two of the most frequently mutated genes, with 27.5% and 16.3% mutation rates. GRM3 mutants selectively mediate the MEK signaling that contributes to tumor growth in melanoma, acting as an indicator for patient stratification and precision medicine [62]. MutSig2CV analysis suggests the three most mutant GPCRs in colon cancer are *GPR98* (21.25%), *TSHR* (13.90%), and *BAI3* (13.62%), while *CELSR1* (11.20%), *EDNRB* (8.14%), and *GPR45* (5.09%) account for the three most frequently mutated GPCRs in GC. However, the functional roles of these mutant GPCRs in GI cancers remain unknown.

Besides mutations, GPCRs, like chemokine and histamine receptors (*HRH2*), exhibit significant copy number variations (CNVs) in tumors. Several broad-type GPCRs are universally overexpressed throughout the GI tract, regulating digestive and pathophysiological processes [54,63,64]. The upregulation of receptors like 5-HTRs, FFARs, HRs, PARs, EPs, and TGRs plays pivotal roles in proliferation, invasion, metastasis, and inflammation in the small intestine and colon. It has been reported that the CNVs of chemokine receptors, LPARs, and ARs contribute to the initiation and progression of hepatocellular carcinoma [65,66]. Early studies have implicated numerous viruses that harbored open reading frames and evolved to take advantage of the signaling network for replicative success by encoding GPCRs [67]. In GC, the Epstein-Barr virus (EBV/HHV-4) encodes a class A GPCR called BILF1, affecting multiple cellular pathways [68].

2.6. Widespread Mutations of G Proteins

As oncogenic drivers in multiple prevalent cancers, many G proteins are considered part of the cancer-associated gene panels routinely employed by a wide range of clinical oncology studies. MutSig2CV analysis indicates that *GNAS* is the most frequently mutated G protein in TCGA cohorts, concordant with the sequence results of the catalog of somatic mutations in the cancer (COSMIC) database. *GNAS* aberrations widely occur in tumors originating from the pituitary (28%), pancreas (12%), thyroid (5%), colon (6%), and a few other locations [69]. Previous studies revealed that the two most frequently mutated residents, Arg 201 [70,71] and Gln 227 [72], might be functionally important. The significance of these two sites has been first confirmed in pituitary tumors [73]. The disease-causing altered resident Arg 201 leads to the constitutive cAMP signaling by reducing the GTP hydrolysis of the active GTP-bound Gαs. However, the conclusion was reshaped by a recent structural study of *GNAS*, indicating that the stabilization of the intramolecular hydrogen bond network (H-bond network) plays a pivotal role in mutation-mediated constitutive activation [74]. These aberrations in *GNAS* are responsible for initiating and progressing multiple types of GI cancer, such as colon neoplasia, GC, and pancreatic adenocarcinomas (PDAs). In colon cancer, the *GNAS* can mediate the tumorigenesis of inflammatory factors by stimulating the Gs-Axin-β-Catenin pathway axis [75]. In the rare gastric adenocarcinoma, *GNAS* mutations were tightly associated with deep submucosal invasion and increased tumor size by activating the Wnt/β-catenin pathway [76]. Besides, at the early onset of invasive PDAs, frequent *GNAS* mutations (~41–75%) suppressed the PKA-mediated SIK and reprogrammed lipid metabolism in the precursor of PDAs [77,78].

Although *GNAQ* and *GNA11* mutations were less studied in tumors than *GNAS*, these mutations were well-established in Sturge-Weber syndrome [79] and leptomeningeal melanocytosis, arising from the central nervous system (~50%) [80], and also in the blue nevi and the primary uveal melanomas (UVM)/uveal melanoma metastases (83%) [81,82]. The somatic mutations are mainly located in the residues Q209 or R183, which are essential for GTP hydrolysis and cause constitutive activation due to loss of GTPase activity. In uveal melanoma (UVM), the more common Q209 mutations were more potent in tumorigenesis assays in nude mice models [82]. Consistently, the mutant GNAQ^{Q209L} contributed to MAPK pathway activation [81] and exhibited more significant activated ERK than the GNAQ^{R183Q} [83,84]. The activated GNAS mutant can also stimulate YAP-dependent transcription through a Trio-Rho/Rac signaling circuitry instead of the canonical Hippo pathway in UVM [85]. Furthermore, GNA13 is upregulated in several solid tumors, such as GC [86], nasopharyngeal cancer [87], breast cancer [88], squamous cell cancers [89], and colorectal cancer [90]. Interestingly, both GNA13 and RhoA have shown relevance to the transformation capacity and metastatic potential in epithelial cancer and fibroblasts, but the axis appears to play a tumor-suppressive role in B-cell lymphomas [91]. Large-scale sequencing of lymphoid and hematopoietic malignancies indicated that the mutant residues could be found throughout the gene [92,93].

The cDNA library screening distinguished the functionally relevant mutations of the Gβ proteins *GNB1* and *GNB2*. The gain-of-function alterations of these proteins can disrupt the interactions of Gα-Gβγ and constitutively stimulate the downstream signaling effectors, conferring resistance to targeted kinase inhibitors [94,95]. Recently, emerging variants in all five Gβ proteins have been reported, such as *GNB2* Arg52Leu in familial cardiac arrhythmia condition, Gly77Arg in neurodevelopmental disorder, and monoallelic missense variants in developmental delay/intellectual disability (DD/ID) [96]. Emerging evidence supports that Gβ mutants also occur in various cancer types and relate to distinct cancer subtypes. GPCRs also transduce the signal through β arrestins instead of G proteins, mediating GC cell invasion, migration, and epithelial-mesenchymal transition (EMT) [24]. For example, the protein kinase AKT exerts its oncogenic function through the signaling complex GPR39/β arrestin1/Src upon obestatin stimulation [97].

Some mutations occur in oncogenic kinase alterations, such as *BCR-ABL* fusion protein, *ETV6-ABL1*, JAK^{V617F}, and BRAF^{V600K}, to enhance the drug resistance of the corresponding kinase inhibitors [95,98]. Nevertheless, further investigations are needed into how these alterations influence tumorigenesis in different contexts. The potential roles of Gγ proteins, the close partners of Gβ proteins, should be clarified.

3. Aberrant GPCR Signaling in GC

GPCRs play hierarchical roles in many signaling networks. Dysregulations of the GPCRs extensively exist in tumor progression, metastasis, and immune response reprogramming. In recent years, aberrant GPCR members have been emerging in GC studies. This section will outline the updated findings of the GPCR signaling pathway in GC (Table 1).

Table 1. The most reported GPCRs in GC.

GPCRs	Ligand	Expression	Mechanisms	Biological Function	References
Class A Receptors					
Peptide/Protein Receptors					
Protease-activated receptors (PAR)	Proteases, such as Thrombin, TFLLRN (synthetic PAR1-targeted peptide)	PAR1/F2R: upregulation PAR2/F2RL1: upregulation PAR3/F2RL2: - PAR4/F2RL3: -	$H.\ pylori \to$ ERK/PI3K-AKT$\to \alpha$-arrestin$\to$PAR1$\to$CXCL2 PAR2$\to$MAPK$\to$VEGF/COX-2	Inflammation, angiogenesis	[99,100]
Angiotensin receptors (ATR)	Angiotensin II	AT_1R: upregulation AT_2R: upregulation	$AT_1R \to$ VEGF	Angiogenesis, metastasis	[101,102]
Endothelin receptors (ETR)	Endothelin-1	ET_AR: upregulation ET_BR: - ET_CR: -	$ET_AR \to$ VEGF $ET_AR \to \beta$ arrestin/Src$\to$EGFR	proliferation, metastasis	[103]
Formyl peptide receptors (FPR)	fMLF, capthespin G	FP_1R: - FP_2R/ALX: upregulation FP_3R: -	$FP_1R \to$ ALOX5/15, SPMs (RvD1 and LXB4), SPM receptors (BLT1, ChemR23, GPR32) $FP_2R \to$ MAPK	FP1R: inhibiting angiogenesis and proliferation FP2R: invasion and metastasis	[104–106]
Cholecystokinin receptors (CCKR)	CCK, gastrin	CCK1R: upregulation CCK2R/GR: upregulation	gastrin/GR$\to$PKC$\to$IκB, NF-κB	proliferation	[107]
Leucine-rich repeat-containing receptors (LGRs) group B	R-spondin1/2/3/4	Lgr4: upregulation Lgr5: upregulation Lgr6	Lgr4/5/6$\to \beta$ catenin Lgr6$\to$PI3K/AKT/mTOR	proliferation, metastasis	[108,109]
Lipid receptors					
Lysophosphatidic acid receptors (LPAR)	Lysophosphatidic acid	LPA1/Edg-2: - LPA2/Edg-4: upregulation LPA3/Edg-7: -	LPAR2$\to$tyrosine phosphorylation of c-Met LPAR2$\to$Gq11$\to$p38	migration	[110–112]
Sphingosine-1 phosphate receptors (S1PR)	Lysophosphatidic acid: S1P	$S1P_1R$/Edg-1 $S1P_2R$/Edg-5 $S1P_3R$/Edg-3 : ubiquitously expressed $S1P_4R$/Edg-6 $S1P_5R$/Edg-8	$S1P_1R \to$ RAC-CDC42$\to$ERK S1P$\to$EGFR, c-Met	$S1P_1R$ & $S1P_3R$: promote proliferation and migration, angiogenesis $S1P_2R$: inhibit migration	[113,114]
Prostaglandin receptors (EPR)	PGE2	EP_1R: - EP_2R: upregulation EP_3R: - EP_4R: -	PGE2$\to$DNMT3B$\to$5mC enrichment (DNA hypermethylation) $H.\ pylori \to$PGE2 upregulation$\to$macrophage infiltration	proliferation, angiogenesis	[115,116]

Table 1. Cont.

GPCRs	Ligand	Expression	Mechanisms	Biological Function	References
Chemokine receptors					
Chemokine CXC receptors (CXCR)	CXCL12-CXCR4/CXCR7CXCL8-CXCR1/CXCR2CXCL16-CXCR6	CXCR1: upregulated CXCR2: upregulated CXCR3 CXCR4: upregulation CXCR5 CXCR6: upregulation CXCR7	CXCL12/CXCR4→PI3K/Akt/mTOR CXCL12/CXCR4→ERK1/2 H. pylori→CXCL8→AKT/ERK/cyclin D1/EGFR/Bcl2/MMP9/MMP2	proliferation, migration, invasion, angiogenesis, metastasis	[117]
Chemokine CC receptors (CCR)	CCL2-CCR2CCL5-CCR5CCL19/CCL21-CCR7	CCR1/3/4/5/6/8/9: - CCR2/7: upregulation	CCR7→TGF-β1/NF-κB	migration, invasion, survival, metastasis	[118]
Aminergic receptors					
Muscarinic acetylcholine receptor	Acetylcholine, carbachol,	M1R: M3R: upregulation M2R/M4R/M5R: -	M1R-TRPC6→PKC M2R/M4R→PKA→ neurotransmitter release M3R→EGFR→MAPK/ERK M3R→Wnt pathway→YAP	proliferation, migration, invasion,	[119]
β-adrenergic receptor (β-AR)	isoproterenol	β1-adrenergic receptor: - β2-adrenergic receptor: upregulation β3-adrenergic receptor: -	ADRB2→NF-κB/AP-1/CREB/STAT3/ERK/JNK/MAPK→VEGF/MMP2/MMP7/MMP9	proliferation, invasion, metastasis	[120]
Nucleotide receptors					
P2Y receptors (P2YR)	ATP	P2Y4: upregulation P2Y6: downregulation P2Y1/2/11-14: -	P2Y6→β catenin→ c-Myc P2Y2→Gαq→p38-MAPK/ERK/JNK	proliferation	[121]
Adenosine receptors (AR)	adenosine	A_1/A_3: - $A_{2a}R$: upregulation $A_{2b}R$: upregulation	$A_{2a}R$→PI3K-AKT-mTOR $A_{2a}R$→PKA/PKC	proliferation, metastasis	[122,123]
Steroid receptors					
Membrane-type bile acid receptor (M-BAR/TGR5)	Deoxyolate, bile acids	TGR5: upregulation	TGR5→EGFR/MAPK	proliferation	[124]
Orphan receptors					
GPR30	G1	GPR30: upregulation	GPR30→cAMP/Ca^{2+} GPR30→EGFR/PI3K/AKT/ERK	invasion, metastasis	[125]
GPR39	Obestatin	GPR39: -	GPR39→EGFR/MMP→AKT GPR39/β-arrestin/Src→EGFR→AKT	proliferation	[97]

Table 1. Cont.

GPCRs	Ligand	Expression	Mechanisms	Biological Function	References
Class B receptors					
Hormone receptors					
Growth hormone-releasing hormone (GHRH) receptor (GHRHR)	GHRH	GHRHR: upregulation	GHRHR→PAK1→ STAT3/NF-κB	proliferation, inflammation	[126]
Class C receptors					
Ion receptors					
Calcium-sensing receptor (CaSR)	calcium ions	CaSR: upregulation	CaSR→Ca^{2+}/TRPV4/ β-Catenin	proliferation, migration, invasion	[127]
Amino Acid receptors					
γ-Aminobutyric acid (GABA) receptor	GABA	$GABA_A$: upregulation $GABA_B$: -	$GABA_A$→ERK1/2	proliferation, invasion	[128]
Metabotropic glutamate receptors (mGluRs)	Glutamate	mGluR5: upregulation mGluR1/5 (group I): - mGluR2/3 (group II): - mGluR4/6/7/8 (group III): -	mGluR5→ERK1/2	proliferation	[129]
Adhesion receptors					
ADGRE5 (CD97)	CD55, α5β1 integrin, CD90	ADGRE5: upregulation	ADGRE5→MAPK	proliferation, metastasis	[130]
Class F receptors					
Fizzled receptors	WNT, lipoglyco-proteins	FZD2/6/7: upregulation FZD1/3/4/5/8/9/10: -	FZDs→Wingless/Int-1 (WNT)	proliferation	[131,132]
Smoothened receptors (SMO)	cholesterol, sterol	Smo: upregulation	SMO→HH	proliferation, invasion	[131,133]
Viral receptors					
EBV-encoded vGPCR	metal ion (Zn^{2+})	BILF1: upregulation	BILF1→MHC class 1	proliferation, immune evasion	[68,134]

3.1. Proliferation and Apoptosis

Mounting evidence has unveiled the multilayered crosstalk between GPCRs and proliferation- and apoptosis-related signaling circuits. The representative ones involve EGFR transactivation, MAPK cascades, the PI3K-AKT-mTOR pathway, and the Hippo signaling pathway [21,135].

3.1.1. Transactivation in the EGFR and MAPK/ERK Pathway

GPCRs share many similarities with the tyrosine kinase receptors, such as EGFR and the MAPK/ERK signaling pathways [136], in regulating cell proliferation. The EGFR-mediated signaling pathway can be ligand-dependent or independent [137–139]. The "three membrane-passing signal (TMPS)" model is an EGFR ligand-dependent route. The activated RTKs are triggered by activated GPCRs and subsequently activate the extracellular signal-regulated kinase (ERK)/mitogen-activated protein kinase (MAPK) cascade. On the other hand, GPCR-mediated Src activation contributed to EGFR phosphorylation more directly. Both modes have been uncovered in GC. S1P could mediate the progression of GC via Gi- and matrix metalloprotease (MMP)-independent c-Met- and EGFR-transactivation [113]. However, the S1P- or LPA-induced transactivation of ERBB2 (also known as HER2) required the activation of MMP and the tyrosine kinase activity of EGFR [110]. In addition, the knockdown of the membrane-type bile acid receptor (M-BAR)/TGR5 suppressed the deoxycholate (DC)-induced phosphorylation of EGFR, and DC transactivates EGFR through M-BAR- and ADAM/HB-EGF-dependent mechanisms [124]. Infection with *H. pylori* boosted the expression of interleukin-8 (IL-8), which promoted cell proliferation by inducing EGFR transactivation [140]. Due to oncogenic activation and PGE_2-EP4 pathway induction, the ubiquitous overexpression of the EGFR ligands and Adams have been identified in mouse gastric tumors [141]. PGE_2-induced uPAR expression has also been implicated in the activation of Src, c-Jun NH_2-terminal kinase (JNK), extracellular signal-regulated kinase (Erk), and p38 mitogen-activated protein kinase (p38 MAPK) [142,143]. GPCRs may also directly trigger MAPK cascades, establishing a connection between the external stimuli and their effect factors. These effectors may be further subdivided into four core categories: ERK1/2, JNK1-3, p38α-δ MAPKs, and ERK5. The LPAR2 inhibitor suppressed the proliferative and migration abilities of GC cell line SGC-7901 through the LPAR2/Gq11/p38 pathway, suggesting that LPAR2 might be a potential target for GC treatment [112]. Protease-activated receptor family (PAR1-4) also exerted pro-carcinogenic effects via the overactivated ERK1/2-MAPK pathway. For example, the reduction of EPCR impeded PAR1 activation, thus resulting in the downregulation of phosphorylated ERK1/2 and the suppression of the proliferation and migration of GC tumor cells [144].

3.1.2. Activation of the PI3K-AKT-mTOR Pathway

PI3K is stimulated by the activated RTKs or GPCRs, ultimately leading to the synthesis of PIP3 and the recruitment of oncogenic effectors such as the serine/threonine kinase AKT. The PH domain in AKT permitted its binding with PIP3, contributing to the membrane accumulation and subsequent phosphorylation at T308 and S473 by PDK1 and mTORC2 [145]. Even though over 100 AKT substrates have been discovered in different settings, the associated mechanisms for most substrates have not been fully delineated [145]. mTOR is one of the AKT subtracts that is well-established to promote biosynthetic processes for cell growth. Since PI3K/AKT/mTOR signaling has also been identified as an ideal drug target for gastric carcinoma, the regulators may have a role in improving treatment design [146]. Indeed, some GPCRs have been proven to influence the activity of AKT in GC cells, such as the leucine-rich repeat-containing receptor Lgr6, adenosine receptor A2a, and the orphan receptor GPR39. Lgr6 was identified to empower GC cell proliferation by activating the PI3K/AKT/mTOR pathway [109]. Another GPCR, adenosine receptor A2a, was engaged in PI3K/AKT-regulated proliferation and migration in GC [123]. Additionally, GPR39 provided GC cells with a growth advantage by boosting the activity of AKT in an EGFR-dependent manner [97].

3.1.3. Regulation of the Hippo Pathway

The canonical Hippo pathway kinase cascade is a critical tumor suppressor pathway, and its dysregulation has been widely implicated in organ size modulation and carcinogenesis [147]. The core components of the Hippo pathway are composed of STE20-like protein kinase 1/2 (MST1/2) and large tumor suppressor 1/2 and the major functional output Yes-associated protein 1 (YAP) and WW domain-containing transcription regulator protein 1 (WWTR1, also known as TAZ). Because there is a lacking DNA binding site in YAP/TAZ, TEF1-4 (TEAD1-4) is characterized as a bona fide transcription enhancer factor [147,148]. GPCRs have been found to control the Hippo pathway positively and negatively as a significant regulator of the intracellular pathway. The initial implication that GPCRs modulate Hippo signaling through LATS1/2 came from the study in serum starvation cells [149]. Two components, LPA and S1P, have been identified as the effective factors in serum that are responsible for YAP/TAZ activation through the recognition of the corresponding

GPCRs. The LPA/S1P-mediated GPCR activation facilitates YAP/TAZ dephosphorylation via the G protein-cytoskeleton circuit. This study has laid the foundation for how YAP/TAZ senses the diffusible extracellular signals. However, several questions have also been raised after the initial discovery. Given that GPCRs constitute ~800 members and each GPCR can be coupled to diverse G proteins [19], the integrated effects on YAP/TAZ modulation remain elusive. The case can be more complicated when the dysregulation of G proteins and GPCRs is frequently determined in cancers [54]. Moreover, different G proteins stimulate the dephosphorylation of YAP/TAZ to various degrees. GPCRs can trigger YAP/TAZ activation by interacting with $G\alpha_{12/13}$, $G\alpha_{i/o}$, and $G\alpha_{q/11}$ or suppress YAP/TAZ by binding with $G\alpha s$. However, which of these that Rho is involved with has not yet been identified; it is also unclear how the actin cytoskeleton regulates Lats1/2 phosphorylation. Emerging findings have revealed how specific GPCRs may fine-tune YAP/TAZ in given cellular surroundings [150]. The triggered LPA receptors have been demonstrated to play crucial roles in activating YAP/TAZ, causing tumor progression in the colon, ovarian, prostate, and breast [151,152]. The S1P-mediated S1P receptors contribute to hepatocellular carcinoma by coupling to $G\alpha_{12/13}$ and stimulating YAP [153], connecting GPCR signaling to the Hippo pathway. Except for LPARs and S1PRs, the other GPCR-initiated signals can influence YAP/TAZ activity, including polypeptides (Angiotensin II, Thrombin, glucagon, etc.) [154,155], metabolites (purines, fatty acids, epinephrine, glutamate, etc.) [156,157], and hormonal factors (estrogen, endothelin-1, etc.) [158,159]. These signals have been widely indicated in human malignancies and are critical cell niche or microenvironment components. Recent studies pointed out that the mesenchymal niche manipulated the initiation of colorectal cancer by the rare peri-cryptal Ptgs2-expressing fibroblasts, and these fibroblasts exhibited paracrine control over tumor-initiating stem cells via the PGE2-EP4-Yap signaling axis [160]. The GPCR-Hippo crosstalk was also identified in GC stem-like cells. PAR1 stimulated the Hippo-YAP pathway and affected invasion, metastasis, and multidrug resistance [161]. As such, the GPCR regulation of YAP/TAZ has emerged as a driver, or as a potential therapeutic target, in gastric neoplasia. However, another study has found that AMOT, rather than Lats1/2, serves as the bridge between GPCR-mediated cytoskeleton changes and YAP/TAZ modulation in uveal melanoma cells, with an activated mutation at Arg183 and Gly209 in *GNAQ* (encoding for $G\alpha q$) and *GNA11* (encoding for $G\alpha_{11}$), respectively [85]. As a result, the findings provide novel explanations for the alternations in actin dynamics induced by GPCR signals, which are somewhat different from previous studies. Therefore, this warrants exploring the interplay between AMOT, Lats1/2, and the actin cytoskeleton in GC, as the mechanic stress is context-dependent.

3.1.4. GPCR-Signaling Integration and Crosstalk with Other Pathways in GC

Besides the above signaling circuits, other pathways have also been linked to GPCR-mediated oncogenicity in GC. These pathways involve the Notch pathway [162], hedgehog (Hh) signaling [163], and the Wnt/β-catenin pathway [164]. The Hh pathway is crucial for GC cell growth and cancer stem cell maintenance, and its activation has been highlighted in diffuse-type GC [165,166]. Smoothened (Smo, a member of class F) and Gpr161 (an orphan member) can function as positive and negative regulators in the Hh pathway, respectively [167]. The Wnt/β-catenin pathway is involved in tissue homeostasis and embryonic development. As Wnt (Wingless/Int1) stimulates the frizzled receptor (FZD, class F GPCRs), both G-protein independent and dependent signaling can be established [164,168].

3.2. GPCRs-Driven Metastasis of GC

Metastasis is how cancer cells establish 'bench-heads' in other organs or anatomical sites instead of the initial lesion, and it is responsible for more than 90% of cancer-related mortality [169] (Figure 3A). The most prevalent sites for GC metastasis are the liver, lung, bone, and lymph nodes [170]. Since Paget's 'seed and soil' hypothesis laid the fundamental basis for metastasis, many investigators have contributed to a better understanding of the process. Several studies have identified the sequential multistep in GC metastasis:

invasion into the surrounding tissue and the degradation of the basement membrane (BM), intravasation into the blood vessels or lymphatic systems, survival and translocation to distant tissues, extravasation into the foreign environment, and finally, colonization to proliferate and form a macroscopic secondary neoplasm [170–172]. As the complexity and relevance of metastasis have previously been widely reviewed, we will focus mainly on the roles and mechanisms of GPCRs during the invasion, BM degradation, and angiogenesis processes in GC.

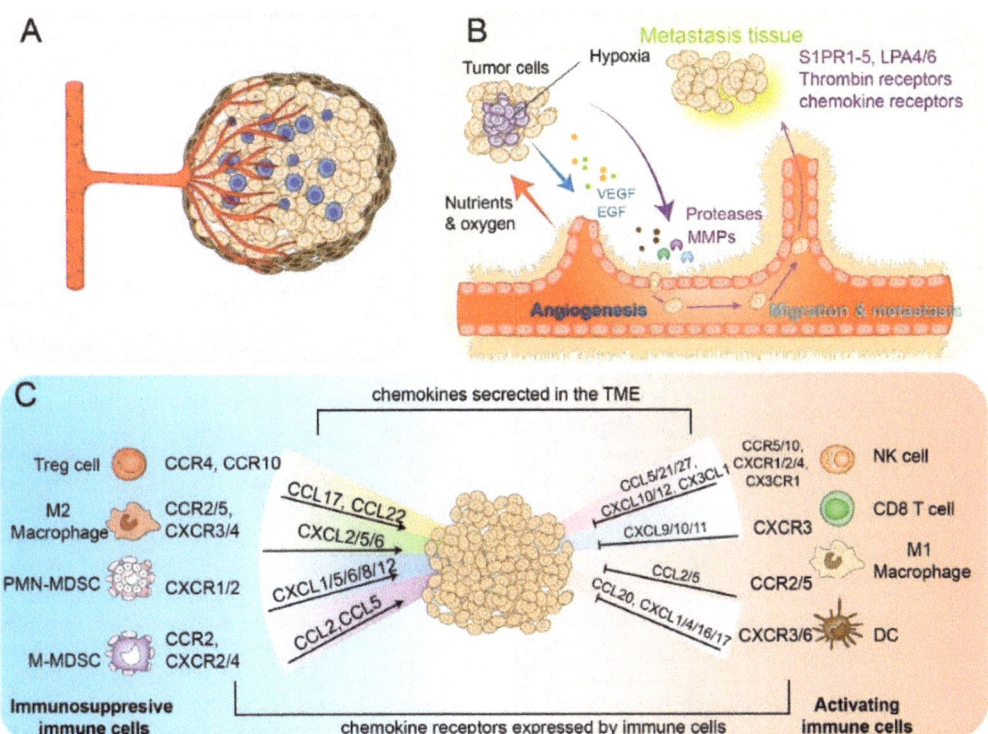

Figure 3. GPCR-mediated metastasis and tumor microenvironment remodeling in GC. (**A**) The TME of GC consists of blood vessels, lymph vessels, immune cells, stromal cells (including fibroblast, pericytes, and adipocytes), extracellular matrix (ECM), and secreted soluble factors, such as proteins, RNAs, and small organelles. (**B**) GPCRs control the process of angiogenesis and metastasis. GPCR activation drives the production of stimulatory angiogenic factors like VEGF and EGF. These factors promote the development of new blood vessels by modulating the mitogenesis, migration, and sprouting of endothelial cells (ECs). Moreover, several GPCRs regulate the metastasis process by influencing ECM, degrading the status of cancer cells (EMT, migration, and invasion), and colonizing foreign sites. (**C**) Chemokine–chemokine receptors modulate immune responses. The chemokines are secreted by tumor cells, immune cells, and stromal cells. The interaction of chemokine and specific chemokine receptors recruits antitumor immune cells and immunosuppressive immune cells into the tumor microenvironment.

3.2.1. Inducing Epithelial-Mesenchymal Transition (EMT), Migration, and Invasion

Epithelial-to-mesenchymal transition (EMT) is regarded as the initial step for invasion, featuring a loss of cell polarity and integrity and the acquisition of motile mesenchymal characteristics. The pathologic activation of the EMT program is primarily executed by transcription factors (including SNAI1/2, TWIST1/2, and ZEB1/2) and microRNAs, ultimately resulting in the accumulation of the genes associated with mesenchymal properties, such as vimentin, fibronectin, and N-cadherin [173]. Though the above-mentioned molecular mechanisms are still lacking regarding GPCR-driven EMT in GC, GPCR signaling dysregulation is still frequently connected to EMT, migration, and invasion processes via dynamically regulating the downstream effectors and downstream cascades [174].

Although chemokines are tiny polypeptides (8-14kDa), they display pleiotropic effects in cancers. The chemokine system comprises nearly 50 chemokines that bind to 20 different chemokine receptors or four atypical chemokine receptors (ACKRs) [175]. This superfamily is distinguished by a substantial degree of redundancy, inferring that these chemokines can bind to different clusters of receptors and vice versa [176]. Intrinsic genetic or epigenetic regulators governed their expression and environmental cues such as hypoxia, microbiota, and metabolic [176]. For example, epigenetic regulator histone deacetylase 1 (HDAC1) suppressed CXCL8 expression by antagonizing the active nuclear transcription factor NF-κB [177]. Hypoxia has been revealed to induce the expression of CXCR4, CXCR7, and CXCL12 in different cancer cells, with the binding sites of hypoxia-inducible factor 1 (HIF1) as the promoters of these genes [178–180]. In GC, the elevated CXCL8 concentration was proved to be tightly correlated with the tumor stage instead of $H.\ pylori$ infection, as shown in previous studies [181–183]. Several clinical investigations have suggested that the upregulation of chemokines and receptors was associated with GC pathogenesis, indicating that the specific chemokines might serve as potential diagnostic and therapeutic targets [184,185].

Chemokine receptors have attracted considerable attention due to their involvement in GC metastasis. A notable correlation was found between CCR7 expression and gastric carcinoma lymph node metastasis via stepwise regression analysis [186,187]. Strikingly, about 67% of primary gastric tumors exhibited CXCR4-positive expression [188]. The high concentration of the CXCR4 ligand CXCL12 has been validated in the malignant ascitic fluids from peritoneal carcinomatosis, and elevated CXCL12 is tightly correlated with the dissemination of GC cells to distant organs [188]. CXCL12 stimulated CXCR4 enhanced NF-κB and STAT3 signaling activation and, in turn, led to its transcriptional upregulation, which formed a positive feedback loop. This loop is linked to EMT, migration, and invasion in GC [189]. In response to CXCL12, CXCR4 also conferred the GC cell EMT and metastasis process via stimulating mTOR and some well-known oncogenic kinases: EGFR, SRC, or c-MET [190,191]. The crosstalk between TGF-β1 and the NF-κB pathway was triggered by the CCL2-CCR2 axis, leading to EMT-related protein upregulation [192]. Besides, chemokine receptors also induced organ-specific metastasis. CXCR4 and CCR7 are the primary receptors guiding the metastasizing cells, including GC cells [193]. Moreover, the high levels of CCR9 in melanoma, breast, and ovarian cancer make them efficiently translocate to the highly CCL25-expressing small intestine [194–196].

Many other GPCRs also govern the development of GC invasion. For example, GPER1 inhibition blocked EMT in GC cells by inhibiting the PI3K/AKT pathway [197]. Similar regulation that is mediated by adenosine receptor 2 ($A_{2a}R$) or GPR30 could also be observed in GC [122,123]. In addition, the MAPK cascades were activated by the formyl peptide receptor 2 (FP_2R), $S1P_2R$, muscarinic acetylcholine receptor 3 (M3R), P2Y receptors (P2YR), and γ-Aminobutyric acid receptor A ($GABA_A$), thus contributing to the invasion and metastasis in GC [104,105,114,119,121,128]. Many other GPCRs have also been linked to GC metastasis, while the underlying mechanisms are unknown. For instance, the angiotensin II receptor type 1/2 (AT1R/AT2R) has been locally upregulated and indicated to carry a much higher risk of nodal spread [101].

3.2.2. Degrading the Barriers to Invasion

BM, a specialized extracellular matrix (ECM), plays a critical role in normal epithelium tissue architecture. BM disruption is a must for cancer cells leaving the primary location, controlled by the balance between the expression of MMPs and their tissue inhibitors (TIMPs) [198,199]. The expression of MMPs was upregulated by a histamine-H2 receptor or Thrombin-PAR1 signaling [143]. $H.\ pylori$ was reported to be crucial during the invasion by upregulating cyclooxygenase-2 (COX-2) through ATF2/MAPK stimulation. The COX-2 inhibitor or EP2 receptor antagonist repressed angiogenesis and tumor invasion via the uPA system, which is a determinant factor in transforming the zymogen plasminogen into plasmin for degrading the ECM constituents [200]. Furthermore, the bacterial pathogen of *H. pylori's* consistent infection manipulated a variety of extracellular proteases [201], but the exact mechanisms need further exploration. The interactions between microbial metabolites and GPCRs may provide new insights into the complicated process [202].

3.2.3. Driving Angiogenesis

Angiogenesis is the process of vessel splitting from pre-existing vessels and is essential for tumorigenesis and progression, especially for those solid tumors exceeding 1–2 mm in diameter, as it provides oxygen and nutrients [203,204]. Many GPCRs exerted pro-tumor effects by promoting tumor-associated angiogenesis, notably Thrombin receptors, S1PRs, lysophosphatidic acid receptors (LPARs), and Prostaglandin receptors (Figure 3B). PAR1 is necessary for physio-pathological angiogenesis since poor vasculature development results in animal embryos dying after PAR1 deprivation. Thrombin-mediated PARs cleavage upregulates the transcription of many proangiogenic genes, such as VEGF and its receptor VEGFR, MMP2, angiopoietin-2 (Ang-2), and others [99,100,205,206]. Moreover, endothelial differentiation gene 1 (Edg1)/S1P$_1$R is the first reported GPCR in blood vessel formation. Furthermore, the intrauterine death of Edg1 ablation mice happened mainly due to abnormal angiogenesis [207,208]. The Gα12/Gα13-coupled receptors LPA4 and LPA6 synergistically regulate endothelial Dll4 expression through YAP/TAZ activation, which mediates sprouting angiogenesis [209]. Moreover, *H. pylori*-induced VEGF upregulation was activated through p38 MAPK COX2-PEG$_2$-EP2/4 signaling [210]. Other orphan receptors are also involved in tumor angiogenesis, such as KSHV-GPCR, GPR124, ELDT1, and GPER [211].

3.3. Remodeling the Tumor Microenvironment (TME) to Promote Immune Escape

TME acts as a unique niche populated by multiple cell types (including cancer cells, immune cells, and stromal cells), ECM, and diverse secreted factors (such as exosomes and microRNAs) [212,213]. The altered TME landscape is related to tumor progression, metastasis, and therapeutic responses [214]. Recently, the sophisticated TME infiltration pattern of GC (termed as TMEscore) has been defined based on the assessment of 22 immune cell types and cancer-associated fibroblasts (CAFs), which were correlated with genomic characteristics and pathologic features [212]. The biology and function of CAFs have emerged as an area of active investigation and have been reviewed elsewhere [215,216]. The compositions of infiltrated immune cells within TME varied greatly, and one of the most important mechanisms involved the chemokines and their receptors [176].

It is noteworthy that chemokines and chemokine receptors can be ubiquitously expressed in tumor cells, immune cells, and stromal cells [217]. Alternations in chemokines and their receptors shaped the TME immune cell constitution and remodeled the immune responses, some of which are hijacked by tumor cells to avoid immune surveillance and elimination [218]. The antitumor immune responses were driven by the recruiting immune cells, mainly including dendritic cells (DCs), CD8+ T cells, natural killer (NK) cells, and M1 macrophages. GC with a high CXCR3 expression level was shown to have increased DC and T cell infiltration. The CXCR3/CXCL4 or CXCR3/CXCL4L1 axis is necessary to recruit DCs as they elicit potent antitumor functions through substantially stimulating T cells and activating the related humoral response [219,220]. Similarly, CXCR3 also plays a vital role in CD8+ T cell infiltration that directly damages the tumor cells after being differentiated into cytotoxic CD8+ T cells [221,222]. In addition, NK cells represent professional killer cells, whose accumulation in the TME is the consequence of upregulated CXCL10 and CXCL12 signaling through CCR7 or CXCR3 [223,224].

On the other hand, chemokine signaling is also involved in the formation of immune-suppressive TME, where tumors evolve to escape recognition and clearance. This process has been largely linked to the infiltration of diverse protumor immune cell populations, such as regulatory T (Treg) cells, the M2 macrophages, monocytic myeloid-derived suppressor cells (M-MDSCs), and granulocytic (or PMN-) MDSCs [176,225]. CCL22, mainly produced by tumor cells (or macrophage-mediated), causes an abundance of Treg cells in TME via interacting with the receptor CCR4 on the surface of Treg. Another receptor, CCR10, in Treg cells also facilitated their migration in response to CCL28 [226,227]. Moreover, the nonpolarized macrophages (M0) originating from the recruited monocytes can be differentiated into two main subtypes, M1 and M2 macrophages, exhibiting extremely distinct functions toward cancers. These transitions depended on a large spectrum of chemokine signals. Active monocyte recruitment required tumor-derived chemokine releases, such as CCL2, CCL3, CCL4, CCL5, CCL20, and CCL18. Additionally, the blockade of the CCL2-CCR2 circuit led to M2 macrophage accumulation, whereas CCL11 skewed macrophages toward an M2 phenotype [228–233]. MDSCs were subdivided into two major groups: polymorphonuclear MDSCs (PMN-MDSCs) and mononuclear MDSCs (M-MDSCs). CXCR2 specifically mediated the migration of PMN-MDSCs to the tumor site by binding with CXCL1/CXCL2/CXCL5, whereas the accumulation of M-MDSCs requires CCL2-CCR2 interaction. Functionally, MDSCs employed diverse mechanisms to suppress T cell functioning, mainly through releasing high levels of arginase 1 (Arg1), reactive oxygen species (ROS), and nitric oxide (NO). Further research also suggested additional mechanisms, including the upregulation of COX2 and PGE2 in these MDSCs [234,235] (Figure 3C).

Other GPCRs are also involved in the regulations of immune responses. For example, prostaglandin (PG) production can mediate inflammation through its cognate GPCR EP1-EP4 (*PTGFR1-4*). PGs, especially the PGE2, were produced by the cyclo-oxygenases COX-1 and COX-2, the inhibitors (nonsteroidal anti-inflammatory drugs (NSAIDs)) of which have been utilized to comfort pains and reduce the incidence of a broad range of cancer types. The role of PGE2 has been extensively studied for inducing inflammation by stimulating other signaling pathways, including the Toll-like receptor (TLR)/MyD88 pathway [236], Wnt, and EGFR signal [237]. PAR1-deficient mice infected with *H. pylori* may suffer from severe gastritis due to lacking suppressing macrophage cytokine secretion and cellular infiltration [238]. In addition, TGR5 antagonized gastric inflammation by inhibiting the transcription activity of NF-κB signaling [239].

The involvement of GPCRs in immune remodeling is summarized in Table 2.

Table 2. Immune cell infiltration induced by chemokine and related receptors.

Cell Type	Receptors	Chemokines	Mechanisms Underlying Recruitment	Effects on Tumor Cells after Recruitment	References
Anti-tumoral immune cells					
Dendritic cell	CXCR3, CXCR6	CXCL4, CXCL1, CXCL16, CXCL17, CCL20	IFN-γ-induced chemokines production, *H. pylori* involvement	The most potent professional antigen-presenting cells, activation of cellular immunity, and T cell-dependent humoral immunity	[219,220]
CD8 T cell	CXCR3	CXCL9, CXCL10, CXCL11	CAFs-mediated IL6 secretion, tumor cell chemokines secretion, adhesion molecules (ICAM-1, VCAM-1)	Differentiated into cytotoxic CD8+ T cells to destroy tumor cells or memory CD8+ T cells to recirculate in the blood	[221,222]

Table 2. Cont.

Cell Type	Receptors	Chemokines	Mechanisms Underlying Recruitment	Effects on Tumor Cells after Recruitment	References
NK cell	CXCR1, CXCR2, CXCR4, CX3CR1, CCR5, CCR10	CXCL10, CXCL12, CCL21, CX3CL1, CCL5, CCL27	Chemokine signaling regulated by HLA-G and CD47; stromal barriers	Cytokine production and cytotoxicity on tumor cells through STAT3; regulating DC maturation; modulating T cell activity	[223,224]
M1 macrophage	CCR2, CCR5	CCL2, CCL5	Disrupting NF-κB signaling or interacting with TNF-α;	High capacity to present antigens; proinflammatory cytokines (IL-1β, IL-1α, IL-12, TNF-α, and GFAP) production; stimulation of type-I T cell responses	[232,233]
Tumor-promoting immune cells					
Treg	CCR4, CCR10	CCL17, CCL22	Stimulation of JAK-STAT3 signaling pathway; remodeling of gastric microbiota by H.pylori.; stimulation of DCs due to H.pylori. infection	Suppressing CD4+ T cells, CD8+ cells, antigen-presenting cell (APC), monocytes, and macrophages; inhibitory cytokines like IL10, IL35, and TGF-β; inducing apoptosis by perforin/ granzyme production	[226,227]
M2 macrophage	CCR2, CCR5, CXCR3, CXCR4	CCL2, CCL5, CXCL9, CXCR12	STAT3 activation; PI3K/AKT/mTOR signaling pathway	Growth factors (FGF, VEGF, and IL-6) production; secreting matrix-degrading enzymes and cytokines	[230,231]
Monocytic MDSC	CCR2, CXCR2, CXCR4	CCL2, CXCL5, CXCL12	IL-6 production, JAK-STAT3 signaling,	High amounts of NO, Arg1, and immune-suppressive cytokines; suppression of nonspecific T cell responses	[234]
Granulocytic (or PMN-) MDSC	CXCR1, CXCR2	CXCL8, CXCL1, CXCL12, CXCL5, CXCL6	HGF/TGF-β/MCP-1 production, JAK-STAT3 signaling, IRF-8, NF-κB pathway, hypoxia	Large amounts of O^{2-}, H_2O_2, and PNT (ROS) production; blocking T cell proliferation; depleting entry of CD8+ T cells to tumors	[234,235]

4. Therapeutic Strategies for Targeting GPCRs in GC

Despite the improving clinical outcomes, advanced GC patients benefit little from traditional surgery or chemotherapy and suffer from painful lives [240]. Personalized medicine and targeted therapy have been introduced to clinical applications for over two decades. For instance, trastuzumab has been integrated into the treatment for HER2-expressing patients, and ramucirumab has been utilized for VEGFR2-positive GC individuals [241]. Immune checkpoint inhibitors (ICI) also have been investigated as a frontline treatment [242–245]. Meanwhile, biomarkers and novel targeted therapies have been intensely investigated for advanced GC [246]. Substantial progress has been made by deciphering the functions of GPCR members in GC progression. However, only a handful of drugs that target GPCRs have been conducted in clinical trials for GC treatment (Table 3).

Table 3. Drugs and antibodies against GPCRs in GC clinical trials [6,135,247].

Drug Name	Targeted GPCRs	Types of Drugs	Tested Cancer Types	Status	NCT
Mogamulizumab	CCR4	mAb	Cataneous/Peripheral T-cell lymphoma; Adult T-cell lymphoma	Phase I: complete	NCT02946671
Vismodegib	SMO	small molecule	Basal-cell carcinoma; Head and neck cancer	Phase II: complete Phase II: complete Phase II: recruiting	NCT03052478 NCT00982592 NCT02465060
Sonidegib	SMO	small molecule	Basal-cell carcinoma	Phase I: recruiting Phase I: complete	NCT04007744 NCT01576666
Lutathera (Lutetium Lu 177 dotatate)	SSTR	peptide	Gastroenteropancreatic neuroendocrine tumors (GEP-NETs)	Most on recruiting	NCT04949282 NCT04727723 NCT04609592 NCT04524442 NCT02736500 NCT02489604 NCT04614766 NCT01860742
Lanreotide	SSTR	peptide	Advanced prostate cancer	Phase III: recruiting	NCT04852679 NCT03043664 NCT03017690 NCT02730104 NCT02736448

In order to accelerate the GPCR-targeted drug development for GC, many groups have identified potent compounds to inhibit or enhance the activity of GPCRs. However, the structures are only available for small partial GPCRs (~50 GPCRs), and 54% of GPCRs are orphans that are under-exploited. Machine learning approaches may be employed for predicting the interaction between immersed compounds and GPCRs based on established high-quality structural models [248]. With the evolving knowledge of GPCR pathways, we will be able to identify more effective drugs in formats, tissue-specific drug delivery systems, and appropriate treatment periods. Small molecules are the most prevalent GPCR modulators, while biologics are receiving more and more attention because of their versatility and specificity [249]. Antibodies, including antibody fragments and variable antibody domains, function with great penetration traits and are attracting considerable interest in drug development. Downregulated targeted GPCRs via RNA interference (RNAi) can represent potential approaches to gene therapy [250].

More efficient drug delivery systems with enhanced solubility and stability, lower dosages, and less toxicity have been developed, such as nanomaterials, nanocarriers, nanoconjugation, and nanoencapsulation techniques [251,252]. Furthermore, several solid tumors have well-established patient-derived xenografts (PDX) and xenograft-derived organoid models. These preclinical platforms recapitulated the genotypic and phenotypic landscape, endowed with a high predictive value for high-throughput drug screening. Nevertheless, they still have limitations, such as intratumor heterogeneity, compromised immune systems, and diverse tumor environments in GC [253].

5. Summary and Future Perspectives

GPCRs govern multiple signaling pathways and regulate GC development in various aspects. The heterogeneous and complicated characteristics of GPCRs contribute to GC heterogeneity and result in the current untimely diagnosis and inefficiency of therapeutic applications. Not only can GPCRs transduce the extracellular changes to the intracellular signaling circuits, but the conformational changes of GPCRs can also continuously influence intracellular events. Aberrant GPCR activation and mutated GPCRs/G proteins can fuel cancer cell proliferation, migration, invasion, angiogenesis, and metastasis. In addition, the dysregulation of GPCRs affords advantages for immunosuppressive TME and drug resistance to malignancies.

Although much progress has been made on novel biomarker identification and molecular mechanism investigation, the current GPCR-based diagnosis and therapy in GC are far from clinically available. It is urgent that GPCR signaling-targeted therapy be developed. In future studies, several issues need to be addressed. First, as the mutation rates of GPCRs and G proteins are prominent in some cancer types, the development of small molecules that target the driver mutations is urgent. Second, because of the heterogeneity of the cancer cells and tumor microenvironment, we need to comprehensively appraise the activation of GPCR signaling and its crosstalk by using cutting-edge techniques such as scRNA-seq or scDNA-seq. Last but not least, more preclinical models based on patient-derived samples, such as organoids or xenografts, need to be developed to evaluate the efficacies and side effects of the screened drugs. With the deep investigation of the molecular mechanisms of GPCR signaling and the multicenter clinical trials, more therapeutic strategies will be delivered for targeting GPCR signaling, which will benefit GC patients.

Author Contributions: W.K. provided direction on this manuscript. H.Y. and J.-L.Z. reviewed the literature and drafted the manuscript. K.-T.L., K.-W.L., J.Y. and K.-F.T. reviewed and revised the manuscript. All authors have read and agreed to the published version of the manuscript.

Funding: This study is supported by the National Natural Science Foundation of China (NSFC) (No. 82272990).

Institutional Review Board Statement: Not applicable.

Informed Consent Statement: Not applicable.

Data Availability Statement: Not applicable.

Acknowledgments: We acknowledge the BioRender for the Figure generation (https://biorender.com/, accessed on 14 December 2022) and the technical support from Core Utilities of Cancer Genomics and Pathobiology, The Chinese University of Hong Kong.

Conflicts of Interest: The authors declare no competing interests.

Abbreviations

$A_{2a}R$, adenosine receptor 2; ACKRs, atypical chemokine receptors; ADAM17, metallopeptidase domain 17; Ang-2, angiopoietin-2; Arg1, arginase 1; AT1R/AT2R, angiotensin II receptor type 1/2; BM, basement membrane; CAFs, cancer-associated fibroblasts; cAMP, cyclic adenosine monophosphate; CHRMs, muscarinic receptors; CIN, chromosomal instability; CNS, central nervous system; CNVs, copy number variations; COX-2, cyclooxygenase-2; cryo-EM, cryo-electron microscopy; DAG, diacylglycerol; DCs, dendritic cells; DD/ID, developmental delay/intellectual disability; EBV, Epstein Barr virus; ECD, extracellular domain; ECL, extracellular loop; ECM, extracellular matrix; Edg1, endothelial differentiation gene 1; EMT, epithelial-mesenchymal transition; ERK1/2, extracellular signal-regulated kinase 1/2; FP2R, formyl peptide receptor 2; GABAB, γ-aminobutyric acid B receptors; GAIN, autoproteolysis-inducing domain; GC, GC; GISTIC, Genomic Identification of Significant Targets in Cancers; GnRH, gonadotrophin-releasing hormone; GPCRs, G protein-coupled receptors; GRKs, GPCR kinases; GRMs, glutamate metabotropic receptors; GS, genomically stable; HDAC1, histone deacetylase 1; Hh, hedgehog; HIF1, hypoxia-inducible factor 1; HRHs, histamine receptors; IBS, irritable bowel syndrome; ICI, Immune checkpoint inhibitors; ICL, intracellular loop; JNK, c-Jun NH2-terminal kinase; LPA, lysophosphatidic acid; M-BAR, membrane-type bile acid receptor; M-MDSCs, monocytic myeloid-derived suppressor cells; M3R, muscarinic acetylcholine receptor 3; MAPK, mitogen-activated protein kinase; mGluRs, metabotropic glutamate receptors; MMP, matrix metalloprotease; MSI, microsatellite instability; MST1/2, STE20-like protein kinase 1/2; NAMs, negative allosteric modulators; NK cells, natural killer cells; NMR, nuclear magnetic resonance; NO, nitric oxide; NSAIDs, nonsteroidal anti-inflammatory drugs; PAMs, positive allosteric modulators; PARs, Protease-activated receptors; PDAs, pancreatic adenocarcinomas; PDEs, phosphodiesterases; PDX, patient-derived xenografts; PIP2, phosphatidylinositol 4,5-biphosphate;

PIP3, inositol 1,4,5-triphosphate; PIs, phospholipases; PKA, activate protein kinase A; PKC, protein kinase C; PMN-MDSCs, polymorphonuclear MDSCs; RNAi, RNA interference; ROS, reactive oxygen species; S1P, sphingosine-1-phosphate; TCGA, The Cancer Genome Atlas; TEAD1-4, TEF1-4; TIMPs, tissue inhibitors of metalloprotease; TME, tumor microenvironment; TMPS, three membrane-passing signal; Treg, regulatory T; UVM, uveal melanoma; WWTR1, WW domain-containing transcription regulator protein 1; XDO, xenograft-derived organoid; YAP, Yes-associated protein 1.

References

1. Bray, F.; Ferlay, J.; Soerjomataram, I.; Siegel, R.L.; Torre, L.A.; Jemal, A. Global cancer statistics 2018: GLOBOCAN estimates of incidence and mortality worldwide for 36 cancers in 185 countries. *CA Cancer J. Clin.* **2018**, *68*, 394–424. [CrossRef]
2. Siegel, R.L.; Miller, K.D.; Jemal, A. Cancer statistics, 2020. *CA Cancer J. Clin.* **2020**, *70*, 7–30. [CrossRef]
3. Lauren, P. The Two Histological Main Types of Gastric Carcinoma: Diffuse and So-Called Intestinal-Type Carcinoma. An Attempt at a Histo-Clinical Classification. *Acta Pathol. Microbiol. Scand.* **1965**, *64*, 31–49. [CrossRef] [PubMed]
4. Cancer Genome Atlas Research, N. Comprehensive molecular characterization of gastric adenocarcinoma. *Nature* **2014**, *513*, 202–209. [CrossRef] [PubMed]
5. Smyth, E.C.; Moehler, M. Late-line treatment in metastatic gastric cancer: Today and tomorrow. *Ther. Adv. Med. Oncol.* **2019**, *11*, 1758835919867522. [CrossRef]
6. Usman, S.; Khawer, M.; Rafique, S.; Naz, Z.; Saleem, K. The current status of anti-GPCR drugs against different cancers. *J. Pharm. Anal.* **2020**, *10*, 517–521. [CrossRef]
7. Abbaszadegan, M.R.; Mojarrad, M.; Moghbeli, M. Role of extra cellular proteins in gastric cancer progression and metastasis: An update. *Genes Environ.* **2020**, *42*, 18. [CrossRef]
8. Fredriksson, R.; Lagerstrom, M.C.; Lundin, L.G.; Schioth, H.B. The G-protein-coupled receptors in the human genome form five main families. Phylogenetic analysis, paralogon groups, and fingerprints. *Mol. Pharmacol.* **2003**, *63*, 1256–1272. [CrossRef]
9. Erlandson, S.C.; McMahon, C.; Kruse, A.C. Structural Basis for G Protein-Coupled Receptor Signaling. *Annu. Rev. Biophys.* **2018**, *47*, 1–18. [CrossRef]
10. Alexander, S.P.H.; Christopoulos, A.; Davenport, A.P.; Kelly, E.; Mathie, A.; Peters, J.A.; Veale, E.L.; Armstrong, J.F.; Faccenda, E.; Harding, S.D.; et al. THE CONCISE GUIDE TO PHARMACOLOGY 2019/20: G protein-coupled receptors. *Br. J. Pharmacol.* **2019**, *176*, S21–S141. [CrossRef]
11. Zhang, Y.; Sun, B.F.; Feng, D.; Hu, H.L.; Chu, M.; Qu, Q.H.; Tarrasch, J.T.; Li, S.; Kobilka, T.S.; Kobilka, B.K.; et al. Cryo-EM structure of the activated GLP-1 receptor in complex with a G protein. *Nature* **2017**, *546*, 248–253. [CrossRef] [PubMed]
12. Liang, Y.L.; Khoshouei, M.; Radjainia, M.; Zhang, Y.; Glukhova, A.; Tarrasch, J.; Thal, D.M.; Furness, S.G.B.; Christopoulos, G.; Coudrat, T.; et al. Phase-plate cryo-EM structure of a class B GPCR-G-protein complex. *Nature* **2017**, *546*, 118–123. [CrossRef] [PubMed]
13. Paek, J.; Kalocsay, M.; Staus, D.P.; Wingler, L.; Pascolutti, R.; Paulo, J.A.; Gygi, S.P.; Kruse, A.C. Multidimensional Tracking of GPCR Signaling via Peroxidase-Catalyzed Proximity Labeling. *Cell* **2017**, *169*, 338–349. [CrossRef]
14. Flock, T.; Hauser, A.S.; Lund, N.; Gloriam, D.E.; Balaji, S.; Babu, M.M. Selectivity determinants of GPCR-G-protein binding. *Nature* **2017**, *545*, 317–322. [CrossRef]
15. Gurevich, V.V.; Gurevich, E.V. GPCR Signaling Regulation: The Role of GRKs and Arrestins. *Front. Pharmacol.* **2019**, *10*, 125. [CrossRef]
16. Gutkind, J.S.; Kostenis, E. Arrestins as rheostats of GPCR signalling. *Nat. Rev. Mol. Cell Biol.* **2018**, *19*, 615–616. [CrossRef] [PubMed]
17. Smith, J.S.; Lefkowitz, R.J.; Rajagopal, S. Biased signalling: From simple switches to allosteric microprocessors. *Nat. Rev. Drug Discov.* **2018**, *17*, 243–260. [CrossRef]
18. Nogues, L.; Palacios-Garcia, J.; Reglero, C.; Rivas, V.; Neves, M.; Ribas, C.; Penela, P.; Mayor, F., Jr. G protein-coupled receptor kinases (GRKs) in tumorigenesis and cancer progression: GPCR regulators and signaling hubs. *Semin. Cancer Biol.* **2018**, *48*, 78–90. [CrossRef] [PubMed]
19. Campbell, A.P.; Smrcka, A.V. Targeting G protein-coupled receptor signalling by blocking G proteins. *Nat. Rev. Drug Discov.* **2018**, *17*, 789–803. [CrossRef]
20. Dorsam, R.T.; Gutkind, J.S. G-protein-coupled receptors and cancer. *Nat. Rev. Cancer* **2007**, *7*, 79–94. [CrossRef] [PubMed]
21. O'Hayre, M.; Degese, M.S.; Gutkind, J.S. Novel insights into G protein and G protein-coupled receptor signaling in cancer. *Curr. Opin. Cell Biol.* **2014**, *27*, 126–135. [CrossRef] [PubMed]
22. Grundmann, M.; Merten, N.; Malfacini, D.; Inoue, A.; Preis, P.; Simon, K.; Ruttiger, N.; Ziegler, N.; Benkel, T.; Schmitt, N.K.; et al. Lack of beta-arrestin signaling in the absence of active G proteins. *Nat. Commun.* **2018**, *9*, 341. [CrossRef]
23. O'Hayre, M.; Eichel, K.; Avino, S.; Zhao, X.F.; Steffen, D.J.; Feng, X.D.; Kawakami, K.; Aoki, J.; Messer, K.; Sunahara, R.; et al. Genetic evidence that β-arrestins are dispensable for the initiation of β_2-adrenergic receptor signaling to ERK. *Sci. Signal.* **2017**, *10*, eaal3395. [CrossRef]

24. Luttrell, L.M.; Wang, J.; Plouffe, B.; Smith, J.S.; Yamani, L.; Kaur, S.; Jean-Charles, P.Y.; Gauthier, C.; Lee, M.H.; Pani, B.; et al. Manifold roles of β-arrestins in GPCR signaling elucidated with siRNA and CRISPR/Cas9. *Sci. Signal.* **2018**, *11*, eaat7650. [CrossRef] [PubMed]
25. Nusse, R.; Clevers, H. Wnt/β-Catenin Signaling, Disease, and Emerging Therapeutic Modalities. *Cell* **2017**, *169*, 985–999. [CrossRef]
26. Lee, R.T.; Zhao, Z.; Ingham, P.W. Hedgehog signalling. *Development* **2016**, *143*, 367–372. [CrossRef]
27. Bray, S.J. Notch signalling in context. *Nat. Rev. Mol. Cell Biol.* **2016**, *17*, 722–735. [CrossRef] [PubMed]
28. May, L.T.; Leach, K.; Sexton, P.M.; Christopoulos, A. Allosteric modulation of G protein-coupled receptors. *Annu. Rev. Pharmacol. Toxicol.* **2007**, *47*, 1–51. [CrossRef]
29. Foster, D.J.; Conn, P.J. Allosteric Modulation of GPCRs: New Insights and Potential Utility for Treatment of Schizophrenia and Other CNS Disorders. *Neuron* **2017**, *94*, 431–446. [CrossRef]
30. Voss, T.; Li, J.; Cummings, J.; Farlow, M.; Assaid, C.; Froman, S.; Leibensperger, H.; Snow-Adami, L.; McMahon, K.B.; Egan, M.; et al. Randomized, controlled, proof-of-concept trial of MK-7622 in Alzheimer's disease. *Alzheimer's Dement.* **2018**, *4*, 173–181. [CrossRef]
31. Uslaner, J.M.; Kuduk, S.D.; Wittmann, M.; Lange, H.S.; Fox, S.V.; Min, C.; Pajkovic, N.; Harris, D.; Cilissen, C.; Mahon, C.; et al. Preclinical to Human Translational Pharmacology of the Novel M_1 Positive Allosteric Modulator MK-7622. *J. Pharmacol. Exp. Ther.* **2018**, *365*, 556–566. [CrossRef]
32. Wootten, D.; Christopoulos, A.; Marti-Solano, M.; Babu, M.M.; Sexton, P.M. Mechanisms of signalling and biased agonism in G protein-coupled receptors. *Nat. Rev. Mol. Cell Biol.* **2018**, *19*, 638–653. [CrossRef]
33. Kohout, T.A.; Nicholas, S.L.; Perry, S.J.; Reinhart, G.; Junger, S.; Struthers, R.S. Differential desensitization, receptor phosphorylation, β-arrestin recruitment, and ERK1/2 activation by the two endogenous ligands for the CC chemokine receptor 7. *J. Biol. Chem.* **2004**, *279*, 23214–23222. [CrossRef] [PubMed]
34. Soergel, D.G.; Subach, R.A.; Burnham, N.; Lark, M.W.; James, I.E.; Sadler, B.M.; Skobieranda, F.; Violin, J.D.; Webster, L.R. Biased agonism of the mu-opioid receptor by TRV130 increases analgesia and reduces on-target adverse effects versus morphine: A randomized, double-blind, placebo-controlled, crossover study in healthy volunteers. *Pain* **2014**, *155*, 1829–1835. [CrossRef] [PubMed]
35. Singla, N.K.; Skobieranda, F.; Soergel, D.G.; Salamea, M.; Burt, D.A.; Demitrack, M.A.; Viscusi, E.R. APOLLO-2: A Randomized, Placebo and Active-Controlled Phase III Study Investigating Oliceridine (TRV130), a G Protein-Biased Ligand at the mu-Opioid Receptor, for Management of Moderate to Severe Acute Pain Following Abdominoplasty. *Pain Pract.* **2019**, *19*, 715–731. [CrossRef]
36. James, J.R.; Oliveira, M.I.; Carmo, A.M.; Iaboni, A.; Davis, S.J. A rigorous experimental framework for detecting protein oligomerization using bioluminescence resonance energy transfer. *Nat. Methods* **2006**, *3*, 1001–1006. [CrossRef] [PubMed]
37. Bouvier, M.; Heveker, N.; Jockers, R.; Marullo, S.; Milligan, G. BRET analysis of GPCR oligomerization: Newer does not mean better. *Nat. Methods* **2007**, *4*, 3–4. [CrossRef]
38. Conn, P.M.; Rogers, D.C.; McNeil, R. Potency enhancement of a GnRH agonist: GnRH-receptor microaggregation stimulates gonadotropin release. *Endocrinology* **1982**, *111*, 335–337. [CrossRef]
39. Conn, P.M.; Rogers, D.C.; Stewart, J.M.; Niedel, J.; Sheffield, T. Conversion of a gonadotropin-releasing hormone antagonist to an agonist. *Nature* **1982**, *296*, 653–655. [CrossRef]
40. Borroto-Escuela, D.O.; Rodriguez, D.; Romero-Fernandez, W.; Kapla, J.; Jaiteh, M.; Ranganathan, A.; Lazarova, T.; Fuxe, K.; Carlsson, J. Mapping the Interface of a GPCR Dimer: A Structural Model of the A_{2A} Adenosine and D_2 Dopamine Receptor Heteromer. *Front. Pharmacol.* **2018**, *9*, 829. [CrossRef]
41. Vischer, H.F.; Watts, A.O.; Nijmeijer, S.; Leurs, R. G protein-coupled receptors: Walking hand-in-hand, talking hand-in-hand? *Br. J. Pharmacol.* **2011**, *163*, 246–260. [CrossRef]
42. Smith, N.J.; Milligan, G. Allostery at G protein-coupled receptor homo- and heteromers: Uncharted pharmacological landscapes. *Pharmacol. Rev.* **2010**, *62*, 701–725. [CrossRef] [PubMed]
43. Hanyaloglu, A.C.; von Zastrow, M. Regulation of GPCRs by endocytic membrane trafficking and its potential implications. *Annu. Rev. Pharmacol. Toxicol.* **2008**, *48*, 537–568. [CrossRef] [PubMed]
44. Shenoy, S.K.; Lefkowitz, R.J. Seven-transmembrane receptor signaling through β-arrestin. *Sci. Signal.* **2005**, *2005*, cm10. [CrossRef]
45. Reiter, E.; Ahn, S.; Shukla, A.K.; Lefkowitz, R.J. Molecular mechanism of β-arrestin-biased agonism at seven-transmembrane receptors. *Annu. Rev. Pharmacol. Toxicol.* **2012**, *52*, 179–197. [CrossRef]
46. Rolland-Fourcade, C.; Denadai-Souza, A.; Cirillo, C.; Lopez, C.; Jaramillo, J.O.; Desormeaux, C.; Cenac, N.; Motta, J.P.; Larauche, M.; Tache, Y.; et al. Epithelial expression and function of trypsin-3 in irritable bowel syndrome. *Gut* **2017**, *66*, 1767–1778. [CrossRef] [PubMed]
47. Du, L.; Long, Y.; Kim, J.J.; Chen, B.; Zhu, Y.; Dai, N. Protease Activated Receptor-2 Induces Immune Activation and Visceral Hypersensitivity in Post-infectious Irritable Bowel Syndrome Mice. *Dig. Dis. Sci.* **2019**, *64*, 729–739. [CrossRef]
48. Jimenez-Vargas, N.N.; Pattison, L.A.; Zhao, P.; Lieu, T.; Latorre, R.; Jensen, D.D.; Castro, J.; Aurelio, L.; Le, G.T.; Flynn, B.; et al. Protease-activated receptor-2 in endosomes signals persistent pain of irritable bowel syndrome. *Proc. Natl. Acad. Sci. USA* **2018**, *115*, E7438–E7447. [CrossRef]
49. Heng, B.C.; Aubel, D.; Fussenegger, M. An overview of the diverse roles of G-protein coupled receptors (GPCRs) in the pathophysiology of various human diseases. *Biotechnol. Adv.* **2013**, *31*, 1676–1694. [CrossRef]

50. Spiegel, A.M.; Weinstein, L.S. Inherited diseases involving g proteins and g protein-coupled receptors. *Annu. Rev. Med.* **2004**, *55*, 27–39. [CrossRef]
51. Raimondi, F.; Inoue, A.; Kadji, F.M.N.; Shuai, N.; Gonzalez, J.C.; Singh, G.; de la Vega, A.A.; Sotillo, R.; Fischer, B.; Aoki, J.; et al. Rare, functional, somatic variants in gene families linked to cancer genes: GPCR signaling as a paradigm. *Oncogene* **2019**, *38*, 6491–6506. [CrossRef] [PubMed]
52. Lawrence, M.S.; Stojanov, P.; Polak, P.; Kryukov, G.V.; Cibulskis, K.; Sivachenko, A.; Carter, S.L.; Stewart, C.; Mermel, C.H.; Roberts, S.A.; et al. Mutational heterogeneity in cancer and the search for new cancer-associated genes. *Nature* **2013**, *499*, 214–218. [CrossRef]
53. Hauser, A.S.; Chavali, S.; Masuho, I.; Jahn, L.J.; Martemyanov, K.A.; Gloriam, D.E.; Babu, M.M. Pharmacogenomics of GPCR Drug Targets. *Cell* **2018**, *172*, 41–54. [CrossRef]
54. Wu, V.; Yeerna, H.; Nohata, N.; Chiou, J.; Harismendy, O.; Raimondi, F.; Inoue, A.; Russell, R.B.; Tamayo, P.; Gutkind, J.S. Illuminating the Onco-GPCRome: Novel G protein-coupled receptor-driven oncocrine networks and targets for cancer immunotherapy. *J. Biol. Chem.* **2019**, *294*, 11062–11086. [CrossRef] [PubMed]
55. O'Hayre, M.; Vazquez-Prado, J.; Kufareva, I.; Stawiski, E.W.; Handel, T.M.; Seshagiri, S.; Gutkind, J.S. The emerging mutational landscape of G proteins and G-protein-coupled receptors in cancer. *Nat. Rev. Cancer* **2013**, *13*, 412–424. [CrossRef]
56. Kan, Z.Y.; Jaiswal, B.S.; Stinson, J.; Janakiraman, V.; Bhatt, D.; Stern, H.M.; Yue, P.; Haverty, P.M.; Bourgon, R.; Zheng, J.B.; et al. Diverse somatic mutation patterns and pathway alterations in human cancers. *Nature* **2010**, *466*, 869–873. [CrossRef] [PubMed]
57. Ragnarsson, L.; Andersson, A.; Thomas, W.G.; Lewis, R.J. Mutations in the NPxxY motif stabilize pharmacologically distinct conformational states of the α_{1B}- and β_2-adrenoceptors. *Sci. Signal.* **2019**, *12*, eaas9485. [CrossRef]
58. Stoy, H.; Gurevich, V.V. How genetic errors in GPCRs affect their function: Possible therapeutic strategies. *Genes Dis.* **2015**, *2*, 108–132. [CrossRef]
59. Teh, J.L.; Chen, S. Glutamatergic signaling in cellular transformation. *Pigment Cell Melanoma Res.* **2012**, *25*, 331–342. [CrossRef]
60. Vizurraga, A.; Adhikari, R.; Yeung, J.; Yu, M.; Tall, G.G. Mechanisms of adhesion G protein-coupled receptor activation. *J. Biol. Chem.* **2020**, *295*, 14065–14083. [CrossRef]
61. Paavola, K.J.; Hall, R.A. Adhesion G protein-coupled receptors: Signaling, pharmacology, and mechanisms of activation. *Mol. Pharmacol.* **2012**, *82*, 777–783. [CrossRef]
62. Prickett, T.D.; Wei, X.; Cardenas-Navia, I.; Teer, J.K.; Lin, J.C.; Walia, V.; Gartner, J.; Jiang, J.; Cherukuri, P.F.; Molinolo, A.; et al. Exon capture analysis of G protein-coupled receptors identifies activating mutations in GRM3 in melanoma. *Nat. Genet.* **2011**, *43*, 1119–1126. [CrossRef] [PubMed]
63. Mermel, C.H.; Schumacher, S.E.; Hill, B.; Meyerson, M.L.; Beroukhim, R.; Getz, G. GISTIC2.0 facilitates sensitive and confident localization of the targets of focal somatic copy-number alteration in human cancers. *Genome Biol.* **2011**, *12*, R41. [CrossRef] [PubMed]
64. Sriram, K.; Moyung, K.; Corriden, R.; Carter, H.; Insel, P.A. GPCRs show widespread differential mRNA expression and frequent mutation and copy number variation in solid tumors. *PLoS Biol.* **2019**, *17*, e3000434. [CrossRef] [PubMed]
65. Canals, M.; Poole, D.P.; Veldhuis, N.A.; Schmidt, B.L.; Bunnett, N.W. G-Protein-Coupled Receptors Are Dynamic Regulators of Digestion and Targets for Digestive Diseases. *Gastroenterology* **2019**, *156*, 1600–1616. [CrossRef] [PubMed]
66. Gottesman-Katz, L.; Latorre, R.; Vanner, S.; Schmidt, B.L.; Bunnett, N.W. Targeting G protein-coupled receptors for the treatment of chronic pain in the digestive system. *Gut* **2020**, *70*, 970–981. [CrossRef] [PubMed]
67. van Senten, J.R.; Fan, T.S.; Siderius, M.; Smit, M.J. Viral G protein-coupled receptors as modulators of cancer hallmarks. *Pharmacol. Res.* **2020**, *156*, 104804. [CrossRef]
68. Knerr, J.M.; Kledal, T.N.; Rosenkilde, M.M. Molecular Properties and Therapeutic Targeting of the EBV-Encoded Receptor BILF1. *Cancers* **2021**, *13*, 4079. [CrossRef]
69. Wood, L.D.; Parsons, D.W.; Jones, S.; Lin, J.; Sjoblom, T.; Leary, R.J.; Shen, D.; Boca, S.M.; Barber, T.; Ptak, J.; et al. The genomic landscapes of human breast and colorectal cancers. *Science* **2007**, *318*, 1108–1113. [CrossRef]
70. Cassel, D.; Selinger, Z. Mechanism of Adenylate-Cyclase Activation by Cholera Toxin: Inhibition of GTP Hydrolysis at Regulatory Site. *Proc. Natl. Acad. Sci. USA* **1977**, *74*, 3307–3311. [CrossRef]
71. Vandop, C.; Tsubokawa, M.; Bourne, H.R.; Ramachandran, J. Amino-Acid-Sequence of Retinal Transducin at the Site Adp-Ribosylated by Cholera-Toxin. *J. Biol. Chem.* **1984**, *259*, 696–698. [CrossRef]
72. Masters, S.B.; Stroud, R.M.; Bourne, H.R. Family of G protein alpha chains: Amphipathic analysis and predicted structure of functional domains. *Protein Eng.* **1986**, *1*, 47–54. [CrossRef] [PubMed]
73. Landis, C.A.; Masters, S.B.; Spada, A.; Pace, A.M.; Bourne, H.R.; Vallar, L. GTPase inhibiting mutations activate the α chain of G_s and stimulate adenylyl cyclase in human pituitary tumours. *Nature* **1989**, *340*, 692–696. [CrossRef] [PubMed]
74. Hu, Q.; Shokat, K.M. Disease-Causing Mutations in the G Protein Gαs Subvert the Roles of GDP and GTP. *Cell* **2018**, *173*, 1254–1264. [CrossRef]
75. Castellone, M.D.; Teramoto, H.; Williams, B.O.; Druey, K.M.; Gutkind, J.S. Prostaglandin E_2 promotes colon cancer cell growth through a G_s-axin-β-catenin signaling axis. *Science* **2005**, *310*, 1504–1510. [CrossRef]
76. Ikuta, K.; Seno, H.; Chiba, T. Molecular changes leading to gastric cancer: A suggestion from rare-type gastric tumors with GNAS mutations. *Gastroenterology* **2014**, *146*, 1417–1418. [CrossRef]
77. Hollstein, P.E.; Shaw, R.J. GNAS shifts metabolism in pancreatic cancer. *Nat Cell Biol* **2018**, *20*, 740–741. [CrossRef]

78. Patra, K.C.; Kato, Y.; Mizukami, Y.; Widholz, S.; Boukhali, M.; Revenco, I.; Grossman, E.A.; Ji, F.; Sadreyev, R.I.; Liss, A.S.; et al. Mutant GNAS drives pancreatic tumourigenesis by inducing PKA-mediated SIK suppression and reprogramming lipid metabolism. *Nat. Cell Biol.* **2018**, *20*, 811–822. [CrossRef]
79. Wu, Y.; Peng, C.; Huang, L.; Xu, L.; Ding, X.; Liu, Y.; Zeng, C.; Sun, H.; Guo, W. Somatic GNAQ R183Q mutation is located within the sclera and episclera in patients with Sturge-Weber syndrome. *Br. J. Ophthalmol.* **2022**, *106*, 1006–1011. [CrossRef]
80. Kusters-Vandevelde, H.V.; Klaasen, A.; Kusters, B.; Groenen, P.J.; van Engen-van Grunsven, I.A.; van Dijk, M.R.; Reifenberger, G.; Wesseling, P.; Blokx, W.A. Activating mutations of the *GNAQ* gene: A frequent event in primary melanocytic neoplasms of the central nervous system. *Acta Neuropathol.* **2010**, *119*, 317–323. [CrossRef]
81. Van Raamsdonk, C.D.; Bezrookove, V.; Green, G.; Bauer, J.; Gaugler, L.; O'Brien, J.M.; Simpson, E.M.; Barsh, G.S.; Bastian, B.C. Frequent somatic mutations of *GNAQ* in uveal melanoma and blue naevi. *Nature* **2009**, *457*, 599–602. [CrossRef] [PubMed]
82. Van Raamsdonk, C.D.; Griewank, K.G.; Crosby, M.B.; Garrido, M.C.; Vemula, S.; Wiesner, T.; Obenauf, A.C.; Wackernagel, W.; Green, G.; Bouvier, N.; et al. Mutations in GNA11 in Uveal Melanoma. *N. Engl. J. Med.* **2010**, *363*, 2191–2199. [CrossRef]
83. Shirley, M.D.; Tang, H.; Gallione, C.J.; Baugher, J.D.; Frelin, L.P.; Cohen, B.; North, P.E.; Marchuk, D.A.; Comi, A.M.; Pevsner, J. Sturge-Weber syndrome and port-wine stains caused by somatic mutation in *GNAQ*. *N. Engl. J. Med.* **2013**, *368*, 1971–1979. [CrossRef] [PubMed]
84. Hubbard, K.B.; Hepler, J.R. Cell signalling diversity of the Gqalpha family of heterotrimeric G proteins. *Cell. Signal.* **2006**, *18*, 135–150. [CrossRef] [PubMed]
85. Feng, X.D.; Degese, M.S.; Iglesias-Bartolome, R.; Vaque, J.P.; Molinolo, A.A.; Rodrigues, M.; Zaidi, M.R.; Ksander, B.R.; Merlino, G.; Sodhi, A.; et al. Hippo-Independent Activation of YAP by the GNAQ Uveal Melanoma Oncogene through a Trio-Regulated Rho GTPase Signaling Circuitry. *Cancer Cell* **2014**, *25*, 831–845. [CrossRef]
86. Zhang, J.X.; Yun, M.; Xu, Y.; Chen, J.W.; Weng, H.W.; Zheng, Z.S.; Chen, C.; Xie, D.; Ye, S. GNA13 as a prognostic factor and mediator of gastric cancer progression. *Oncotarget* **2016**, *7*, 4414–4427. [CrossRef]
87. Liu, S.C.; Jen, Y.M.; Jiang, S.S.; Chang, J.L.; Hsiung, C.A.; Wang, C.H.; Juang, J.L. Gα_{12}-mediated pathway promotes invasiveness of nasopharyngeal carcinoma by modulating actin cytoskeleton reorganization. *Cancer Res.* **2009**, *69*, 6122–6130. [CrossRef]
88. Yagi, H.; Tan, W.; Dillenburg-Pilla, P.; Armando, S.; Amornphimoltham, P.; Simaan, M.; Weigert, R.; Molinolo, A.A.; Bouvier, M.; Gutkind, J.S. A synthetic biology approach reveals a CXCR4-G$_{13}$-Rho signaling axis driving transendothelial migration of metastatic breast cancer cells. *Sci. Signal.* **2011**, *4*, ra60. [CrossRef]
89. Rasheed, S.A.K.; Leong, H.S.; Lakshmanan, M.; Raju, A.; Dadlani, D.; Chong, F.T.; Shannon, N.B.; Rajarethinam, R.; Skanthakumar, T.; Tan, E.Y.; et al. GNA13 expression promotes drug resistance and tumor-initiating phenotypes in squamous cell cancers. *Oncogene* **2018**, *37*, 1340–1353. [CrossRef]
90. Zhang, Z.; Tan, X.; Luo, J.; Cui, B.; Lei, S.; Si, Z.; Shen, L.; Yao, H. GNA13 promotes tumor growth and angiogenesis by upregulating CXC chemokines via the NF-kappaB signaling pathway in colorectal cancer cells. *Cancer Med.* **2018**, *7*, 5611–5620. [CrossRef]
91. O'Hayre, M.; Inoue, A.; Kufareva, I.; Wang, Z.; Mikelis, C.M.; Drummond, R.A.; Avino, S.; Finkel, K.; Kalim, K.W.; DiPasquale, G.; et al. Inactivating mutations in GNA13 and RHOA in Burkitt's lymphoma and diffuse large B-cell lymphoma: A tumor suppressor function for the Gα_{13}/RhoA axis in B cells. *Oncogene* **2016**, *35*, 3771–3780. [CrossRef]
92. Lohr, J.G.; Stojanov, P.; Lawrence, M.S.; Auclair, D.; Chapuy, B.; Sougnez, C.; Cruz-Gordillo, P.; Knoechel, B.; Asmann, Y.W.; Slager, S.L.; et al. Discovery and prioritization of somatic mutations in diffuse large B-cell lymphoma (DLBCL) by whole-exome sequencing. *Proc. Natl. Acad. Sci. USA* **2012**, *109*, 3879–3884. [CrossRef] [PubMed]
93. Love, C.; Sun, Z.; Jima, D.; Li, G.; Zhang, J.; Miles, R.; Richards, K.L.; Dunphy, C.H.; Choi, W.W.; Srivastava, G.; et al. The genetic landscape of mutations in Burkitt lymphoma. *Nat. Genet.* **2012**, *44*, 1321–1325. [CrossRef] [PubMed]
94. Ford, C.E.; Skiba, N.P.; Bae, H.S.; Daaka, Y.H.; Reuveny, E.; Shekter, L.R.; Rosal, R.; Weng, G.Z.; Yang, C.S.; Iyengar, R.; et al. Molecular basis for interactions of G protein βγ subunits with effectors. *Science* **1998**, *280*, 1271–1274. [CrossRef] [PubMed]
95. Yoda, A.; Adelmant, G.; Tamburini, J.; Chapuy, B.; Shindoh, N.; Yoda, Y.; Weigert, O.; Kopp, N.; Wu, S.C.; Kim, S.S.; et al. Mutations in G protein β subunits promote transformation and kinase inhibitor resistance. *Nat. Med.* **2015**, *21*, 71–75. [CrossRef]
96. Tan, N.B.; Pagnamenta, A.T.; Ferla, M.P.; Gadian, J.; Chung, B.H.; Chan, M.C.; Fung, J.L.; Cook, E.; Guter, S.; Boschann, F.; et al. Recurrent de novo missense variants in GNB2 can cause syndromic intellectual disability. *J. Med. Genet.* **2022**, *59*, 511–516. [CrossRef]
97. Alvarez, C.J.P.; Lodeiro, M.; Theodoropoulou, M.; Camina, J.P.; Casanueva, F.F.; Pazos, Y. Obestatin stimulates Akt signalling in gastric cancer cells through beta-arrestin-mediated epidermal growth factor receptor transactivation. *Endocr. Relat. Cancer* **2009**, *16*, 599–611. [CrossRef]
98. Zimmermannova, O.; Doktorova, E.; Stuchly, J.; Kanderova, V.; Kuzilkova, D.; Strnad, H.; Starkova, J.; Alberich-Jorda, M.; Falkenburg, J.H.F.; Trka, J.; et al. An activating mutation of GNB1 is associated with resistance to tyrosine kinase inhibitors in ETV6-ABL1-positive leukemia. *Oncogene* **2017**, *36*, 5985–5994. [CrossRef]
99. Liu, Y.G.; Teng, Y.S.; Shan, Z.G.; Cheng, P.; Hao, C.J.; Lv, Y.P.; Mao, F.Y.; Yang, S.M.; Chen, W.; Zhao, Y.L.; et al. Arrestin domain containing 3 promotes *Helicobacter pylori*-associated gastritis by regulating protease-activated receptor 1. *JCI Insight* **2020**, *5*, e135849. [CrossRef]

100. Zhang, C.; Gao, G.R.; Lv, C.G.; Zhang, B.L.; Zhang, Z.L.; Zhang, X.F. Protease-activated receptor-2 induces expression of vascular endothelial growth factor and cyclooxygenase-2 via the mitogen-activated protein kinase pathway in gastric cancer cells. *Oncol. Rep.* **2012**, *28*, 1917–1923. [CrossRef]
101. Rocken, C.; Rohl, F.W.; Diebler, E.; Lendeckel, U.; Pross, M.; Carl-McGrath, S.; Ebert, M.P. The angiotensin II/angiotensin II receptor system correlates with nodal spread in intestinal type gastric cancer. *Cancer Epidemiol. Biomark. Prev.* **2007**, *16*, 1206–1212. [CrossRef] [PubMed]
102. Huang, W.; Wu, Y.L.; Zhong, J.; Jiang, F.X.; Tian, X.L.; Yu, L.F. Angiotensin II type 1 receptor antagonist suppress angiogenesis and growth of gastric cancer xenografts. *Dig. Dis. Sci.* **2008**, *53*, 1206–1210. [CrossRef]
103. Shen, W.; Xi, H.; Li, C.; Bian, S.; Cheng, H.; Cui, J.; Wang, N.; Wei, B.; Huang, X.; Chen, L. Endothelin-A receptor in gastric cancer and enhanced antitumor activity of trastuzumab in combination with the endothelin-A receptor antagonist ZD4054. *Ann. N. Y. Acad. Sci.* **2019**, *1448*, 30–41. [CrossRef] [PubMed]
104. Prevete, N.; Liotti, F.; Visciano, C.; Marone, G.; Melillo, R.M.; de Paulis, A. The formyl peptide receptor 1 exerts a tumor suppressor function in human gastric cancer by inhibiting angiogenesis. *Oncogene* **2015**, *34*, 3826–3838. [CrossRef]
105. Hou, X.L.; Ji, C.D.; Tang, J.; Wang, Y.X.; Xiang, D.F.; Li, H.Q.; Liu, W.W.; Wang, J.X.; Yan, H.Z.; Wang, Y.; et al. FPR2 promotes invasion and metastasis of gastric cancer cells and predicts the prognosis of patients. *Sci. Rep.* **2017**, *7*, 3153. [CrossRef] [PubMed]
106. Prevete, N.; Liotti, F.; Illiano, A.; Amoresano, A.; Pucci, P.; de Paulis, A.; Melillo, R.M. Formyl peptide receptor 1 suppresses gastric cancer angiogenesis and growth by exploiting inflammation resolution pathways. *Oncoimmunology* **2017**, *6*, e1293213. [CrossRef] [PubMed]
107. Ogasa, M.; Miyazaki, Y.; Hiraoka, S.; Kitamura, S.; Nagasawa, Y.; Kishida, O.; Miyazaki, T.; Kiyohara, T.; Shinomura, Y.; Matsuzawa, Y. Gastrin activates nuclear factor κB (NFκB) through a protein kinase C dependent pathway involving NFκB inducing kinase, inhibitor κB (IκB) kinase, and tumour necrosis factor receptor associated factor 6 (TRAF6) in MKN-28 cells transfected with gastrin receptor. *Gut* **2003**, *52*, 813–819. [CrossRef] [PubMed]
108. de Lau, W.; Barker, N.; Low, T.Y.; Koo, B.K.; Li, V.S.W.; Teunissen, H.; Kujala, P.; Haegebarth, A.; Peters, P.J.; van de Wetering, M.; et al. Lgr5 homologues associate with Wnt receptors and mediate R-spondin signalling. *Nature* **2011**, *476*, 293–297. [CrossRef]
109. Ke, J.; Ma, P.; Chen, J.P.; Qin, J.; Qian, H.X. LGR6 promotes the progression of gastric cancer through PI3K/AKT/mTOR pathway. *OncoTargets Ther.* **2018**, *11*, 3025–3033. [CrossRef]
110. Shida, D.; Kitayama, J.; Yamaguchi, H.; Yamashita, H.; Mori, K.; Watanabe, T.; Nagawa, H. Lysophospholipids transactivate HER2/neu (erbB-2) in human gastric cancer cells. *BBRC* **2005**, *327*, 907–911. [CrossRef]
111. Shida, D.; Kitayama, J.; Yamaguchi, H.; Hama, K.; Aoki, J.; Arai, H.; Yamashita, H.; Mori, K.; Sako, A.; Konishi, T.; et al. Dual mode regulation of migration by lysophosphatidic acid in human gastric cancer cells. *Exp. Cell Res.* **2004**, *301*, 168–178. [CrossRef] [PubMed]
112. Yang, D.Z.; Yang, W.H.; Zhang, Q.; Hu, Y.; Bao, L.; Damirin, A. Migration of gastric cancer cells in response to lysophosphatidic acid is mediated by LPA receptor 2. *Oncol. Lett.* **2013**, *5*, 1048–1052. [CrossRef]
113. Shida, D.; Kitayama, J.; Yamaguchi, H.; Yamashita, H.; Mori, K.; Watanabe, T.; Yatomi, Y.; Nagawa, H. Sphingosine 1-phosphate transactivates c-Met as well as epidermal growth factor receptor (EGFR) in human gastric cancer cells. *FEBS Lett.* **2004**, *577*, 333–338. [CrossRef] [PubMed]
114. Yamashita, H.; Kitayama, J.; Shida, D.; Yamaguchi, H.; Mori, K.; Osada, M.; Aoki, S.; Yatomi, Y.; Takuwa, Y.; Nagawa, H. Sphingosine 1-phosphate receptor expression profile in human gastric cancer cells: Differential regulation on the migration and proliferation. *J. Surg. Res.* **2006**, *130*, 80–87. [CrossRef]
115. Wong, C.C.; Kang, W.; Xu, J.; Qian, Y.; Luk, S.T.Y.; Chen, H.; Li, W.; Zhao, L.; Zhang, X.; Chiu, P.W.; et al. Prostaglandin E2 induces DNA hypermethylation in gastric cancer in vitro and in vivo. *Theranostics* **2019**, *9*, 6256–6268. [CrossRef] [PubMed]
116. Oshima, H.; Oshima, M.; Inaba, K.; Taketo, M.M. Hyperplastic gastric tumors induced by activated macrophages in COX-2/mPGES-1 transgenic mice. *EMBO J.* **2004**, *23*, 1669–1678. [CrossRef]
117. Lee, H.J.; Song, I.C.; Yun, H.J.; Jo, D.Y.; Kim, S. CXC chemokines and chemokine receptors in gastric cancer: From basic findings towards therapeutic targeting. *World J. Gastroenterol.* **2014**, *20*, 1681–1693. [CrossRef]
118. Xu, M.; Wang, Y.; Xia, R.; Wei, Y.; Wei, X. Role of the CCL2-CCR2 signalling axis in cancer: Mechanisms and therapeutic targeting. *Cell Prolif.* **2021**, *54*, e13115. [CrossRef]
119. Shah, N.; Khurana, S.; Cheng, K.R.; Raufman, J.P. Muscarinic receptors and ligands in cancer. *Am. J. Physiol. Cell Physiol.* **2009**, *296*, C221–C232. [CrossRef]
120. Zhang, X.; Zhang, Y.; He, Z.; Yin, K.; Li, B.; Zhang, L.; Xu, Z. Chronic stress promotes gastric cancer progression and metastasis: An essential role for ADRB2. *Cell Death Dis.* **2019**, *10*, 788. [CrossRef]
121. Bellefeuille, S.D.; Molle, C.M.; Gendron, F.P. Reviewing the role of P2Y receptors in specific gastrointestinal cancers. *Purinergic Signal.* **2019**, *15*, 451–463. [CrossRef] [PubMed]
122. Tsuchiya, A.; Nishizaki, T. Anticancer effect of adenosine on gastric cancer via diverse signaling pathways. *World J. Gastroenterol.* **2015**, *21*, 10931–10935. [CrossRef] [PubMed]
123. Shi, L.; Wu, Z.; Miao, J.; Du, S.; Ai, S.; Xu, E.; Feng, M.; Song, J.; Guan, W. Adenosine interaction with adenosine receptor A2a promotes gastric cancer metastasis by enhancing PI3K-AKT-mTOR signaling. *Mol. Biol. Cell* **2019**, *30*, 2527–2534. [CrossRef] [PubMed]

124. Yasuda, H.; Hirata, S.; Inoue, K.; Mashima, H.; Ohnishi, H.; Yoshiba, M. Involvement of membrane-type bile acid receptor M-BAR/TGR5 in bile acid-induced activation of epidermal growth factor receptor and mitogen-activated protein kinases in gastric carcinoma cells. *Biochem. Biophys. Res. Commun.* 2007, *354*, 154–159. [CrossRef]
125. Wang, X.F.; Xu, Z.Y.; Sun, J.C.; Lv, H.; Wang, Y.P.; Ni, Y.X.; Chen, S.Q.; Hu, C.; Wang, L.J.; Chen, W.; et al. Cisplatin resistance in gastric cancer cells is involved with GPR30-mediated epithelial-mesenchymal transition. *J. Cell. Mol. Med.* 2020, *24*, 3625–3633. [CrossRef]
126. Gan, J.F.; Ke, X.R.; Jiang, J.L.; Dong, H.M.; Yao, Z.M.; Lin, Y.S.; Lin, W.; Wu, X.; Yan, S.M.; Zhuang, Y.X.; et al. Growth hormone-releasing hormone receptor antagonists inhibit human gastric cancer through downregulation of PAK1-STAT3/NF-κB signaling. *Proc. Natl. Acad. Sci. USA* 2016, *113*, 14745–14750. [CrossRef]
127. Xie, R.; Xu, J.Y.; Xiao, Y.F.; Wu, J.L.; Wan, H.X.; Tang, B.; Liu, J.J.; Fan, Y.H.; Wang, S.M.; Wu, Y.Y.; et al. Calcium Promotes Human Gastric Cancer via a Novel Coupling of Calcium-Sensing Receptor and TRPV4 Channel. *Cancer Res.* 2017, *77*, 6499–6512. [CrossRef]
128. Wang, K.; Zhao, X.H.; Liu, J.; Zhang, R.; Li, J.P. Nervous system and gastric cancer. *BBA-Rev. Cancer* 2020, *1873*, 188313. [CrossRef]
129. Ferrigno, A.; Berardo, C.; Di Pasqua, L.G.; Siciliano, V.; Richelmi, P.; Vairetti, M. Localization and role of metabotropic glutamate receptors subtype 5 in the gastrointestinal tract. *World J. Gastroenterol.* 2017, *23*, 4500–4507. [CrossRef]
130. Gad, A.A.; Balenga, N. The Emerging Role of Adhesion GPCRs in Cancer. *ACS Pharmacol. Transl. Sci.* 2020, *3*, 29–42. [CrossRef]
131. Kozielewicz, P.; Turku, A.; Schulte, G. Molecular Pharmacology of Class F Receptor Activation. *Mol. Pharmacol.* 2020, *97*, 62–71. [CrossRef] [PubMed]
132. Flanagan, D.J.; Barker, N.; Di Costanzo, N.S.; Mason, E.A.; Gurney, A.; Meniel, V.S.; Koushyar, S.; Austin, C.R.; Ernst, M.; Pearson, H.B.; et al. Frizzled-7 Is Required for Wnt Signaling in Gastric Tumors with and Without Apc Mutations. *Cancer Res.* 2019, *79*, 970–981. [CrossRef] [PubMed]
133. Xin, L.; Liu, L.; Liu, C.; Zhou, L.Q.; Zhou, Q.; Yuan, Y.W.; Li, S.H.; Zhang, H.T. DNA-methylation-mediated silencing of miR-7-5p promotes gastric cancer stem cell invasion via increasing Smo and Hes1. *J. Cell. Physiol.* 2020, *235*, 2643–2654. [CrossRef]
134. Fares, S.; Spiess, K.; Olesen, E.T.B.; Zuo, J.M.; Jackson, S.; Kledal, T.N.; Wills, M.R.; Rosenkilde, M.M. Distinct Roles of Extracellular Domains in the Epstein-Barr Virus-Encoded BILF1 Receptor for Signaling and Major Histocompatibility Complex Class I Downregulation. *mBio* 2019, *10*, e01707-18. [CrossRef] [PubMed]
135. Perez Almeria, C.V.; Setiawan, I.M.; Siderius, M.; Smit, M.J. G protein-coupled receptors as promising targets in cancer. *Curr. Opin. Endocr. Metab. Res.* 2021, *16*, 119–127. [CrossRef]
136. Warren, C.M.; Landgraf, R. Signaling through ERBB receptors: Multiple layers of diversity and control. *Cell. Signal.* 2006, *18*, 923–933. [CrossRef] [PubMed]
137. Wetzker, R.; Bohmer, F.D. Transactivation joins multiple tracks to the ERK/MAPK cascade. *Nat. Rev. Mol. Cell Biol.* 2003, *4*, 651–657. [CrossRef]
138. Kose, M. GPCRs and EGFR—Cross-talk of membrane receptors in cancer. *Bioorg. Med. Chem. Lett.* 2017, *27*, 3611–3620. [CrossRef]
139. Kilpatrick, L.E.; Hill, S.J. Transactivation of G protein-coupled receptors (GPCRs) and receptor tyrosine kinases (RTKs): Recent insights using luminescence and fluorescence technologies. *Curr. Opin. Endocr. Metab. Res.* 2021, *16*, 102–112. [CrossRef]
140. Joh, T.; Kataoka, H.; Tanida, S.; Watanabe, K.; Ohshima, T.; Sasaki, M.; Nakao, H.; Ohhara, H.; Higashiyama, S.; Itoh, M. Helicobacter pylori-stimulated interleukin-8 (IL-8) promotes cell proliferation through transactivation of epidermal growth factor receptor (EGFR) by disintegrin and metalloproteinase (ADAM) activation. *Dig. Dis. Sci.* 2005, *50*, 2081–2089. [CrossRef]
141. Oshima, H.; Popivanova, B.K.; Oguma, K.; Kong, D.; Ishikawa, T.O.; Oshima, M. Activation of epidermal growth factor receptor signaling by the prostaglandin E_2 receptor EP4 pathway during gastric tumorigenesis. *Cancer Sci.* 2011, *102*, 713–719. [CrossRef] [PubMed]
142. Lian, S.; Xia, Y.; Ung, T.T.; Khoi, P.N.; Yoon, H.J.; Lee, S.G.; Kim, K.K.; Jung, Y.D. Prostaglandin E2 stimulates urokinase-type plasminogen activator receptor via EP2 receptor-dependent signaling pathways in human AGS gastric cancer cells. *Mol. Carcinog.* 2017, *56*, 664–680. [CrossRef] [PubMed]
143. Ancha, H.R.; Kurella, R.R.; Stewart, C.A.; Damera, G.; Ceresa, B.P.; Harty, R.F. Histamine stimulation of MMP-1(collagenase-1) secretion and gene expression in gastric epithelial cells: Role of EGFR transactivation and the MAP kinase pathway. *Int. J. Biochem. Cell Biol.* 2007, *39*, 2143–2152. [CrossRef] [PubMed]
144. Wang, Q.; Yang, H.; Zhuo, Q.; Xu, Y.; Zhang, P. Knockdown of EPCR inhibits the proliferation and migration of human gastric cancer cells via the ERK1/2 pathway in a PAR-1-dependent manner. *Oncol. Rep.* 2018, *39*, 1843–1852. [CrossRef] [PubMed]
145. Manning, B.D.; Toker, A. AKT/PKB Signaling: Navigating the Network. *Cell* 2017, *169*, 381–405. [CrossRef] [PubMed]
146. Matsuoka, T.; Yashiro, M. The Role of PI3K/Akt/mTOR Signaling in Gastric Carcinoma. *Cancers* 2014, *6*, 1441–1463. [CrossRef]
147. Li, F.L.; Guan, K.L. The two sides of Hippo pathway in cancer. *Semin. Cancer Biol.* 2021, *85*, 33–42. [CrossRef]
148. Zhao, B.; Ye, X.; Yu, J.; Li, L.; Li, W.; Li, S.; Yu, J.; Lin, J.D.; Wang, C.Y.; Chinnaiyan, A.M.; et al. TEAD mediates YAP-dependent gene induction and growth control. *Genes Dev.* 2008, *22*, 1962–1971. [CrossRef]
149. Yu, F.X.; Zhao, B.; Panupinthu, N.; Jewell, J.L.; Lian, I.; Wang, L.H.; Zhao, J.; Yuan, H.; Tumaneng, K.; Li, H.; et al. Regulation of the Hippo-YAP pathway by G-protein-coupled receptor signaling. *Cell* 2012, *150*, 780–791. [CrossRef]
150. Luo, J.; Yu, F.X. GPCR-Hippo Signaling in Cancer. *Cells* 2019, *8*, 426. [CrossRef]
151. Cai, H.; Xu, Y. The role of LPA and YAP signaling in long-term migration of human ovarian cancer cells. *Cell Commun. Signal.* 2013, *11*, 31. [CrossRef] [PubMed]

152. Yung, Y.C.; Stoddard, N.C.; Chun, J. LPA receptor signaling: Pharmacology, physiology, and pathophysiology. *J. Lipid Res.* **2014**, *55*, 1192–1214. [CrossRef]
153. Cheng, J.C.; Wang, E.Y.; Yi, Y.; Thakur, A.; Tsai, S.H.; Hoodless, P.A. S1P Stimulates Proliferation by Upregulating CTGF Expression through S1PR2-Mediated YAP Activation. *Mol. Cancer Res.* **2018**, *16*, 1543–1555. [CrossRef] [PubMed]
154. Saikawa, S.; Kaji, K.; Nishimura, N.; Seki, K.; Sato, S.; Nakanishi, K.; Kitagawa, K.; Kawaratani, H.; Kitade, M.; Moriya, K.; et al. Angiotensin receptor blockade attenuates cholangiocarcinoma cell growth by inhibiting the oncogenic activity of Yes-associated protein. *Cancer Lett.* **2018**, *434*, 120–129. [CrossRef] [PubMed]
155. Mo, J.S.; Yu, F.X.; Gong, R.; Brown, J.H.; Guan, K.L. Regulation of the Hippo-YAP pathway by protease-activated receptors (PARs). *Genes Dev.* **2012**, *26*, 2138–2143. [CrossRef]
156. Anakk, S.; Bhosale, M.; Schmidt, V.A.; Johnson, R.L.; Finegold, M.J.; Moore, D.D. Bile acids activate YAP to promote liver carcinogenesis. *Cell Rep.* **2013**, *5*, 1060–1069. [CrossRef]
157. Thirunavukkarasan, M.; Wang, C.; Rao, A.; Hind, T.; Teo, Y.R.; Siddiquee, A.A.; Goghari, M.A.I.; Kumar, A.P.; Herr, D.R. Short-chain fatty acid receptors inhibit invasive phenotypes in breast cancer cells. *PLoS ONE* **2017**, *12*, e0186334. [CrossRef] [PubMed]
158. Zhou, X.; Wang, S.; Wang, Z.; Feng, X.; Liu, P.; Lv, X.B.; Li, F.; Yu, F.X.; Sun, Y.; Yuan, H.; et al. Estrogen regulates Hippo signaling via GPER in breast cancer. *J. Clin. Investig.* **2015**, *125*, 2123–2135. [CrossRef]
159. Wang, Z.; Liu, P.; Zhou, X.; Wang, T.; Feng, X.; Sun, Y.P.; Xiong, Y.; Yuan, H.X.; Guan, K.L. Endothelin Promotes Colorectal Tumorigenesis by Activating YAP/TAZ. *Cancer Res.* **2017**, *77*, 2413–2423. [CrossRef]
160. Roulis, M.; Kaklamanos, A.; Schernthanner, M.; Bielecki, P.; Zhao, J.; Kaffe, E.; Frommelt, L.S.; Qu, R.; Knapp, M.S.; Henriques, A.; et al. Paracrine orchestration of intestinal tumorigenesis by a mesenchymal niche. *Nature* **2020**, *580*, 524–529. [CrossRef]
161. Fujimoto, D.; Ueda, Y.; Hirono, Y.; Goi, T.; Yamaguchi, A. PAR1 participates in the ability of multidrug resistance and tumorigenesis by controlling Hippo-YAP pathway. *Oncotarget* **2015**, *6*, 34788–34799. [CrossRef]
162. Katoh, M.; Katoh, M. Precision medicine for human cancers with Notch signaling dysregulation (Review). *Int. J. Mol. Med.* **2020**, *45*, 279–297. [CrossRef]
163. Merchant, J.L. Hedgehog signalling in gut development, physiology and cancer. *J. Physiol.* **2012**, *590*, 421–432. [CrossRef] [PubMed]
164. Chiurillo, M.A. Role of the Wnt/β-catenin pathway in gastric cancer: An in-depth literature review. *World J. Exp. Med.* **2015**, *5*, 84–102. [CrossRef] [PubMed]
165. Liang, Y.; Yang, L.; Xie, J. The Role of the Hedgehog Pathway in Chemoresistance of Gastrointestinal Cancers. *Cells* **2021**, *10*, 2030. [CrossRef]
166. Fukaya, M.; Isohata, N.; Ohta, H.; Aoyagi, K.; Ochiya, T.; Saeki, N.; Yanagihara, K.; Nakanishi, Y.; Taniguchi, H.; Sakamoto, H.; et al. Hedgehog signal activation in gastric pit cell and in diffuse-type gastric cancer. *Gastroenterology* **2006**, *131*, 14–29. [CrossRef] [PubMed]
167. Mukhopadhyay, S.; Rohatgi, R. G-protein-coupled receptors, Hedgehog signaling and primary cilia. *Semin. Cell Dev. Biol.* **2014**, *33*, 63–72. [CrossRef]
168. Koushyar, S.; Powell, A.G.; Vincan, E.; Phesse, T.J. Targeting Wnt Signaling for the Treatment of Gastric Cancer. *Int. J. Mol. Sci.* **2020**, *21*, 3927. [CrossRef]
169. Valastyan, S.; Weinberg, R.A. Tumor metastasis: Molecular insights and evolving paradigms. *Cell* **2011**, *147*, 275–292. [CrossRef]
170. Li, W.; Ng, J.M.; Wong, C.C.; Ng, E.K.W.; Yu, J. Molecular alterations of cancer cell and tumour microenvironment in metastatic gastric cancer. *Oncogene* **2018**, *37*, 4903–4920. [CrossRef]
171. Gupta, G.P.; Massagué, J. Cancer metastasis: Building a framework. *Cell* **2006**, *127*, 679–695. [CrossRef] [PubMed]
172. Chaffer, C.L.; Weinberg, R.A. A perspective on cancer cell metastasis. *Science* **2011**, *331*, 1559–1564. [CrossRef]
173. Brabletz, S.; Schuhwerk, H.; Brabletz, T.; Stemmler, M.P. Dynamic EMT: A multi-tool for tumor progression. *EMBO J.* **2021**, *40*, e108647. [CrossRef] [PubMed]
174. Lombardi, L.; Tavano, F.; Morelli, F.; Latiano, T.P.; Di Sebastiano, P.; Maiello, E. Chemokine receptor CXCR4: Role in gastrointestinal cancer. *Crit. Rev. Oncol. Hematol.* **2013**, *88*, 696–705. [CrossRef]
175. Marcuzzi, E.; Angioni, R.; Molon, B.; Cali, B. Chemokines and Chemokine Receptors: Orchestrating Tumor Metastasization. *Int. J. Mol. Sci.* **2018**, *20*, 96. [CrossRef]
176. Nagarsheth, N.; Wicha, M.S.; Zou, W.P. Chemokines in the cancer microenvironment and their relevance in cancer immunotherapy. *Nat. Rev. Immunol.* **2017**, *17*, 559–572. [CrossRef]
177. Ashburner, B.P.; Westerheide, S.D.; Baldwin, A.S., Jr. The p65 (RelA) subunit of NF-kappaB interacts with the histone deacetylase (HDAC) corepressors HDAC1 and HDAC2 to negatively regulate gene expression. *Mol. Cell. Biol.* **2001**, *21*, 7065–7077. [CrossRef]
178. Tarnowski, M.; Grymula, K.; Reca, R.; Jankowski, K.; Maksym, R.; Tarnowska, J.; Przybylski, G.; Barr, F.G.; Kucia, M.; Ratajczak, M.Z. Regulation of expression of stromal-derived factor-1 receptors: CXCR4 and CXCR7 in human rhabdomyosarcomas. *Mol. Cancer Res.* **2010**, *8*, 1–14. [CrossRef] [PubMed]
179. Schioppa, T.; Uranchimeg, B.; Saccani, A.; Biswas, S.K.; Doni, A.; Rapisarda, A.; Bernasconi, S.; Saccani, S.; Nebuloni, M.; Vago, L.; et al. Regulation of the chemokine receptor CXCR4 by hypoxia. *J. Exp. Med.* **2003**, *198*, 1391–1402. [CrossRef] [PubMed]
180. Hitchon, C.; Wong, K.; Ma, G.; Reed, J.; Lyttle, D.; El-Gabalawy, H. Hypoxia-induced production of stromal cell-derived factor 1 (CXCL12) and vascular endothelial growth factor by synovial fibroblasts. *Arthritis Rheum.* **2002**, *46*, 2587–2597. [CrossRef]

181. Lee, K.E.; Khoi, P.N.; Xia, Y.; Park, J.S.; Joo, Y.E.; Kim, K.K.; Choi, S.Y.; Jung, Y.D. Helicobacter pylori and interleukin-8 in gastric cancer. *World J. Gastroenterol.* **2013**, *19*, 8192–8202. [CrossRef] [PubMed]
182. Haghazali, M.; Molaei, M.; Mashayekhi, R.; Zojaji, H.; Pourhoseingholi, M.A.; Shooshtarizadeh, T.; Mirsattari, D.; Zali, M.R. Proinflammatory cytokines and thrombomodulin in patients with peptic ulcer disease and gastric cancer, infected with *Helicobacter pylori*. *Indian J. Pathol. Microbiol.* **2011**, *54*, 103–106. [CrossRef] [PubMed]
183. Jafarzadeh, A.; Nemati, M.; Jafarzadeh, S. The important role played by chemokines influence the clinical outcome of *Helicobacter pylori* infection. *Life Sci.* **2019**, *231*, 116688. [CrossRef]
184. Pawluczuk, E.; Lukaszewicz-Zajac, M.; Mroczko, B. The Role of Chemokines in the Development of Gastric Cancer—Diagnostic and Therapeutic Implications. *Int. J. Mol. Sci.* **2020**, *21*, 8456. [CrossRef]
185. Baj-Krzyworzeka, M.; Weglarczyk, K.; Baran, J.; Szczepanik, A.; Szura, M.; Siedlar, M. Elevated level of some chemokines in plasma of gastric cancer patients. *Cent. Eur. J. Immunol.* **2016**, *41*, 358–362. [CrossRef]
186. Mashino, K.; Sadanaga, N.; Yamaguchi, H.; Tanaka, F.; Ohta, M.; Shibuta, K.; Inoue, H.; Mori, M. Expression of chemokine receptor CCR7 is associated with lymph node metastasis of gastric carcinoma. *Cancer Res.* **2002**, *62*, 2937–2941. [PubMed]
187. Ying, J.E.; Xu, Q.; Zhang, G.; Liu, B.X.; Zhu, L.M. The expression of CXCL12 and CXCR4 in gastric cancer and their correlation to lymph node metastasis. *Med. Oncol.* **2012**, *29*, 1716–1722. [CrossRef] [PubMed]
188. Yasumoto, K.; Koizumi, K.; Kawashima, A.; Saitoh, Y.; Arita, Y.; Shinohara, K.; Minami, T.; Nakayama, T.; Sakurai, H.; Takahashi, Y.; et al. Role of the CXCL12/CXCR4 axis in peritoneal carcinomatosis of gastric cancer. *Cancer Res.* **2006**, *66*, 2181–2187. [CrossRef]
189. Xiang, Z.; Zhou, Z.J.; Xia, G.K.; Zhang, X.H.; Wei, Z.W.; Zhu, J.T.; Yu, J.; Chen, W.; He, Y.; Schwarz, R.E.; et al. A positive crosstalk between CXCR4 and CXCR2 promotes gastric cancer metastasis. *Oncogene* **2017**, *36*, 5122–5133. [CrossRef]
190. Cheng, Y.; Qu, J.L.; Che, X.F.; Xu, L.; Song, N.; Ma, Y.J.; Gong, J.; Qu, X.J.; Liu, Y.P. CXCL12/SDF-1 alpha induces migration via SRC-mediated CXCR4-EGFR cross-talk in gastric cancer cells. *Oncol. Lett.* **2017**, *14*, 2103–2110. [CrossRef]
191. Chen, G.; Chen, S.M.; Wang, X.; Ding, X.F.; Ding, J.; Meng, L.H. Inhibition of chemokine (CXC motif) ligand 12/chemokine (CXC motif) receptor 4 axis (CXCL12/CXCR4)-mediated cell migration by targeting mammalian target of rapamycin (mTOR) pathway in human gastric carcinoma cells. *J. Biol. Chem.* **2012**, *287*, 12132–12141. [CrossRef]
192. Ma, H.Y.; Gao, L.L.; Li, S.C.; Qin, J.; Chen, L.; Liu, X.Z.; Xu, P.P.; Wang, F.; Xiao, H.L.; Zhou, S.; et al. CCR7 enhances TGF-β 1-induced epithelial-mesenchymal transition and is associated with lymph node metastasis and poor overall survival in gastric cancer. *Oncotarget* **2015**, *6*, 24348–24360. [CrossRef]
193. Zlotnik, A.; Burkhardt, A.M.; Homey, B. Homeostatic chemokine receptors and organ-specific metastasis. *Nat. Rev. Immunol.* **2011**, *11*, 597–606. [CrossRef] [PubMed]
194. Amersi, F.F.; Terando, A.M.; Goto, Y.; Scolyer, R.A.; Thompson, J.F.; Tran, A.N.; Faries, M.B.; Morton, D.L.; Hoon, D.S.B. Activation of CCR9/CCL25 in cutaneous melanoma mediates preferential metastasis to the small intestine. *Clin. Cancer. Res.* **2008**, *14*, 638–645. [CrossRef] [PubMed]
195. Johnson-Holiday, C.; Singh, R.; Johnson, E.; Singh, S.; Stockard, C.R.; Grizzle, W.E.; Lillard, J.W. CCL25 mediates migration, invasion and matrix metalloproteinase expression by breast cancer cells in a CCR9-dependent fashion. *Int. J. Oncol.* **2011**, *38*, 1279–1285.
196. Singh, R.; Stockard, C.R.; Grizzle, W.E.; Lillard, J.W.; Singh, S. Expression and histopathological correlation of CCR9 and CCL25 in ovarian cancer. *Int. J. Oncol.* **2011**, *39*, 373–381. [CrossRef]
197. Xu, E.; Xia, X.; Jiang, C.; Li, Z.; Yang, Z.; Zheng, C.; Wang, X.; Du, S.; Miao, J.; Wang, F.; et al. GPER1 Silencing Suppresses the Proliferation, Migration, and Invasion of Gastric Cancer Cells by Inhibiting PI3K/AKT-Mediated EMT. *Front. Cell Dev. Biol.* **2020**, *8*, 591239. [CrossRef] [PubMed]
198. Kessenbrock, K.; Plaks, V.; Werb, Z. Matrix metalloproteinases: Regulators of the tumor microenvironment. *Cell* **2010**, *141*, 52–67. [CrossRef]
199. Jackson, H.W.; Defamie, V.; Waterhouse, P.; Khokha, R. TIMPs: Versatile extracellular regulators in cancer. *Nat. Rev. Cancer* **2017**, *17*, 38–53. [CrossRef]
200. Abdi, E.; Latifi-Navid, S.; Sarvestani, F.A.; Esmailnejad, M.H. Emerging therapeutic targets for gastric cancer from a host-Helicobacter pylori interaction perspective. *Expert Opin. Ther. Targets* **2021**, *25*, 685–699. [CrossRef]
201. Posselt, G.; Crabtree, J.E.; Wessler, S. Proteolysis in Helicobacter pylori-Induced Gastric Cancer. *Toxins* **2017**, *9*, 134. [CrossRef]
202. Colosimo, D.A.; Kohn, J.A.; Luo, P.M.; Piscotta, F.J.; Han, S.M.; Pickard, A.J.; Rao, A.; Cross, J.R.; Cohen, L.J.; Brady, S.F. Mapping Interactions of Microbial Metabolites with Human G-Protein-Coupled Receptors. *Cell Host Microbe* **2019**, *26*, 273–282. [CrossRef]
203. Ziyad, S.; Iruela-Arispe, M.L. Molecular mechanisms of tumor angiogenesis. *Genes Cancer* **2011**, *2*, 1085–1096. [CrossRef]
204. Fidler, I.J. Timeline—The pathogenesis of cancer metastasis: The 'seed and soil' hypothesis revisited. *Nat. Rev. Cancer* **2003**, *3*, 453–458. [CrossRef]
205. Nierodzik, M.L.; Karpatkin, S. Thrombin induces tumor growth, metastasis, and angiogenesis: Evidence for a thrombin-regulated dormant tumor phenotype. *Cancer Cell* **2006**, *10*, 355–362. [CrossRef]
206. Wojtukiewicz, M.Z.; Hempel, D.; Sierko, E.; Tucker, S.C.; Honn, K.V. Protease-activated receptors (PARs)—Biology and role in cancer invasion and metastasis. *Cancer Metastasis Rev.* **2015**, *34*, 775–796. [CrossRef]
207. Liu, Y.J.; Wada, R.; Yamashita, T.; Mi, Y.D.; Deng, C.X.; Hobson, J.P.; Rosenfeldt, H.M.; Nava, V.E.; Chae, S.S.; Lee, M.J.; et al. Edg-1, the G protein-coupled receptor for sphingosine-1-phosphate, is essential for vascular maturation. *J. Clin. Investig.* **2000**, *106*, 951–961. [CrossRef] [PubMed]

208. Cartier, A.; Leigh, T.; Liu, C.H.; Hla, T. Endothelial sphingosine 1-phosphate receptors promote vascular normalization and antitumor therapy. *Proc. Natl. Acad. Sci. USA* **2020**, *117*, 3157–3166. [CrossRef]
209. Yasuda, D.; Kobayashi, D.; Akahoshi, N.; Ohto-Nakanishi, T.; Yoshioka, K.; Takuwa, Y.; Mizuno, S.; Takahashi, S.; Ishii, S. Lysophosphatidic acid-induced YAP/TAZ activation promotes developmental angiogenesis by repressing Notch ligand Dll4. *J. Clin. Investig.* **2019**, *129*, 4332–4349. [CrossRef] [PubMed]
210. Liu, N.; Wu, Q.; Wang, Y.; Sui, H.; Liu, X.; Zhou, N.; Zhou, L.; Wang, Y.; Ye, N.; Fu, X.; et al. Helicobacter pylori promotes VEGF expression via the p38 MAPKmediated COX2PGE2 pathway in MKN45 cells. *Mol. Med. Rep.* **2014**, *10*, 2123–2129. [CrossRef] [PubMed]
211. De Francesco, E.M.; Sotgia, F.; Clarke, R.B.; Lisanti, M.P.; Maggiolini, M. G Protein-Coupled Receptors at the Crossroad between Physiologic and Pathologic Angiogenesis: Old Paradigms and Emerging Concepts. *Int. J. Mol. Sci.* **2017**, *18*, 2713. [CrossRef]
212. Zeng, D.Q.; Li, M.Y.; Zhou, R.; Zhang, J.W.; Sun, H.Y.; Shi, M.; Bin, J.P.; Liao, Y.L.; Rao, J.J.; Liao, W.J. Tumor Microenvironment Characterization in Gastric Cancer Identifies Prognostic and Immunotherapeutically Relevant Gene Signatures. *Cancer Immunol. Res.* **2019**, *7*, 737–750. [CrossRef] [PubMed]
213. Rojas, A.; Araya, P.; Gonzalez, I.; Morales, E. Gastric Tumor Microenvironment. *Adv. Exp. Med. Biol.* **2020**, *1226*, 23–35. [CrossRef]
214. Quail, D.F.; Joyce, J.A. Microenvironmental regulation of tumor progression and metastasis. *Nat. Med.* **2013**, *19*, 1423–1437. [CrossRef]
215. Chiu, K.J.; Chiou, H.C.; Huang, C.H.; Lu, P.C.; Kuo, H.R.; Wang, J.W.; Lin, M.H. Natural Compounds Targeting Cancer-Associated Fibroblasts against Digestive System Tumor Progression: Therapeutic Insights. *Biomedicines* **2022**, *10*, 713. [CrossRef] [PubMed]
216. Peltier, A.; Seban, R.D.; Buvat, I.; Bidard, F.C.; Mechta-Grigoriou, F. Fibroblast heterogeneity in solid tumors: From single cell analysis to whole-body imaging. *Semin. Cancer Biol.* **2022**, *86*, 262–272. [CrossRef]
217. Gorbachev, A.V.; Fairchild, R.L. Regulation of chemokine expression in the tumor microenvironment. *Crit. Rev. Immunol.* **2014**, *34*, 103–120. [CrossRef] [PubMed]
218. Rihawi, K.; Ricci, A.D.; Rizzo, A.; Brocchi, S.; Marasco, G.; Pastore, L.V.; Llimpe, F.L.R.; Golfieri, R.; Renzulli, M. Tumor-Associated Macrophages and Inflammatory Microenvironment in Gastric Cancer: Novel Translational Implications. *Int. J. Mol. Sci.* **2021**, *22*, 3805. [CrossRef]
219. Chen, F.F.; Yin, S.; Niu, L.; Luo, J.; Wang, B.C.; Xu, Z.G.; Yang, G.F. Expression of the Chemokine Receptor CXCR3 Correlates with Dendritic Cell Recruitment and Prognosis in Gastric Cancer. *Genet. Test. Mol. Biomark.* **2018**, *22*, 35–42. [CrossRef]
220. Sehrell, T.A.; Hashimi, M.; Sidar, B.; Wilkinson, R.A.; Kirpotina, L.; Quinn, M.T.; Malkoc, Z.; Taylor, P.J.; Wilking, J.N.; Bimczok, D. A Novel Gastric Spheroid Co-culture Model Reveals Chemokine-Dependent Recruitment of Human Dendritic Cells to the Gastric Epithelium. *Cell. Mol. Gastroenterol. Hepatol.* **2019**, *8*, 157–171. [CrossRef]
221. Maimela, N.R.; Liu, S.; Zhang, Y. Fates of CD8+ T cells in Tumor Microenvironment. *Comput. Struct. Biotechnol. J.* **2019**, *17*, 1–13. [CrossRef] [PubMed]
222. Slaney, C.Y.; Kershaw, M.H.; Darcy, P.K. Trafficking of T Cells into Tumors. *Cancer Res.* **2014**, *74*, 7168–7174. [CrossRef] [PubMed]
223. Ishigami, S.; Natsugoe, S.; Tokuda, K.; Nakajo, A.; Che, X.M.; Iwashige, H.; Aridome, K.; Hokita, S.; Aikou, T. Prognostic value of intratumoral natural killer cells in gastric carcinoma. *Cancer* **2000**, *88*, 577–583. [CrossRef]
224. Bald, T.; Krummel, M.F.; Smyth, M.J.; Barry, K.C. The NK cell-cancer cycle: Advances and new challenges in NK cell-based immunotherapies. *Nat. Immunol.* **2020**, *21*, 835–847. [CrossRef]
225. Bikfalvi, A.; Billottet, C. The CC and CXC chemokines: Major regulators of tumor progression and the tumor microenvironment. *Am. J. Physiol. Cell Physiol.* **2020**, *318*, C542–C554. [CrossRef]
226. Mizukami, Y.; Kono, K.; Kawaguchi, Y.; Akaike, H.; Kamimura, K.; Sugai, H.; Fujii, H. CCL17 and CCL22 chemokines within tumor microenvironment are related to accumulation of Foxp3+ regulatory T cells in gastric cancer. *Int. J. Cancer* **2008**, *122*, 2286–2293. [CrossRef]
227. Liu, X.; Zhang, Z.Z.; Zhao, G. Recent advances in the study of regulatory T cells in gastric cancer. *Int. Immunopharmacol.* **2019**, *73*, 560–567. [CrossRef]
228. Parisi, L.; Gini, E.; Baci, D.; Tremolati, M.; Fanuli, M.; Bassani, B.; Farronato, G.; Bruno, A.; Mortara, L. Macrophage Polarization in Chronic Inflammatory Diseases: Killers or Builders? *J. Immunol. Res.* **2018**, *2018*, 8917804. [CrossRef]
229. Ruytinx, P.; Proost, P.; Van Damme, J.; Struyf, S. Chemokine-Induced Macrophage Polarization in Inflammatory Conditions. *Front. Immunol.* **2018**, *9*, 1930. [CrossRef]
230. Reyes, M.E.; de La Fuente, M.; Hermoso, M.; Ili, C.G.; Brebi, P. Role of CC Chemokines Subfamily in the Platinum Drugs Resistance Promotion in Cancer. *Front. Immunol.* **2020**, *11*, 901. [CrossRef]
231. Chen, F.F.; Yuan, J.P.; Yan, H.L.; Liu, H.; Yin, S. Chemokine Receptor CXCR3 Correlates with Decreased M2 Macrophage Infiltration and Favorable Prognosis in Gastric Cancer. *BioMed Res. Int.* **2019**, *2019*, 6832867. [CrossRef] [PubMed]
232. Gambardella, V.; Castillo, J.; Tarazona, N.; Gimeno-Valiente, F.; Martinez-Ciarpaglini, C.; Cabeza-Segura, M.; Rosello, S.; Roda, D.; Huerta, M.; Cervantes, A.; et al. The role of tumor-associated macrophages in gastric cancer development and their potential as a therapeutic target. *Cancer Treat. Rev.* **2020**, *86*, 102015. [CrossRef]
233. Boutilier, A.J.; Elsawa, S.F. Macrophage Polarization States in the Tumor Microenvironment. *Int. J. Mol. Sci.* **2021**, *22*, 6995. [CrossRef]
234. Condamine, T.; Ramachandran, I.; Youn, J.I.; Gabrilovich, D.I. Regulation of Tumor Metastasis by Myeloid-Derived Suppressor Cells. *Annu. Rev. Med.* **2015**, *66*, 97–110. [CrossRef] [PubMed]

235. Kramer, E.D.; Abrams, S.I. Granulocytic Myeloid-Derived Suppressor Cells as Negative Regulators of Anticancer Immunity. *Front. Immunol.* **2020**, *11*, 1963. [CrossRef]
236. Echizen, K.; Hirose, O.; Maeda, Y.; Oshima, M. Inflammation in gastric cancer: Interplay of the COX-2/prostaglandin E2 and Toll-like receptor/MyD88 pathways. *Cancer Sci.* **2016**, *107*, 391–397. [CrossRef]
237. Oshima, H.; Oshima, M. The role of PGE2-associated inflammatory responses in gastric cancer development. *Semin. Immunopathol.* **2013**, *35*, 139–150. [CrossRef]
238. Chionh, Y.T.; Ng, G.Z.; Ong, L.; Arulmuruganar, A.; Stent, A.; Saeed, M.A.; Wee, J.L.; Sutton, P. Protease-activated receptor 1 suppresses Helicobacter pylori gastritis via the inhibition of macrophage cytokine secretion and interferon regulatory factor 5. *Mucosal Immunol.* **2015**, *8*, 68–79. [CrossRef]
239. Guo, C.; Qi, H.; Yu, Y.J.; Zhang, Q.Q.; Su, J.; Yu, D.; Huang, W.D.; Chen, W.D.; Wang, Y.D. The G-Protein-Coupled Bile Acid Receptor Gpbar1 (TGR5) Inhibits Gastric Inflammation Through Antagonizing NF-kappa B Signaling Pathway. *Front. Pharmacol.* **2015**, *6*, 287. [CrossRef]
240. Ajani, J.A.; D'Amico, T.A.; Bentrem, D.J.; Chao, J.; Cooke, D.; Corvera, C.; Das, P.; Enzinger, P.C.; Enzler, T.; Fanta, P.; et al. Gastric Cancer, Version 2.2022, NCCN Clinical Practice Guidelines in Oncology. *J. Natl. Compr. Canc. Netw.* **2022**, *20*, 167–192. [CrossRef]
241. Van Cutsem, E.; Sagaert, X.; Topal, B.; Haustermans, K.; Prenen, H. Gastric cancer. *Lancet* **2016**, *388*, 2654–2664. [CrossRef] [PubMed]
242. Shitara, K.; Van Cutsem, E.; Bang, Y.J.; Fuchs, C.; Wyrwicz, L.; Lee, K.W.; Kudaba, I.; Garrido, M.; Chung, H.C.; Lee, J.; et al. Efficacy and Safety of Pembrolizumab or Pembrolizumab Plus Chemotherapy vs Chemotherapy Alone for Patients With First-line, Advanced Gastric Cancer: The KEYNOTE-062 Phase 3 Randomized Clinical Trial. *JAMA Oncol.* **2020**, *6*, 1571–1580. [CrossRef] [PubMed]
243. Hindson, J. Nivolumab plus chemotherapy for advanced gastric cancer and oesophageal adenocarcinoma. *Nat. Rev. Gastroenterol. Hepatol.* **2021**, *18*, 523. [CrossRef] [PubMed]
244. Kubota, Y.; Kawazoe, A.; Sasaki, A.; Mishima, S.; Sawada, K.; Nakamura, Y.; Kotani, D.; Kuboki, Y.; Taniguchi, H.; Kojima, T.; et al. The Impact of Molecular Subtype on Efficacy of Chemotherapy and Checkpoint Inhibition in Advanced Gastric Cancer. *Clin. Cancer. Res.* **2020**, *26*, 3784–3790. [CrossRef]
245. Smyth, E.C.; Gambardella, V.; Cervantes, A.; Fleitas, T. Checkpoint inhibitors for gastroesophageal cancers: Dissecting heterogeneity to better understand their role in first-line and adjuvant therapy. *Ann. Oncol.* **2021**, *32*, 590–599. [CrossRef]
246. Chen, Y.; Wei, K.; Liu, D.; Xiang, J.; Wang, G.; Meng, X.; Peng, J. A Machine Learning Model for Predicting a Major Response to Neoadjuvant Chemotherapy in Advanced Gastric Cancer. *Front. Oncol.* **2021**, *11*, 675458. [CrossRef]
247. Congreve, M.; de Graaf, C.; Swain, N.A.; Tate, C.G. Impact of GPCR Structures on Drug Discovery. *Cell* **2020**, *181*, 81–91. [CrossRef]
248. Jabeen, A.; Ranganathan, S. Applications of machine learning in GPCR bioactive ligand discovery. *Curr. Opin. Struct. Biol.* **2019**, *55*, 66–76. [CrossRef]
249. Hauser, A.S.; Attwood, M.M.; Rask-Andersen, M.; Schioth, H.B.; Gloriam, D.E. Trends in GPCR drug discovery: New agents, targets and indications. *Nat. Rev. Drug Discov.* **2017**, *16*, 829–842. [CrossRef]
250. Ceylan, S.; Bahadori, F.; Akbas, F. Engineering of siRNA loaded PLGA Nano-Particles for highly efficient silencing of GPR87 gene as a target for pancreatic cancer treatment. *Pharm. Dev. Technol.* **2020**, *25*, 855–864. [CrossRef]
251. Ma, X.; Xiong, Y.; Lee, L.T.O. Application of Nanoparticles for Targeting G Protein-Coupled Receptors. *Int. J. Mol. Sci.* **2018**, *19*, 2006. [CrossRef] [PubMed]
252. Banerjee, D.; Harfouche, R.; Sengupta, S. Nanotechnology-mediated targeting of tumor angiogenesis. *Vasc. Cell* **2011**, *3*, 3. [CrossRef] [PubMed]
253. Corso, S.; Isella, C.; Bellomo, S.E.; Apicella, M.; Durando, S.; Migliore, C.; Ughetto, S.; D'Errico, L.; Menegon, S.; Moya-Rull, D.; et al. A Comprehensive PDX Gastric Cancer Collection Captures Cancer Cell-Intrinsic Transcriptional MSI Traits. *Cancer Res.* **2019**, *79*, 5884–5896. [CrossRef] [PubMed]

Disclaimer/Publisher's Note: The statements, opinions and data contained in all publications are solely those of the individual author(s) and contributor(s) and not of MDPI and/or the editor(s). MDPI and/or the editor(s) disclaim responsibility for any injury to people or property resulting from any ideas, methods, instructions or products referred to in the content.

Article

Modulatory Properties of *Aloe secundiflora's* Methanolic Extracts on Targeted Genes in Colorectal Cancer Management

John M. Macharia [1,*], Timea Varjas [2], Ruth W. Mwangi [3,4], Zsolt Káposztás [5], Nóra Rozmann [1], Márton Pintér [1], Isabel N. Wagara [4] and Bence L. Raposa [5]

1. Doctoral School of Health Sciences, Faculty of Health Science, University of Pécs, 7621 Pécs, Hungary
2. Department of Public Health Medicine, Medical School, University of Pécs, 7621 Pécs, Hungary
3. Department of Vegetable and Mushroom Growing, Hungarian University of Agriculture and Life Sciences, 1118 Budapest, Hungary
4. Department of Biological Sciences, Egerton University, Nakuru P.O. Box 3366-20100, Kenya
5. Faculty of Health Sciences, University of Pécs, 7621 Pécs, Hungary; bence.raposa@etk.pte.hu
* Correspondence: johnmacharia@rocketmail.com

Simple Summary: This study focused on understanding the potential use of *Aloe secundiflora* (AS) extracts in managing colorectal cancer (CRC). As colon tumors present complex challenges, the research aimed to assess how the AS methanolic extracts impacted the expression of specific genes related to CRC, namely *CASPS9*, *5-LOX*, *Bcl2*, *Bcl-xL*, and *COX-2*. The results demonstrated that the AS extracts, when applied to CRC cell lines, effectively upregulated *CASPS9* expression, promoting apoptosis in a dose-dependent manner. Simultaneously, the extracts downregulated the expressions of *5-LOX*, *Bcl2*, and *Bcl-xL*, crucial in curbing cancer progression. The study suggests that using AS extracts and methanol as an extraction solvent could be beneficial in managing CRC. Furthermore, the researchers recommend exploring the specific metabolites in AS involved in these pathways to better understand how they impede the development and spread of CRC. This research provides promising insights into potential natural treatments for colorectal cancer, offering hope for improved therapies in the future.

Citation: Macharia, J.M.; Varjas, T.; Mwangi, R.W.; Káposztás, Z.; Rozmann, N.; Pintér, M.; Wagara, I.N.; Raposa, B.L. Modulatory Properties of *Aloe secundiflora's* Methanolic Extracts on Targeted Genes in Colorectal Cancer Management. *Cancers* 2023, 15, 5002. https://doi.org/10.3390/cancers 15205002

Academic Editor: Shihori Tanabe

Received: 20 September 2023
Revised: 8 October 2023
Accepted: 12 October 2023
Published: 16 October 2023

Copyright: © 2023 by the authors. Licensee MDPI, Basel, Switzerland. This article is an open access article distributed under the terms and conditions of the Creative Commons Attribution (CC BY) license (https:// creativecommons.org/licenses/by/ 4.0/).

Abstract: Colon tumors have a very complicated and poorly understood pathogenesis. Plant-based organic compounds might provide a novel source for cancer treatment with a sufficient novel mode of action. The objective of this study was to analyze and evaluate the efficacy of *Aloe secundiflora's* (AS) methanolic extracts on the expression of *CASPS9*, *5-LOX*, *Bcl2*, *Bcl-xL*, and *COX-2* in colorectal cancer (CRC) management. Caco-2 cell lines were used in the experimental study. In the serial exhaustive extraction (SEE) method, methanol was utilized as the extraction solvent. Upon treatment of *CASPS9* with the methanolic extracts, the expression of the genes was progressively upregulated, thus, dose-dependently increasing the rate of apoptosis. On the other hand, the expressions of *5-LOX*, *Bcl2*, and *Bcl-xL* were variably downregulated in a dose-dependent manner. This is a unique novel study that evaluated the effects of AS methanolic extracts in vitro on CRC cell lines using different dosage concentrations. We, therefore, recommend the utilization of AS and the application of methanol as the extraction solvent of choice for maximum modulatory benefits in CRC management. In addition, we suggest research on the specific metabolites in AS involved in the modulatory pathways that suppress the development of CRC and potential metastases.

Keywords: molecular mechanism; colorectal cancer; genome modulatory pathways; tumor microenvironment; phytotherapeutic effects

1. Introduction

1.1. Background Information on Colon Cancer and Prevention Approaches

In the developed world, colon cancer is currently the third leading cause of cancer-related fatalities [1]. In general, cancer is the second leading cause of death, behind cardiovascular disorders [2]. The pathophysiology of colon tumors is very complex and poorly understood. On the other hand, the process of its onset has been linked to the interactions between risk factors such as lifestyle, inheritance, and environmental factors, among other identified causes [3,4]. Understanding the processes that swiftly proliferating malignant cells utilize to regulate their metabolism can help scientists create more effective cancer treatments [5].

To investigate and identify efficient bioactive compounds that can destroy malignant cells without harming or killing healthy cells has a huge impact in human medicine [6]. Due to this, management utilizing plant-based dietary supplements is beginning to receive attention as the most efficient way to lower the burden of colon-cancer-associated mortality [7]. Organic components found in plants may offer a fresh source for cancer treatment with a sufficiently revolutionary method of action [8]. Plant extracts have been shown to exhibit astounding therapeutic activities to treat a variety of infectious diseases, in contrast to synthetic pharmaceuticals, which are frequently observed as being associated with serious drawbacks [9].

Because of Africa's diversity and abundance of aloe species, these plants are commonly used as a source of phytotherapeutic medication to improve human health and welfare. There are 500 species of *Aloe* L. (Asphodeloideae), a genus of flowering succulents that includes trees, shrubs, and perennials [10]. The largest genus in the Asphodelaceae family, *Aloe*, has over 400 species that range in size from tiny shrubs to enormous trees and are distributed throughout dry regions of Africa, India, and other places [11]. In South Africa, there is the greatest variety of *Aloe*. The two primary ingredients are *Aloe* exudates and *Aloe* leaf gel. Exudates come from the inner epidermal layers, while parenchymatous cells are the source of the gel. The exudates are primarily a combination of phenolic chemicals, whereas the gel is primarily composed of polysaccharides [12]. In nations such as Ethiopia, Sudan, Kenya, and Tanzania, wide grasslands and bushlands are home to numerous AS (Asphodelaceae) bushes. *Aloe engleri*, *Aloe marsabitensis*, and *Aloe floramaculata* are all synonyms of AS [12]. In our previous research [10], we postulated the potential effects of *A. secundiflora's* active metabolites in colorectal cancer management. In the current study, we substantiate the claims postulated using specific genes present in human cells.

1.2. Phytoconstituent Biomolecules Present in A. secundiflora

Studies on the phytochemical and pharmacological activities of *Aloe* species have led to the discovery of various active compounds. In past decades, herbalists have used them to treat a range of diseases [11]. Most of the biologically active substances found in *Aloe* originate as anthraquinones naturally [13], as we have previously reported [10]. Terpenes, flavonoids, and tannins have been identified in the leaves (Table 1), according to preliminary phytochemical analysis [14], while naphthoquinones have been identified in the roots [15]. Anthraquinones, which are structurally related to anthracene [16], are primarily composed of Anthracenedione (9,10-anthracenedione). They are sometimes referred to as 9,10-dioxoanthracene. Anthraquinones typically occur in their glycosidic state [10]. These elements constitute the pigmentation that gives plants their hue, and they are commonly employed as natural dyes [16]. Aloin (AL) is employed in pharmacotherapy for a variety of purposes, one of which is as a laxative [17]. In both in vivo and in vitro experimental settings, AL was shown to be beneficial in lowering tumor angiogenesis and development by blocking STAT3 activation in CRC cells [18].

In the current study, we demonstrate the activity of AS in regulating the expression of *CASPS9*, *5-LOX*, *Bcl2*, *Bcl-xL*, and *COX-2* genes associated with colon carcinogenesis. In addition, *HPRT1* was employed as a housing keeping gene for internal control. *CASPS9*, the primary enzyme in the mitochondrial caspase pathway, plays a critical role in mediating apoptosis control [22,23]. The relationship between CASPS9 and CRC is currently undetermined in a tangible and substantial manner. Its correlation with clinicopathological characteristics and longevity may provide insightful data that can be used to estimate survival and choose additional treatment options [24]. While the exact sequence of events relating cancer development to *5-LOX* gene expression is equally unknown, it is clear that *5-LOX* expression is occasionally increased during neoplastic transformation [25]. *LOX* inhibitors decrease cancer cell proliferation in both in vivo and in vitro experiments and cause death through mitochondrial induction [26,27]. Targeted modulatory strategies call for knowledge of *CASPS9, 5-LOX, Bcl2, Bcl-xL*, and *COX-2* for effective CRC medical intervention. This is a unique novel study that evaluated the effects of AS methanolic extracts in vitro on CRC cell lines at different dosage/concentration levels. We do, however, support additional research into the specific AS metabolites implicated in the modulatory pathways that prevent the growth of CRC and potential metastases.

Table 1. Phytoconstituent biomolecules present in A. secundiflora [10].

Plant	Phytoconstituents Present in Roots	Phytoconstituents Present in Leaves	Ref.
Aloe secundiflora	Anthraquinones (Chrysophanol, Helminthosporin, Aloe-emodin, Aloesaponarin II, and Aloesaponarin I), laccaic acid D, methyl ester, and asphodelin. Naphthoquinones (5-hydroxy-3,6-dimethoxy-2-methylnaphthalene-1,4-dione and 5,8-dihydroxy-3-methoxy-2-methylnaphthalene-1,4-dione)	Phenols such as anthrones (aloenin, aloenin B, isobarbaloin, barbaloin, and other aloin derivatives), chromones and phenylpyrones, Alkaloids, Saponin, Tannins, Flavonoids (nthoxanthins, flavanones, flavanols, flavans, and anthocyanidin), Steroids, Cardiac Glycosides, Aloeresin, Anthraquinones Aloin, Hydro-xyaloins, Polyphenols, and Terpenoids	[14–16,19–21]

2. Materials and Methods

2.1. Colorectal Cancer Cell Lines (Caco-2 Cell Lines)

The Caco-2 cell lines were purchased from American Type Culture Collection (ATCC) and directly delivered to our laboratory (Department of Public Health) by the University of Pecs' Department of Biochemistry and Medical Chemistry. This type of cell line is ideally suited to be employed in studies on cancer and cytotoxicity and serves as a superb transfection carrier. In conformance with the supplier's guidelines, the Caco-2 cell lines were preserved until use [28].

2.2. Plant Extract Extraction Using Methanol and Region of Acquisition

Shade-dried leaves were pulverized into a fine powder. Methanol was used in the serial exhaustive extraction (SEE). One thousand grams of AS plant organs were placed in a flask and extracted for 3 days with methanol with frequent shaking to adequately extract phytoconstituents. The crude solvent extracts were dechlorophyllated after being filtered via Whatman filter paper with varying pore diameters (Nos. 4 and 1). This technique was performed three times until all soluble elements were extracted completely [29]. The filtrate was concentrated using a rotavapor apparatus paired to a vacuum pump, a condenser apparatus to recover the solvent, and a round-bottomed flask at a temperature of 50 °C. In order to allow the samples to air dry, the concentrated solution was placed in small glass universal bottles and covered with perforated aluminum foil [12]. The dried solvent-free metabolites were stored in tightly sealed sample bottles with parafilm tape in a desiccator at 4 °C in a fridge until use [30]. The identified plant organs were collected from Kampi ya Moto in Rongai constituency, Nakuru County, Kenya and located at 0.1244° S, 35.9431° E, GPS coordinates. The annual temperature in the district is 18.75 °C (65.75 °F), which is

−3.75% lower than the national average for Kenya. Furthermore, 118.62 mm (4.67 inches) of precipitation and 221.53 days of rain are typical yearly totals for Kampi Ya Moto.

2.3. Dissipation of Plant Extracts in DMSO

Dimethyl sulfoxide (DMSO) is a multipurpose solvent that is used in toxicology and pharmacology to improve medication delivery, dissolve various pharmaceuticals, and dissolve plant extracts [31]. It served as a suspending medium for water-insoluble crude plant extracts, as well as an inert diluent. A stock solution of 30 mg/mL was created using 0.5% DMSO and double-distilled phosphate buffer saline (ddPBS) as the dissolving and diluent solvents. The final Caco-2 cell line concentrations of 2 mg/mL, 1 mg/mL, and 0.5 mg/mL were then made using the stock solution.

2.4. Passaging Caco-2 Cell Lines for Bioassays

The lamina hood, which was kept sterile, was carefully placed over the petri dish or flask holding the Caco-2 cells. The utilized medium was sucked out after the dish or flask was opened. PBS was used twice to wash it. After pouring PBS-EDTA over the area, it was set aside for a while. It was then cautiously and softly sucked. To separate and detach the Caco-2 cells from clumps and the surface, respectively, 2 mL of trypsin was applied. Trypsin was applied to the surface by gently swiping the dish side to side across it. For five minutes, the dish was placed in the thermostat. The dish was removed after 5 min, and following a successful visible detachment, the Caco-2 medium was carefully added to it. The Caco-2 cells are adherent and have a propensity to stick to surfaces. The dish's whole contents were sucked into a tube and centrifuged for five minutes at 125 rpm. The Caco-2 cells stayed at the tube's bottom as the supernatant was sucked away. A pipette was used to gently disrupt the cells by sucking up and down while adding fresh media into the tube. The suspension was divided into fresh dishes, the medium was added, and the thermostat for growth was then turned on. The confluence was watched until it reached 70–80% so that it could be treated.

2.5. Exposure of Caco-2 Cell Lines to Solvent Extracts

Various quantities of the extract solutions (0.5 mg/mL, 1 mg/mL, and 2 mg/mL) were applied to 200 µL of passed Caco-2 cell lines that were refilled with new Caco-2 medium. Following that, the treated cells were incubated at 37 °C for 36 h. The condition of the cells was examined under a light microscope following the incubation period. The typical cell-doubling time of cancer cell lines, which is between 36 and 48 h, was used for the estimation of the duration of the exposure (36 h). Treatments were administered to the cells at concentrations ranging from 0.5 mg/mL at the lowest point to 2 mg/mL at the highest point to test the cells' dose and time responsiveness. This allowed us to identify and analyze the pharmacological modulatory effects of the treatments with the greatest accuracy. The growth and potential biological impacts of exposure were assessed every 12 h.

2.6. Isolation of RNA

The media was removed from the cell cultures, trypsin-EDTA was applied, and they were then washed twice with PBS. After centrifuging, the cell suspension was pipetted into a 4 cm^3 centrifuge tube. ExtraZol Tri-reagent solution diluted to 1 cm^3 was added and allowed to sit at room temperature for 5 min. Chloroform 0.2 cm^3 was added. The sample was centrifuged at 12,000× g for 10 min at 2–8 °C after a 2–3 min incubation period. A clean tube was used to transfer the aqueous phase. Isopropyl alcohol (0.2 cm^3) was added. The sample was centrifuged once more at 12,000× g for 10 min at 2–8 °C after being incubated for 10 min.

The RNA precipitate is frequently invisible prior to centrifugation but produces a gel-like pellet at the tube's bottom following centrifugation. The supernatant was removed before washing the RNA pellet with 1 cm^3 of 75% alcohol. The pellet was then centrifuged at 7500× g for 5 min at 2–8 °C after vortexing. The pellet was dried after the supernatant was drained off. It was then dissolved in 50–100 L of DEPC water, which is RN-ase free. After being vortexed, the sample was incubated for 10 min at 55 °C. Before usage, the isolated total RNA was kept at −80 °C.

2.7. Protocol and Equipment Used for qRT-PCR (SYBR Green Protocol)

Nucleic acid quantification is sensitive, specific, and repeatable with real-time quantitative PCR. According to the manufacturer's instructions, one-step PCR was carried out using the One-Step Detect SyGreen Lo-ROX one-step RT-PCR kit (Nucleotest Bio Ltd. PB25.11–12) on a 96-well plate using a LightCycler 480 qPCR platform. The thermal program was set up as follows: 45 cycles of 95 °C for 5 s, 56 °C for 15 s, and 72 °C for 5 s, with a fluorescence readout being obtained at the conclusion of each cycle. The sample was incubated at 42 °C for 5 min, followed by incubation at 95 °C for 3 min. A melting curve analysis (95 °C—5 s, 65 °C—60 s, 97 °C) was performed after each run to validate amplification specificity. The reaction mixture was as follows: 10 l Master Mix, 0.4 l RT Mix, 0.4 l dUTP, 0.4 l primers, and 5 l mRNA template added to a total volume of 20 l of sterile double-distilled water. Integrated DNA Technologies (Bio-Sciences) created the primers, and Primer Express™ Software v3.0.1 was used to create the sequences (Table 2).

Table 2. Forward and reverse primer sequences adopted and applied in the experimental study.

Primer ID	Forward Primer	Reverse Primer
COX-2	CGGTGAAACTCTGGCTAGACAG	GCAAACCGTAGATGCTCAGGGA
5-LOX	GGAGAACCTGTTCATCAACCGC	CAGGTCTTCCTGCCAGTGATTC
Bcl2	ATCGCCCTGTGGATGACTGAGT	GCCAGGAGAAATCAAACAGAGGC
Bcl xL	GCCACTTACCTGAATGACCACC	AACCAGCGGTTGAAGCGTTCCT
Casp9	GTTTGAGGACCTTCGACCAGCT	CAACGTACCAGGAGCCACTCTT
HPRT1	TGCTTCTCCTCAGCTTCA	CTCAGGAGGAGGAAGCC

2.8. qRT-PCR Result Analysis

The Cp numbers used to express the PCR findings represent the intersection of the amplification curve and the threshold value. Using the 2-Cp (Livak method) and the Cp values, the fold changes of the target genes from the control sample were calculated [32].

2.9. Data Analysis

IBM SPSS 26.0 (IBM Corp. Released 2019. IBM SPSS Statistics for Windows, Version 26.0. Armonk, NY, USA: IBM Corp) and MS Excel (Microsoft Corp., released in 2013, Redmond, WA, USA) were used to calculate the statistical analysis. IBM SPSS Statistics for Windows, version 26.0.3. From the collected data, the normality analysis was conducted using the Kolmogorov–Smirnov test, and the means of the relevant variables were then compared using the ANOVA test. If the results were significant at the 95% confidence level, the p value was $p \leq 0.05$.

3. Results

3.1. Upregulatory Effects of AS Extracts on CASPS9 Expression

When the *CASPS9* genes were treated with AS's leaf extracts, apoptosis was observed to gradually occur upon exposure. Of interest, the increased optimal activity of apoptotic expressions was observed at a dosage concentration of 1 mg/mL. However, the apoptotic activity diminished and was at minimum in a concentration of 2 mg/mL. Observably, the expressions of *CASPS9* were progressively upregulated in a dose-dependent manner on exposure to the treatment extracts. In addition, the effects were statistically significant ($p < 0.001$), (Figure 1, Tables 3 and 4).

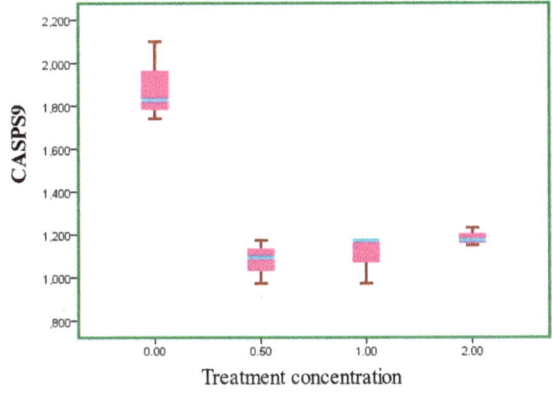
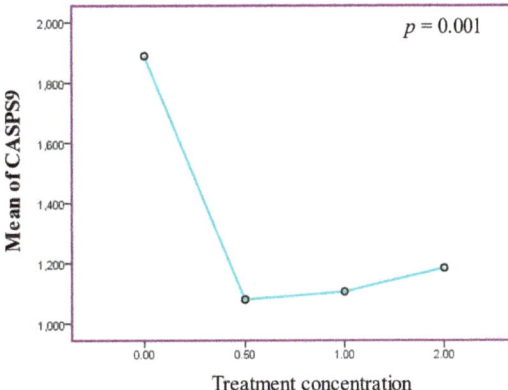

a) Upregulated *CASPS9* dose-dependently

b) Mean plot clearly showing the upregulated nature of *CASPS9* in a dose-dependently

Figure 1. Gradual upregulatory activities were observed in *CASPS9* after treatment with AS leaf extracts, as clearly shown in sub-figures (**a**,**b**). The regulatory characteristics were statistically significant.

Table 3. Post hoc computed test analysis of multiple relationships showing the targeted gene expressions at increasing concentrations (Tukey HSD, multiple comparisons [a]).

Dependent Variable	(I) Conc	(J) Conc	Mean Difference (I−J)	Std. Error	Sig.	95% Confidence Interval	
						Lower Bound	Upper Bound
COX2	0.00	0.50	−0.0011000	0.0050780	0.996	−0.017362	0.015162
		1.00	−0.0030333	0.0050780	0.930	−0.019295	0.013228
		2.00	−0.0018333	0.0050780	0.983	−0.018095	0.014428
	0.50	0.00	0.0011000	0.0050780	0.996	−0.015162	0.017362
		1.00	−0.0019333	0.0050780	0.980	−0.018195	0.014328
		2.00	−0.0007333	0.0050780	0.999	−0.016995	0.015528
	1.00	0.00	0.0030333	0.0050780	0.930	−0.013228	0.019295
		0.50	0.0019333	0.0050780	0.980	−0.014328	0.018195
		2.00	0.0012000	0.0050780	0.995	−0.015062	0.017462
	2.00	0.00	0.0018333	0.0050780	0.983	−0.014428	0.018095
		0.50	0.0007333	0.0050780	0.999	−0.015528	0.016995
		1.00	−0.0012000	0.0050780	0.995	−0.017462	0.015062
Lox5	0.00	0.50	0.2052000 *	0.0596828	0.036	0.014075	0.396325
		1.00	0.2887667 *	0.0596828	0.006	0.097641	0.479892
		2.00	0.3763333 *	0.0596828	0.001	0.185208	0.567459
	0.50	0.00	−0.2052000 *	0.0596828	0.036	−0.396325	−0.014075
		1.00	0.0835667	0.0596828	0.533	−0.107559	0.274692
		2.00	0.1711333	0.0596828	0.080	−0.019992	0.362259
	1.00	0.00	−0.2887667 *	0.0596828	0.006	−0.479892	−0.097641
		0.50	−0.0835667	0.0596828	0.533	−0.274692	0.107559
		2.00	0.0875667	0.0596828	0.497	−0.103559	0.278692
	2.00	0.00	−0.3763333 *	0.0596828	0.001	−0.567459	−0.185208
		0.50	−0.1711333	0.0596828	0.080	−0.362259	0.019992
		1.00	−0.0875667	0.0596828	0.497	−0.278692	0.103559

Table 3. Cont.

Dependent Variable	(I) Conc	(J) Conc	Mean Difference (I−J)	Std. Error	Sig.	95% Confidence Interval Lower Bound	Upper Bound
Bcl2	0.00	0.50	−0.0001333	0.0007605	0.998	−0.002569	0.002302
		1.00	0.0003333	0.0007605	0.970	−0.002102	0.002769
		2.00	0.0014333	0.0007605	0.306	−0.001002	0.003869
	0.50	0.00	0.0001333	0.0007605	0.998	−0.002302	0.002569
		1.00	0.0004667	0.0007605	0.925	−0.001969	0.002902
		2.00	0.0015667	0.0007605	0.244	−0.000869	0.004002
	1.00	0.00	−0.0003333	0.0007605	0.970	−0.002769	0.002102
		0.50	−0.0004667	0.0007605	0.925	−0.002902	0.001969
		2.00	0.0011000	0.0007605	0.508	−0.001335	0.003535
	2.00	0.00	−0.0014333	0.0007605	0.306	−0.003869	0.001002
		0.50	−0.0015667	0.0007605	0.244	−0.004002	0.000869
		1.00	−0.0011000	0.0007605	0.508	−0.003535	0.001335
Bcl-xL	0.00	0.50	−0.7222000	0.4281958	0.389	−2.093434	0.649034
		1.00	−0.2432333	0.4281958	0.939	−1.614467	1.128000
		2.00	0.0371000	0.4281958	1.000	−1.334134	1.408334
	0.50	0.00	0.7222000	0.4281958	0.389	−0.649034	2.093434
		1.00	0.4789667	0.4281958	0.689	−0.892267	1.850200
		2.00	0.7593000	0.4281958	0.351	−0.611934	2.130534
	1.00	0.00	0.2432333	0.4281958	0.939	−1.128000	1.614467
		0.50	−0.4789667	0.4281958	0.689	−1.850200	0.892267
		2.00	0.2803333	0.4281958	0.911	−1.090900	1.651567
	2.00	0.00	−0.0371000	0.4281958	1.000	−1.408334	1.334134
		0.50	−0.7593000	0.4281958	0.351	−2.130534	0.611934
		1.00	−0.2803333	0.4281958	0.911	−1.651567	1.090900
CASPS9	0.00	0.50	0.8094667 *	0.1002133	0.000	.488548	1.130385
		1.00	0.7833000 *	0.1002133	0.000	.462382	1.104218
		2.00	0.7052000 *	0.1002133	0.000	.384282	1.026118
	0.50	0.00	−0.8094667 *	0.1002133	0.000	−1.130385	−0.488548
		1.00	−0.0261667	0.1002133	0.993	−0.347085	0.294752
		2.00	−0.1042667	0.1002133	0.732	−0.425185	0.216652
	1.00	0.00	−0.7833000 *	0.1002133	0.000	−1.104218	−0.462382
		0.50	0.0261667	0.1002133	0.993	−0.294752	0.347085
		2.00	−0.0781000	0.1002133	0.862	−0.399018	0.242818
	2.00	0.00	−0.7052000 *	0.1002133	0.000	−1.026118	−0.384282
		0.50	0.1042667	0.1002133	0.732	−0.216652	0.425185
		1.00	0.0781000	0.1002133	0.862	−0.242818	0.399018

* The mean difference is significant at the 0.05 level; [a] treatment = methanolic leaf extracts.

Table 4. Statistical analysis of the effects of methanolic leaf extracts on the targeted genes.

Target Genes		Sum of Squares	df	Mean Square	F	Sig.
COX2	Between Groups	0.000	3	0.000	0.126	0.942
	Within Groups	0.000	8	0.000		
	Total	0.000	11			
Lox5	Between Groups	0.233	3	0.078	14.554	0.001
	Within Groups	0.043	8	0.005		
	Total	0.276	11			
Bcl2	Between Groups	0.000	3	0.000	1.748	0.235
	Within Groups	0.000	8	0.000		
	Total	0.000	11			
Bcl-xL	Between Groups	1.100	3	0.367	1.333	0.330
	Within Groups	2.200	8	0.275		
	Total	3.300	11			
Caspase9	Between Groups	1.338	3	0.446	29.603	0.000
	Within Groups	0.121	8	0.015		
	Total	1.458	11			

3.2. Downregulatory Effects of AS Extracts on 5-LOX Expression

After the *5-LOX* genes were treated with AS extract, a characteristic decrease in expression was gradually observed across the varying concentrations of exposure. An optimal mechanism of action resulting in a high decrease in expression was seen to occur at a concentration of 0.5 mg/mL. While the mechanism of action steadily increased at a dose of 2 mg/mL, very little transcriptional activity was seen at a concentration of 1 mg/mL. It is appropriate to state that the downregulatory effects in this investigation were dose dependent. The analysis yielded statistically significant findings (Figure 2, Tables 3 and 4).

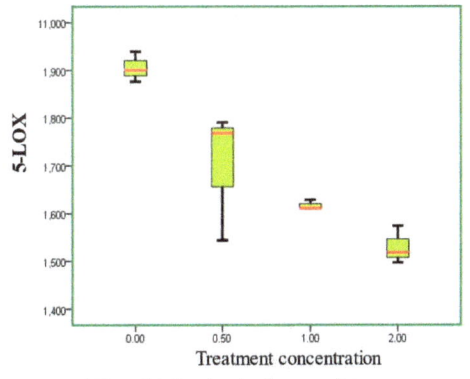
a) Box plot showing the downregulatory properties of the extract on *5-LOX*

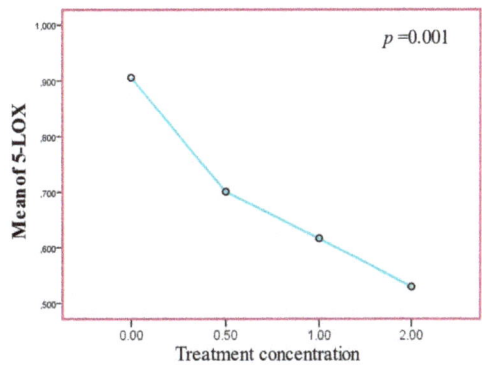
b) Mean plot showing the expression of *5-LOX*

Figure 2. Downregulatory effects were observed in *5-LOX* upon treatment with AS leaf extracts, as clearly shown in sub-figures (**a**,**b**). The modulatory expressions were statistically significant. (**a**) Box plot, (**b**) mean plot.

3.3. Downregulatory Effects of AS Extracts on Bcl2 Expressions

In all treatment concentrations, the decreased expression of *Bcl2* was observed to be dose-dependent. Even though the statistical output was not statistically significant ($p = 0.235$), the extract treatments exhibited significant downregulatory effects required to stimulate inhibitory cellular growth effects, as shown in Figure 3, Tables 3 and 4.

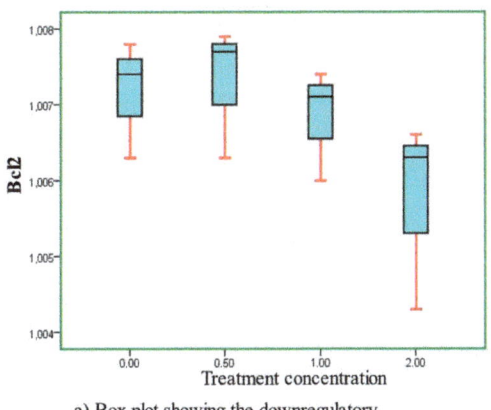
a) Box plot showing the downregulatory properties of the extract on *Bcl2*

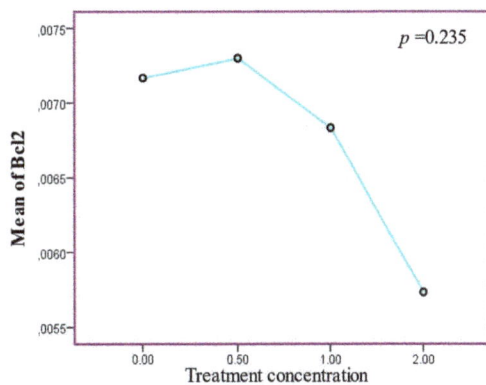
b) Mean plot showing the statistical expression of *Bcl2*

Figure 3. Downregulatory properties recorded in *Bcl2* after treatment with AS leaf extracts. Sub-figures (**a**,**b**) represent the box and mean lots, respectively.

3.4. Downregulatory Effects of AS Extracts on Bcl-xL Gene Expressions

The downregulatory effects were progressively dose-dependent with optimal activities observed at 0.50 mg/mL (Figure 4). The treatments exhibited significant modulatory and beneficial properties as required to stimulate downregulation. The downregulatory effects were sufficient to elicit beneficial inhibitive properties (Tables 3 and 4).

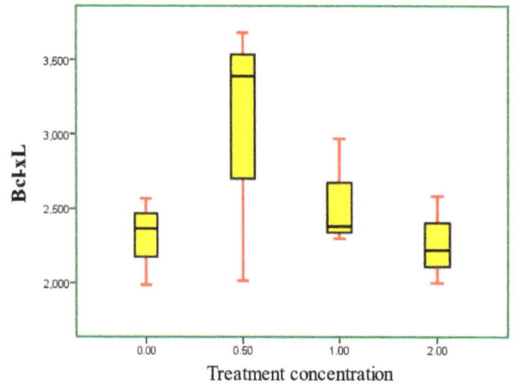
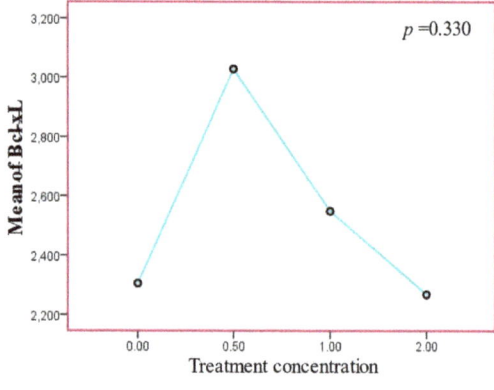

a) Box plot showing the downregulatory properties of the extract on *Bcl-xL*

b) Mean plot showing the statistical expression of *Bcl-xL*

Figure 4. Downregulatory properties recorded in *Bcl-xL* after treatment with AS leaf extracts. Sub-figures (**a,b**) represent the box and mean lots, respectively.

3.5. Downregulatory Effects of AS Extracts on COX-2 Expression at High Concentrations

When the Caco-2 cell lines were treated with methanolic leaf extracts, the *COX-2* genes were modulated variably with an increasing dosage across the concentration gradients. Notably, at a low concentration, the extract products stimulated upregulation of *COX-2* genome expression. However, downregulatory effects were observed at high concentrations (2.0 mg/mL), indicating potential optimal benefits at a high dosage concentration, as exhibited in (Figure 5). In summary, the concentration of AS extract treatment poses a direct impact on molecular activities, resulting in relative changes in gene expression (Figure 6, Tables 3 and 4).

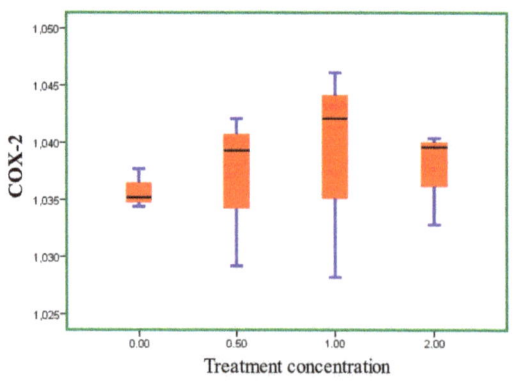
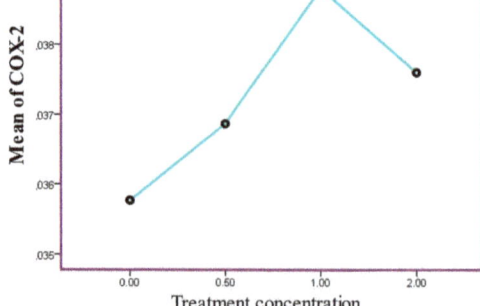

a) Box plot showing the regulatory properties of the extract on *COX-2*

b) Mean plot showing the statistical expression of *COX-2*

Figure 5. Varied regulatory properties observed in *COX-2* after treatment with AS leaf extracts. Sub-figures (**a,b**) represent the box and mean lots, respectively.

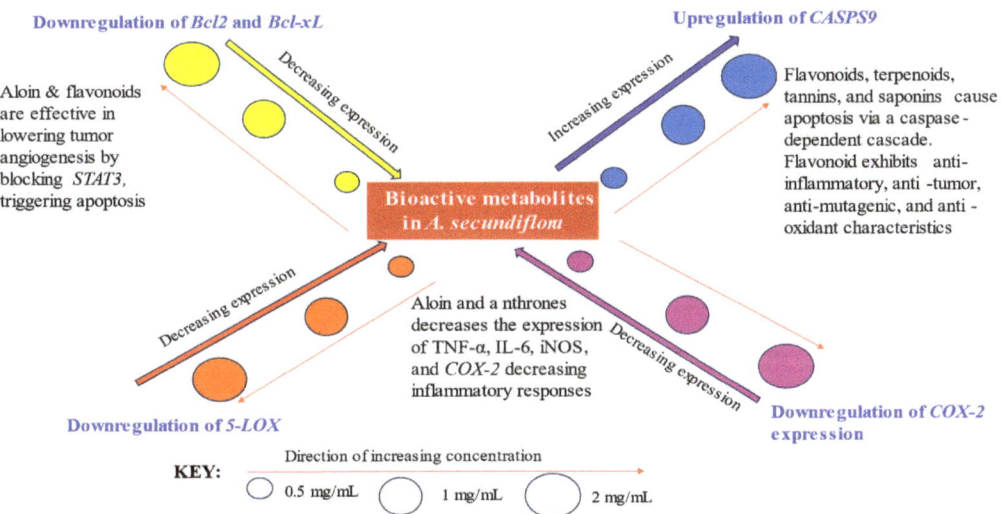

Figure 6. A summarized mechanistic schematic presentation of CASPS9, 5-LOX, Bcl2/Bcl-xL, and COX-2 relative expressions with increasing AS extract concentration.

4. Discussion

4.1. Phytotherapeutic Effects of AS Leaf Extracts on CASPS9 Expression

An essential mediator in the regulation of apoptosis is CASPS9, the starter of the mitochondrial caspase cascade. These enzymes (Caspases) are a component of a cascade that is triggered by pro-apoptotic cues and causes the fragmentation of cells, as well as the dissociation of many peptides. For the aim of medical treatment, it is essential to selectively modulate apoptosis [22,23]. In a variety of biological events, apoptosis is a crucial physiological process that involves the selective elimination of cells [33]. According to some of the research, inhibiting spontaneous apoptosis raises the risk of developing cancer [34,35]. Similarly, it has been noted that a lower rate of apoptosis is strongly associated with a higher frequency of colorectal adenoma [36]. Aloin was shown to significantly decrease tumor sizes and weight in mice xenografts while inducing apoptosis and lowering tumor cell viability in vitro [18]. It was demonstrated that natural plant-based constituents found in AS, such as flavonoids [37], terpenoids, tannins [38], and saponins [39–41], can cause and enhance apoptosis via a caspase-dependent cascade. In addition to their powerful anti-inflammatory, anti-tumor, anti-mutagenic, and antioxidant characteristics, flavonoids also have the capacity to regulate key cellular enzyme functions [42]. The findings of this study showed that methanolic extracts demonstrated high efficacy with significant activity in the upregulation of CASPS9 in a dose-dependent manner. The increased upregulation of CASPS9 resulted in decreased tumor development and, thus, an important breakthrough in CRC growth inhibition. The capacity to cause apoptosis in gastrointestinal epithelial cells is one potential chemo-prevention technique [35]. Cancer incidence has been postulated to increase when spontaneous apoptosis is suppressed [34,35]. Similar findings have been made regarding the relationship between colorectal adenoma prevalence and apoptosis [36]. One of the prospective chemoprevention techniques is the ability to cause apoptosis in epithelial cells with gastrointestinal origin [35]. For this reason, a promising direction for colorectal cancer research is the study of the apoptotic process. Therefore, the level of expression of the apoptosis-associated CASPS9 genes may be helpful in predicting the prognosis for people with stage II colorectal cancer. From our findings, methanol is, thus, highly recommended for use as an efficient extraction solvent for bioactive ingredients in AS with demonstrated upregulatory effects of CASPS9 in CRC treatment. Of great significance, the dose-dependent variation of these genes under the influence of AS leaf extracts has not been investigated in the previous literature. Thus, highlighting the importance of this research, which evaluates and documents the activity of *A. secundiflora's* active metabolites in modulation of the CRC *CASPS9* genes responsible for programmed cell death (apoptosis), for deployment in CRC treatment.

4.2. Phytotherapeutic Effects of AS Leaf Extracts on 5-LOX Expression

Inhibiting the expression of *5-LOX*, which is upregulated in colorectal cancer, could be helpful in both the prevention and therapy of the disease [43]. It has been demonstrated that when arachidonic acid, a polyunsaturated fatty acid, is metabolized by either the COX pathway or the LOX mechanism, eicosanoids such as prostaglandins, thromboxanes, and leukotrienes, among others, operate as potent autocrine and paracrine regulators of cell biology. Only a few of the physiological and pathological responses that these compounds are known to alter include the growth and invasiveness of tumor cells, as well as the suppression of immune surveillance [44]. Targeting arachidonic acid pathways may be useful in delaying the progression of CRC and other types of malignancies, since LOXs enzymes produce metabolites in the arachidonic acid pathway that seem to promote carcinogenesis.

The results of this study demonstrated that downregulatory properties were observed to occur in a dose-dependent manner, with the expressions of *5-LOX* being suppressed with increasing concentration. Although prostaglandins (PGs) and other COX-derived metabolites have received most of the attention, the new research indicates that leukotrienes (LTs) and hydroxyeicosatetraenoic acids (HETEs), two products catalyzed by LOX enzymes, also have a significant biological impact on the initiation and progression of human cancers. In several human cancer cell lines and tissues, including those of the colon [43,45,46], an increase in the expression of *5-LOX* and their metabolites has been found. This over expression was reported to be significantly linked to tumor cell proliferation, resistance to apoptosis, and angiogenesis [47,48]. Additionally, it was discovered that the direct suppression of *5-LOX* or *12-LOX* significantly reduced the development of tumor cells [44,47–49].

Our most important finding was the substantial correlation between extract dose concentration and *5-LOX* expression level. This finding is consistent with other research, which showed that gastric or colorectal [47] tumors with higher *5-LOX* and *COX-2* levels grew deeper and larger. Additionally, it lends credence to the idea that *5-LOX* and *COX-2* share features of expression and function that are proangiogenic and anti-apoptotic [50] as well as substrate preference in human cancer. Like COX-2 and PGs, 5-LOX enzymes and their products may operate on tumor cells by inhibiting apoptosis, increasing cell proliferation, and stimulating angiogenesis, according to a number of experimental investigations [48,50]. The degree of *5-LOX* expression and LTB4 synthesis in cancer cell lines was recently found to be correlated with the dose- and time-dependent reduction in cell viability and induction of apoptosis caused by *5-LOX* inhibitors [48]. A key mechanism of 5-LOX activities on proliferation and apoptosis was the release of endothelial growth factor (VEGF) and mRNA levels by *5-LOX* activity in malignant mesothelial cells [48].

The abundance of biomolecules found in AS, as we have previously described [10], suggests the value of using this plant as a potential anti-CRC agent with the assurance of phytotherapeutic effects. The overwhelming evidence is in congruent support of our findings with regards to the benefits of suppressing the expression of *5-LOX*. *A. secundiflora* (AS) is a promising phytotherapeutic plant, and based on our thorough research, this is among the first studies to assess the levels of expression and clinicopathologic significance of the *5-LOX* gene on human CRC using AS. AS leaf extracts are, therefore, unreservedly perfect natural *5-LOX* inhibitors whose exploitation for therapy is of paramount importance in CRC management. Thus, our findings imply that *5-LOX* overexpression may have a significant impact on the emergence of CRC. Therefore, blocking this metabolic pathway might be a timely and useful therapeutic strategy for both the prevention and treatment of CRC.

4.3. Phytotherapeutic Effects of AS Leaf Extracts on Bcl2 and Bcl-xL Expression

The best-defined protein family involved in the regulation of apoptotic cell death is the Bcl-2 protein family, which comprises members that are both anti- and pro-apoptotic. The anti-apoptotic members of this family include, among others, *Bcl2* and *Bcl-xL*. In the present study, it was exclusively demonstrated that in all treatments, the downregulatory effects of the extracts on *Bcl-xL* and *Bcl2* were progressively influenced dose-dependently. All of the extract treatments exhibited significant downregulatory and beneficial properties as required to stimulate inhibitory properties on genome expression. Other investigators have also asserted that aloin is effective in lowering tumor angiogenesis and growth by blocking *STAT3* activation in CRC cells, which, in turn, controls the expression of the antiapoptotic protein *Bcl-xL* gene [18].

One of the latent self-signaling transcription factors in the cytoplasm is *STAT3* (e.g., VEGF), which is activated by cytokines (such as IL-6) and progenitor cells. The stimulation of STAT3 homodimerization and nuclear translocation modulates the transcription of responsive genes encoding apoptotic-cell-death inhibitors (e.g., *Bcl-xL*, *Bcl2*) and inducers of angiogenesis (e.g., VEGF) [51]. These genes play roles in human defense evasion, angiogenesis, metastatic spread, cell survival, differentiation, and programmed cell death [52]. In recent years, there has been an abundance of research demonstrating that blocking constitutive STAT3 signaling substantially inhibits tumor development and triggers apoptosis [51,53]. By downregulating the biosynthesis of *Bcl2* and *Bcl-xL* through the p53 signaling pathway, a notable apoptotic target in many cancer types, flavonoids diminish a significant dysregulated pathway in cancer [54]. Therefore, it is crucial to realize that Bcl2 and Bcl-xL suppression causes advantageous apoptotic effects, which subsequently reduce CRC cancer growth. Pharmacological manipulation of the Bcl-2 family activities will be limited until a deeper scientific knowledge of how Bcl-2 family proteins control apoptosis in cells is established [55]. This is, therefore, the first study that has successfully evaluated the downregulatory effects of *A. secundiflora*'s extracts on the expression of *Bcl2* and *Bcl-xL* in colorectal cancer cell lines. In line with these fundamental findings, the methanolic leaf extracts of AS are recommended for considerable deployment for further in vivo and subsequent clinical trials for therapeutic management of CRC in humans, with substantial beneficial effects postulated.

4.4. Phytotherapeutic Effects of AS Leaf Extracts on COX-2 Expression

With regards to *COX-2* expression, cancer risk is increased by persistent inflammation [56], while strong induction of *COX-2* occurs during inflammation. This study demonstrates that downregulatory effects were quite minimal across all doses applied but sufficient to elicit beneficial properties. Of significance to note is that one of the crucial enzymes involved in the manufacture of prostaglandins, which is linked to inflammatory processes, is the COX-2 enzyme. Relief from these inflammatory conditions may result from inhibition of the COX-2 enzyme, and consequently, of the prostaglandin biosynthetic pathway [57]. Tumor necrosis factor alpha (TNF-α), interleukin 6 (IL-6), inducible nitric oxide synthase (iNOS), and cyclooxygenase-2 (COX-2) expression are all downregulated by aloin (highly abundant in the *Aloe* species), which has been shown to suppress lipopolysaccharide-induced pro-inflammatory cytokine secretion and nitric oxide production [58]. Phenols such as anthrones that are also abundant in AS were reported to strongly inhibit the expression of *COX-2* [59]. Aloin and aloe-emodin are said to limit the expression of iNOS and COX-2 mRNA, hence, reducing inflammatory responses. The authors further assert that aloe-emodin may be a crucial component behind aloe's anti-inflammatory properties [60]. The phenolic components of the methanolic extract may have contributed to the downregulatory effects of the extracts. Therefore, it was presupposed that the radical scavenging capabilities of naphthalene analogs and the inhibition of *COX-2* by anthraquinones and naphthalene derivatives were the mechanisms by which the inhibitory properties were mediated. Remarkably, this is one of the first studies that has demonstrated an explicit link between AS leaf extracts and *COX-2* downregulation in CRC cell lines, with prospective application for CRC management.

5. Conclusions

In this remarkable study, CRC cell lines were exposed to various AS methanolic extract doses in vitro. In our investigation, AS's leaf extracts were arduously proven and listed as being incredibly effective at modulating the activities of *CASPS9, 5-LOX, Bcl2, Bcl-xL,* and *COX-2* genes for pharmacotherapeutic beneficial effects. Methanol was also demonstrated to be a promising extraction solvent for the effective metabolites responsible for the modulatory properties discussed in our study. Given the copious amounts of phytoconstituent biomolecules present in AS, it is beneficial to use the plant as a prospective CRC inhibiting agent with potent phytotherapeutic benefits. To treat CRC, targeted apoptotic modulation necessitates an understanding of genome programming. The regulatory effects exhibited in this research establish AS as an important plant of choice, one whose extracts have demonstrated significant CRC inhibitory properties. This is one of the initial studies that explicitly hyperlinks the positive effects of AS leaf extracts on the targeted genes in our study for potential applications for therapeutic management of CRC. We, therefore, strongly recommend the utilization of AS for further in vivo and subsequent clinical trials for CRC management. In addition, we suggest the application of methanol as a promising AS extraction solvent for maximum regulatory benefits towards *CASPS9, 5-LOX, Bcl2, Bcl-xL,* and *COX-2* expression. Finally, we also encourage further investigation into the particular AS metabolites involved in the modulatory pathways that inhibit the development of CRC, as well as potential metastases.

Author Contributions: Conceptualization, J.M.M.; data curation, R.W.M. and Z.K.; formal analysis, J.M.M., N.R. and M.P.; funding acquisition, J.M.M. and B.L.R.; investigation, R.W.M., I.N.W. and Z.K.; methodology, J.M.M., T.V. and M.P.; project administration, B.L.R.; supervision, Z.K. and B.L.R.; validation, N.R., I.N.W. and B.L.R.; visualization, T.V., R.W.M., I.N.W. and M.P.; writing—original draft, J.M.M., T.V. and M.P.; writing—review and editing, T.V., Z.K., N.R. and B.L.R. All authors have read and agreed to the published version of the manuscript.

Funding: Our work was funded by the Tempus Public Foundation, through the Doctoral School of Health Sciences, University of Pecs. In addition, the APC was fully paid by the University of Pécs.

Institutional Review Board Statement: Not applicable.

Informed Consent Statement: Not applicable.

Data Availability Statement: The datasets generated and/or analyzed during the current study are available from the corresponding author on reasonable request.

Acknowledgments: We express our gratitude to Stipendium Hungaricum Programme/Tempus Public Foundation for the study and research grant awarded to John M. Macharia, Doctoral School of Health Sciences, University of Pecs, Hungary. We are also grateful to Ruth W. Mwangi, Hungarian University of Agriculture and Life Sciences, Institute of Horticultural Sciences, Budapest, Hungary for collection and identification of the plant species.

Conflicts of Interest: The authors declare no conflict of interest.

References

1. Arnold, C.N.; Goel, A.; Blum, H.E.; Boland, C.R. Molecular pathogenesis of colorectal cancer: Implications for molecular diagnosis. *Cancer* **2005**, *104*, 2035–2047. [CrossRef] [PubMed]
2. Halder, S.; Modak, P.; Sarkar, B.K.; Das, A.; Sarkar, A.P.; Chowdhury, A.R.; Kundu, S.K. Traditionally Used Medicinal Plants with Anticancer Effect: A Review. *Int. J. Pharm. Sci. Rev. Res.* **2020**, *65*, 1–13. [CrossRef]
3. Colussi, D.; Brandi, G.; Bazzoli, F.; Ricciardiello, L. Molecular Pathways Involved in Colorectal Cancer: Implications for Disease Behavior and Prevention. *Int. J. Mol. Sci.* **2013**, *14*, 16365–16385. [CrossRef]
4. Carvalho, C.; Marinho, A.; Leal, B.; Bettencourt, A.; Boleixa, D.; Almeida, I.; Farinha, F.; Costa, P.P.; Vasconcelos, C.; Silva, B.M. Association between vitamin D receptor (*VDR*) gene polymorphisms and systemic lupus erythematosus in Portuguese patients. *Lupus* **2015**, *24*, 846–853. [CrossRef] [PubMed]

5. Macharia, J.M.; Kaposztas, Z.; Varjas, T.; Budán, F.; Zand, A.; Bodnar, I.; Bence, R.L. Targeted lactate dehydrogenase genes silencing in probiotic lactic acid bacteria: A possible paradigm shift in colorectal cancer treatment? *Biomed. Pharmacother.* **2023**, *160*, 114371. [CrossRef] [PubMed]
6. Hashemzaei, M.; Delarami Far, A.; Yari, A.; Heravi, R.E.; Tabrizian, K.; Taghdisi, S.M.; Sadegh, S.E.; Tsarouhas, K.; Kouretas, D.; Tzanakakis, G.; et al. Anticancer and apoptosis-inducing effects of quercetin in vitro and in vivo. *Oncol. Rep.* **2017**, *38*, 819–828. [CrossRef] [PubMed]
7. Macharia, J.M.; Mwangi, R.W.; Rozmann, N.; Zsolt, K.; Varjas, T.; Uchechukwu, P.O.; Wagara, I.N.; Raposa, B.L. Medicinal plants with anti-colorectal cancer bioactive compounds: Potential game-changers in colorectal cancer management. *Biomed. Pharmacother.* **2022**, *153*, 113383. [CrossRef]
8. Macharia, J.M.; Káposztás, Z.; Bence, R.L. Medicinal Characteristics of *Withania somnifera* L. in Colorectal Cancer Management. *Pharmaceuticals* **2023**, *16*, 915. [CrossRef]
9. Kaur, R.; Kaur, H. The Antimicrobial activity of essential oil and plant extracts of *Woodfordia fruticosa*. *Arch. Appl. Sci. Res.* **2011**, *2*, 373–383.
10. Macharia, J.M.; Ngure, V.; Emődy, B.; Király, B.; Káposztás, Z.; Rozmann, N.; Erdélyi, A.; Raposa, B. Pharmacotherapeutic Potential of *Aloe secundiflora* against Colorectal Cancer Growth and Proliferation. *Pharmaceutics* **2023**, *15*, 1558. [CrossRef]
11. Diriba, T.F.; Deresa, E.M. Botanical description, ethnomedicinal uses, phytochemistry, and pharmacological activities of genus Kniphofia and Aloe: A review. *Arab. J. Chem.* **2022**, *15*, 104111. [CrossRef]
12. Kaingu, F.; Kibor, A.; Waihenya, R.; Shivairo, R.; Mungai, L. Efficacy of Aloe secundiflora Crude Extracts on *Ascaridia galli* in Vitro. *Sustain. Agric. Res.* **2012**, *2*, 49. [CrossRef]
13. Puia, A.; Puia, C.; Moiş, E.; Graur, F.; Fetti, A.; Florea, M. The phytochemical constituents and therapeutic uses of genus Aloe: A review. *Not. Bot. Horti Agrobot. Cluj-Napoca* **2021**, *49*, 12332. [CrossRef]
14. Okemo, P.; Kirimuhuzya, C.; Otieno, J.N.; Magadula, J.J.; Mariita, R.M.; Orodho, J. Methanolic extracts of Aloe secundiflora Engl. inhibits in vitro growth of tuberculosis and diarrhea-causing bacteria. *Pharmacogn. Res.* **2011**, *3*, 95–99. [CrossRef] [PubMed]
15. Induli, M.; Cheloti, M.; Wasuna, A.; Wekesa, I.; Wanjohi, J.M.; Byamukama, R.; Heydenrich, M.; Makayoto, M.; Yenesew, A. Naphthoquinones from the roots of Aloe secundiflora. *Phytochem. Lett.* **2012**, *5*, 506–509. [CrossRef]
16. Simpson, D.; Amos, S. Other Plant Metabolites. *Pharmacognosy* **2017**, 267–280. [CrossRef]
17. Patel, K.; Patel, D.K. Medicinal importance, pharmacological activities, and analytical aspects of aloin: A concise report. *J. Acute Dis.* **2013**, *2*, 262–269. [CrossRef]
18. Pan, Q.; Pan, H.; Lou, H.; Xu, Y.; Tian, L. Inhibition of the angiogenesis and growth of Aloin in human colorectal cancer in vitro and in vivo. *Cancer Cell Int.* **2013**, *13*, 1–9. [CrossRef]
19. Cardarelli, M.; Rouphael, Y.; Pellizzoni, M.; Colla, G.; Lucini, L. Profile of bioactive secondary metabolites and antioxidant capacity of leaf exudates from eighteen Aloe species. *Ind. Crop. Prod.* **2017**, *108*, 44–51. [CrossRef]
20. Rachuonyo, H.O.; Ogola, P.E.; Arika, W.M.; Wambani, J.R. Efficacy of Crude Leaf Extracts of Aloe secundiflora on Selected Enteric Bacterial Pathogens and Candida albicans. *J. Antimicrob. Agents* **2016**, *2*, 2. [CrossRef]
21. Waithaka, P.N.; Gathuru, E.M.; Githaiga, B.M.; Kazungu, R.Z. Antimicrobial Properties of Aloe vera, Aloe volkensii and Aloe secundiflora from Egerton University. *Acta Sci. Microbiol.* **2018**, *1*, 6–10.
22. Thornberry, N.A.; Lazebnik, Y. Caspases: Enemies within. *Science* **1998**, *281*, 1312–1316. [CrossRef] [PubMed]
23. Raff, M. Cell suicide for beginners. *Nature* **1998**, *396*, 119–122. [CrossRef] [PubMed]
24. Prabhu, A. Anti-angiogenic, apoptotic and matrix metalloproteinase inhibitory activity of Withania somnifera (ashwagandha) on lung adenocarcinoma cells. *Phytomedicine* **2021**, *90*, 153639. [CrossRef] [PubMed]
25. Ye, Y.N.; Wu, W.K.K.; Shin, V.Y.; Bruce, I.C.; Wong, B.C.Y.; Cho, C.H. Dual inhibition of 5-LOX and COX-2 suppresses colon cancer formation promoted by cigarette smoke. *Carcinogenesis* **2005**, *26*, 827–834. [CrossRef] [PubMed]
26. Ding, X.; Zhu, C.; Qiang, H.; Zhou, X.; Zhou, G. Enhancing antitumor effects in pancreatic cancer cells by combined use of COX-2 and 5-LOX inhibitors. *Biomed. Pharmacother.* **2011**, *65*, 486–490. [CrossRef]
27. Rådmark, O.; Shimizu, T.; Jörnvall, H.; Samuelsson, B. Leukotriene A4 hydrolase in human leukocytes. Purification and properties. *J. Biol. Chem.* **1984**, *259*, 12339–12345. [CrossRef]
28. American Type Culture Collection. Caco-2 [Caco2]—HTB-37 | ATCC 2022. Available online: https://www.atcc.org/products/htb-37 (accessed on 17 May 2023).
29. Nelson, V.K.; Sahoo, N.K.; Sahu, M.; Sudhan, H.H.; Pullaiah, C.P.; Muralikrishna, K.S. In vitro anticancer activity of Eclipta alba whole plant extract on colon cancer cell HCT-116. *BMC Complement. Med. Ther.* **2020**, *20*, 1–8. [CrossRef]
30. Saravanakumar, D.S.D.; Karthiba, L.K.L.; Ramjegathesh, R.R.R.; Prabakar, K.P.K.; Raguchander, T.R.T. Characterization of Bioactive Compounds from Botanicals for the Management of Plant Diseases. In *Sustainable Crop Disease Management Using Natural Products*; CABI: Wallingford, UK, 2015; pp. 1–18. [CrossRef]
31. Sangweni, N.F.; Dludla, P.V.; Chellan, N.; Mabasa, L.; Sharma, J.R.; Johnson, R. The Implication of Low Dose Dimethyl Sulfoxide on Mitochondrial Function and Oxidative Damage in Cultured Cardiac and Cancer Cells. *Molecules* **2021**, *26*, 7305. [CrossRef]
32. Doak, S.H.; Zaïr, Z.M. Real-Time Reverse-Transcription Polymerase Chain Reaction: Technical Considerations for Gene Expression Analysis. *Methods Mol. Biol.* **2012**, *817*, 251–270. [CrossRef]

33. Morimoto, Y.; Takada, K.; Takeuchi, O.; Watanabe, K.; Hirohara, M.; Hamamoto, T.; Masuda, Y. Bcl-2/Bcl-xL inhibitor navitoclax increases the antitumor effect of Chk1 inhibitor prexasertib by inducing apoptosis in pancreatic cancer cells via inhibition of Bcl-xL but not Bcl-2. *Mol. Cell. Biochem.* **2020**, *472*, 187–198. [CrossRef] [PubMed]
34. Sjöström, J.; Bergh, J. How apoptosis is regulated, and what goes wrong in cancer. *BMJ* **2001**, *322*, 1538–1539. [CrossRef] [PubMed]
35. Wong, B.C.Y.; Zhu, G.H.; Lam, S.K. Aspirin induced apoptosis in gastric cancer cells. *Biomed. Pharmacother.* **1999**, *53*, 315–318. [CrossRef] [PubMed]
36. Martin, C.; Connelly, A.; Keku, T.O.; Mountcastle, S.B.; Galanko, J.; Woosley, J.T.; Schliebe, B.; Lund, P.; Sandler, R.S. Nonsteroidal anti-inflammatory drugs, apoptosis, and colorectal adenomas. *Gastroenterology* **2002**, *123*, 1770–1777. [CrossRef]
37. Yang, H.; Dou, Q.P. Targeting apoptosis pathway with natural terpenoids: Implications for treatment of breast and prostate cancer. *Curr. Drug Targets* **2010**, *11*, 733–744. [CrossRef]
38. Larrosa, M.; Tomás-Barberán, F.A.; Espín, J.C. The dietary hydrolysable tannin punicalagin releases ellagic acid that induces apoptosis in human colon adenocarcinoma Caco-2 cells by using the mitochondrial pathway. *J. Nutr. Biochem.* **2006**, *17*, 611–625. [CrossRef]
39. Jiang, X.-P.; Jin, S.; Shao, W.; Zhu, L.; Yan, S.; Lu, J. Saponins of Marsdenia Tenacissima promotes apoptosis of hepatocellular carcinoma cells through damaging mitochondria then activating cytochrome C/Caspase-9/Caspase-3 pathway. *J. Cancer* **2022**, *13*, 2855–2862. [CrossRef]
40. Elekofehinti, O.O.; Iwaloye, O.; Olawale, F.; Ariyo, E.O. Saponins in Cancer Treatment: Current Progress and Future Prospects. *Pathophysiology* **2021**, *28*, 250–272. [CrossRef]
41. Cheng, Y.; He, W.; He, Y. Gleditsia Saponin C Induces A549 Cell Apoptosis via Caspase-Dependent Cascade and Suppresses Tumor Growth on Xenografts Tumor Animal Model. *Front. Pharmacol.* **2018**, *8*, 988. [CrossRef]
42. Vetrivel, P.; Kim, S.M.; Saralamma, V.V.G.; Ha, S.E.; Kim, E.H.; Min, T.S.; Kim, G.S. Function of flavonoids on different types of programmed cell death and its mechanism: A review. *J. Biomed. Res.* **2019**, *33*, 363–370. [CrossRef]
43. Soumaoro, L.T.; Iida, S.; Uetake, H.; Ishiguro, M.; Takagi, Y.; Higuchi, T.; Yasuno, M.; Enomoto, M.; Sugihara, K. Expression of 5-Lipoxygenase in human colorectal cancer. *World J. Gastroenterol.* **2006**, *12*, 6355–6360. [CrossRef] [PubMed]
44. Ghosh, J.; Myers, C.E. Inhibition of arachidonate 5-lipoxygenase triggers massive apoptosis in human prostate cancer cells. *Proc. Natl. Acad. Sci. USA* **1998**, *95*, 13182–13187. [CrossRef]
45. Öhd, J.F.; Nielsen, C.K.; Campbell, J.; Landberg, G.; Löfberg, H.; Sjölander, A. Expression of the leukotriene D4 receptor CysLT1, COX-2, and other cell survival factors in colorectal adenocarcinomas. *Gastroenterology* **2003**, *124*, 57–70. [CrossRef] [PubMed]
46. Gupta, S.; Srivastava, M.; Ahmad, N.; Sakamoto, K.; Bostwick, D.G.; Mukhtar, H. Lipoxygenase-5 is overexpressed in prostate adenocarcinoma. *Cancer* **2001**, *91*, 737–743. [CrossRef] [PubMed]
47. Avis, I.; Hong, S.H.; Martínez, A.; Moody, T.; Choi, Y.H.; Trepel, J.; Das, R.; Jett, M.; Mulshine, J.L. Five-lipoxygenase inhibitors can mediate apoptosis in human breast cancer cell lines through complex eicosanoid interactions. *FASEB J.* **2001**, *15*, 2007–2009. [CrossRef]
48. Hoque, A.; Lippman, S.M.; Wu, T.T.; Xu, Y.; Liang, Z.D.; Swisher, S. Increased 5-lipoxygenase expression and induction of apoptosis by its inhibitors in esophageal cancer: A potential target for prevention. *Carcinogenesis* **2005**, *26*, 785–791. [CrossRef]
49. Romano, M.; Clària, J. Cyclooxygenase-2 and 5-lipoxygenase converging functions on cell proliferation and tumor angiogenesis: Implications for cancer therapy. *FASEB J.* **2003**, *17*, 1986–1995. [CrossRef]
50. Romano, M.; Catalano, A.; Nutini, M.; D'Urbano, E.; Crescenzi, C.; Claria, J.; Libner, R.; Davi, G.; Procopio, A. 5-Lipoxygenase regulates malignant mesothelial cell survival: Involvement of vascular endothelial growth factor. *FASEB J.* **2001**, *15*, 2326–2336. [CrossRef]
51. Gamero, A.M.; Young, H.A.; Wiltrout, R.H. Inactivation of Stat3 in tumor cells: Releasing a brake on immune responses against cancer? *Cancer Cell* **2004**, *5*, 111–112. [CrossRef]
52. Johnston, P.A.; Grandis, J.R. STAT3 SIGNALING: Anticancer Strategies and Challenges. *Mol. Interv.* **2011**, *11*, 18–26. [CrossRef]
53. Chen, J.; Wang, J.; Lin, L.; He, L.; Wu, Y.; Zhang, L.; Yi, Z.; Chen, Y.; Pang, X.; Liu, M. Inhibition of STAT3 Signaling Pathway by Nitidine Chloride Suppressed the Angiogenesis and Growth of Human Gastric Cancer. *Mol. Cancer Ther.* **2012**, *11*, 277–287. [CrossRef] [PubMed]
54. Rahman, N.; Khan, H.; Zia, A.; Khan, A.; Fakhri, S.; Aschner, M.; Gul, K.; Saso, L. Bcl-2 Modulation in p53 Signaling Pathway by Flavonoids: A Potential Strategy towards the Treatment of Cancer. *Int. J. Mol. Sci.* **2021**, *22*, 11315. [CrossRef] [PubMed]
55. Macharia, J.M.; Mwangi, R.W.; Szabó, I.; Zand, A.; Kaposztas, Z.; Varjas, T.; Rozmann, N.; Raposa, B.L. Regulatory activities of Warbugia ugandensis ethanolic extracts on colorectal cancer-specific genome expression dose-dependently. *Biomed. Pharmacother.* **2023**, *166*, 115325. [CrossRef] [PubMed]
56. Ezziyyani, M. Advances in Intelligent Systems and Computing 1103 Advanced Intelligent Systems for Sustainable Development (AI2SD'2019) Volume 2-Advanced Intelligent Systems for Sustainable Development Applied to Agriculture and Health, Conference Proceeding Held in Marrakech, Morocco, from 8 to 11 July 2019. Available online: https://link.springer.com/conference/aisd (accessed on 18 May 2023).
57. Lindsey, K.L.; Jäger, A.K.; Viljoen, A.M.; van Wyk, B.-E. Cyclooxygenase inhibitory activity of Aloe species. *S. Afr. J. Bot.* **2002**, *68*, 47–50. [CrossRef]
58. Luo, X.; Zhang, H.; Wei, X.; Shi, M.; Fan, P.; Xie, W.; Zhang, Y.; Xu, N. Aloin Suppresses Lipopolysaccharide-Induced Inflammatory Response and Apoptosis by Inhibiting the Activation of NF-κB. *Molecules* **2018**, *23*, 517. [CrossRef]

59. Capes, I.C.; Universidade, U.; Lib, D. Copyrighted I. Inhibitory, and Free Radical Scavenging Effects of Rumex nepalensis. *Planta Medicaica* **2010**, 1564–1569.
60. Park, M.-Y.; Kwon, H.-J.; Sung, M.-K. Evaluation of Aloin and Aloe-Emodin as Anti-Inflammatory Agents in Aloe by Using Murine Macrophages. *Biosci. Biotechnol. Biochem.* **2009**, *73*, 828–832. [CrossRef]

Disclaimer/Publisher's Note: The statements, opinions and data contained in all publications are solely those of the individual author(s) and contributor(s) and not of MDPI and/or the editor(s). MDPI and/or the editor(s) disclaim responsibility for any injury to people or property resulting from any ideas, methods, instructions or products referred to in the content.

Article

New Insights on the Progesterone (P4) and PGRMC1/NENF Complex Interactions in Colorectal Cancer Progression

Joanna Kamińska [1,*,†], Olga Martyna Koper-Lenkiewicz [1], Donata Ponikwicka-Tyszko [2], Weronika Lebiedzińska [3], Ewelina Palak [2], Maria Sztachelska [2], Piotr Bernaczyk [4], Justyna Dorf [1], Katarzyna Guzińska-Ustymowicz [5], Konrad Zaręba [6], Sławomir Wołczyński [3], Nafis Ahmed Rahman [3,7] and Violetta Dymicka-Piekarska [1,*,†]

[1] Department of Clinical Laboratory Diagnostics, Medical University of Bialystok, Waszyngtona 15A, 15-269 Bialystok, Poland; o.koper@wp.pl (O.M.K.-L.); justyna.dorf@umb.edu.pl (J.D.)
[2] Department of Biology and Pathology of Human Reproduction, Institute of Animal Reproduction and Food Research, Polish Academy of Sciences, 10-748 Olsztyn, Poland; d.ponikwicka-tyszko@pan.olsztyn.pl (D.P.-T.); e.palak@pan.olsztyn.pl (E.P.); maria.sztachelska@gmail.com (M.S.)
[3] Department of Reproduction and Gynecological Endocrinology, Medical University of Bialystok, 15-269 Bialystok, Poland; weronika.lebiedzinska@umb.edu.pl (W.L.); slawek.wolczynski@gmail.com (S.W.)
[4] Department of Medical Pathomorphology, Medical University of Bialystok, 15-269 Bialystok, Poland; piotr.bernaczyk@umwb.edu.pl
[5] Department of General Pathomorphology, Medical University of Bialystok, 15-269 Bialystok, Poland; katarzyna.guzinska-ustymowicz@umb.edu.pl
[6] 2nd Clinical Department of General and Gastroenterological Surgery, Medical University of Bialystok, 15-094 Bialystok, Poland; konrad.zareba@umb.edu.pl
[7] Institute of Biomedicine, University of Turku, 20014 Turku, Finland; nafis.rahman@utu.fi
* Correspondence: joanna.kaminska@umb.edu.pl (J.K.); piekarskav@yahoo.com (V.D.-P.)
† These authors contributed equally to this work.

Simple Summary: Progesterone (P4) via PGRMC1/NENF may stimulate the proliferation and invasion of colorectal cancer DLD-1 and HT-29 cells. PGRMC1 inhibition abolishes the effect of P4, suggesting that P4 in advanced colorectal cancer may act primarily through PGRMC1. Our data may provide the novel insights into the action of P4, PGRMC1, and NENF in colorectal cancer. It seems that PGRMC1 and NENF may interact as possible cofactors in non-classical P4 signaling. Targeting the PGRMC1/NENF complex may open-up new therapeutic possibilities for patients with advanced colorectal cancer. Therefore, future studies aimed at developing treatment strategies for colorectal cancer could consider simultaneous PGRMC1 inhibition along with a blockage of NENF production and secretion.

Abstract: The literature data regarding the risk of colorectal cancer (CRC) in the context of hormone therapy (HT), including both estrogen–progestogen combinations and estrogen alone, are inconclusive. The precise relationship underlying the action of progesterone (P4) and progesterone receptors in CRC has yet to be determined. We characterized the expression profiles of both nuclear and membrane progesterone receptors and their potential cofactors in CRC tissues. Additionally, we analyzed the P4 and NENF treatment effects on the cell proliferation and invasion of DLD-1 and HT-29 colorectal cancer cells. We observed a weak expression of the nuclear P4 receptor (PGR), but an abundant expression of the P4 receptor membrane component 1 (PGRMC1) and neuron-derived neurotrophic factor (NENF) in the CRC tissues. P4 treatment stimulated the proliferation of the DLD-1 and HT-29 CRC cells. The co-treatment of P4 and NENF significantly increased the invasiveness of the DLD-1 and HT-29 cells. A functional analysis revealed that these effects were dependent on PGRMC1. AN immunocytochemical analysis demonstrated a cytoplasmic co-localization of PGRMC1 and NENF in the CRC cells. Moreover, the concentration of serum NENF was significantly higher in CRC patients, and P4 treatment significantly increased the release of NENF in the DLD-1 cells. P4 or NENF treatment also significantly increased the IL-8 release in the DLD-1 cells. Our data may provide novel insights into the action of P4 and PGRMC1/NENF in CRC progression, where NENF may act as a potential PGRMC1 co-activator in non-classical P4 signaling. Furthermore, NENF, as a secreted

Citation: Kamińska, J.; Koper-Lenkiewicz, O.M.; Ponikwicka-Tyszko, D.; Lebiedzińska, W.; Palak, E.; Sztachelska, M.; Bernaczyk, P.; Dorf, J.; Guzińska-Ustymowicz, K.; Zaręba, K.; et al. New Insights on the Progesterone (P4) and PGRMC1/NENF Complex Interactions in Colorectal Cancer Progression. *Cancers* **2023**, *15*, 5074. https://doi.org/10.3390/cancers15205074

Academic Editors: Shihori Tanabe and Susanne Merkel

Received: 10 August 2023
Revised: 12 October 2023
Accepted: 16 October 2023
Published: 20 October 2023

Copyright: © 2023 by the authors. Licensee MDPI, Basel, Switzerland. This article is an open access article distributed under the terms and conditions of the Creative Commons Attribution (CC BY) license (https://creativecommons.org/licenses/by/4.0/).

protein, potentially could serve as a promising circulating biomarker candidate for distinguishing between colorectal cancer patients and healthy individuals, although large-scale extensive studies are needed to establish this.

Keywords: biomarkers; colorectal cancer; neudesin (NENF); progesterone (P4); progesterone receptor membrane components 1 (PGRMC1)

1. Introduction

Colorectal cancer (CRC) is the third most common cancer worldwide, at an advanced stage with a 25% higher incidence rate in males than females [1–3]. Due to recurrence and distant metastasis [2], the mortality rate for colorectal cancer patients is very high [2,3], and tends to 30% and 40% in females and males, respectively [1]. Therefore, there is still a need to better investigate potential biomarkers for CRC diagnosis, as well as for the evaluation of the disease advancement, prognosis, and choice of rational therapeutic targets for personalized cancer treatment [4].

Progesterone (P4), an endogenous 21-carbon steroid hormone synthesized from cholesterol, is mainly produced by the corpus luteum and by the placenta during pregnancy. To a lesser extent, progesterone is also produced by the adrenal cortex, Leydig cells of the testes in men, adipose, and other tissues [5]. In addition to its reproductive importance in females, progesterone acts through multiple pathways, regulating important processes, e.g., brain development in fetuses, neuroprotection and myelin regeneration, immune response, and the proliferation and migration of various cancer cells in both genders [3,6,7]. P4 signals may be mediated by classical genomic or non-genomic action [7]. The classical P4 effect is dependent on the P4 interaction with the specific nuclear progesterone receptor (PGR) [8]. Rapid non-classical signaling is mediated by membrane P4 receptors (mPRα, mPRβ, mPRγ, mPRδ, and mPRε) and membrane-associated P4 receptors (MAPR), progesterone receptor membrane components 1 and 2 (PGRMC1 and PGRMC2) [9–11]. Among other MAPR proteins, important ones include neudesin (NENF, neuron-derived neurotrophic factor) and neuferricin (CYB5D2) [12]. *NENF* expression has been demonstrated in neurons and various peripheral tissues, such as the lungs, kidneys, and heart. NENF potentially promotes neuronal survival and differentiation by activating the MAPK (mitogen-activated protein kinase) and PI3K/AKT (phosphatidylinositol 3-kinase/protein kinase) pathways [13]. However, the distinct role of NENF in peripheral tissues remains unclear [14]. Recently, NENF has also been investigated as a molecule involved in the tumorigenesis of primary breast tumors, as well as in other human carcinomas of the uterine cervix, malignant lymphoma, colon, lung, skin, liver, and leukemia [15–17]. Our previous study showed elevated concentrations of NENF in the cerebrospinal fluid of patients with astrocytic brain tumors compared to non-tumoral controls, suggesting NENF as a circulating biomarker for brain tumors [17].

It is widely known that P4 plays a pivotal role in the development of breast, ovarian, and brain cancer [18]. Recent studies have suggested that steroid hormones may also affect CRC development, prognosis, and treatment [2]. Due to the increased morbidity and mortality rates for CRC and lack of specific CRC biomarkers [4], it seems crucial to identify the molecular mechanisms that promote CRC growth and metastasis. In recent years, targeted therapies for CRC seem to be promising treatment options [19]. Previous studies have confirmed the effectiveness of anti-progesterone receptor drugs in the treatment of breast, ovarian, lung, and head and neck cancers [20–22]. However, the mechanism of P4 action on progesterone receptors in CRC has not been well studied [2,3,23–28]. In the present study, we characterized the expression profiles of the nuclear and membrane P4 receptors and their cofactors in advanced colorectal cancers and investigated the potential molecular mechanism underlying the P4 action on CRC cell tissues and cell lines. Moreover, we evaluated NENF as a potential CRC biomarker.

2. Materials and Methods

2.1. Human Samples

All the samples were obtained from patients with primary colorectal cancer, who underwent surgical treatment at the 2nd Clinical Department of General and Gastroenterological Surgery of the Medical Clinical Hospital in Bialystok, Poland. Tissue samples for immunohistochemistry (IHC) were preserved in 4% formalin and, for a gene expression analysis, were preserved in snap frozen and stored at $-80\,°C$. Human CRC tissues ($n = 20$; 14 males, 6 females, median age 68) and normal mucosa tissues ($n = 10$; 6 males, 4 females, median age 66) were histologically examined to prove the tumor grade at the Department of Medical Pathomorphology of the Medical University of Bialystok, Bialystok, Poland. Based on their symptoms, medical history, radiological, colonoscopy, and histological examination results, the CRC patients were retrospectively included. Histologically, CRC was classified as grade G2, intermediate grade ($n = 8$; 5 males, 3 females), and G3 high grade ($n = 12$; 9 males, 3 females). Blood samples from the CRC patients ($n = 41$, 27 males, 14 females, median age 69) were collected 1 day before surgery. The control group was composed of healthy volunteers, age- and sex-matched to the study group ($n = 15$; 11 males, 4 females, median age 67. The exclusion criteria encompassed other neoplasia and receiving chemo- or radiotherapy before surgery.

2.2. Cell Cultures

Colorectal adenocarcinoma cell lines DLD-1 (CCL-221) and HT-29 (HTB-38), which differ in their resistance to anticancer treatment, were purchased from American Type Culture Collection, ATCC (Rockville, MD, USA). The DLD-1 cells were cultured in RPMI medium (RPMI 1640 Medium, no phenol red, Gibco™, catalog #: 11835030, Life Technologies Corporation, Grand Island, NE, USA) and the HT-29 cells were cultured in McCoy medium (McCoy's 5A (Modified) Medium, GlutaMAX™ Supplement, Gibco™, catalog #: 36600021, Life Technologies Ltd., Paisley, UK), supplemented with 10% fetal bovine serum (FBS; Biochrom, Berlin, Germany), 100 units/mL of penicillin, and 100 µg/mL of streptomycin (P/S solution; Sigma-Aldrich) at 37 °C in a humidified atmosphere in the presence of 5% CO_2. The cells were treated with P4 (1 µM) and AG-205 (1 µM). The dose of AG-205 was determined based on our previous studies [29,30] and NENF (1 ng/mL) in stimulation medium. Three independent experiments per cell line were run, and each performed cell plating was performed in triplicates for RNA isolation and medium collection.

2.3. Drugs and Inhibitors

Progesterone (P4) and PGRMC1 inhibitor (AG-205) were obtained from Sigma-Aldrich (Saint Louis, MO, USA; catalog #: P8783-25G and A1487, respectively). Recombinant human Neudesin (NENF) was obtained from R&D Systems Europe Ltd., (Abingdon, UK; catalog #: 6714-ND-050).

2.4. Immunohistochemical Staining

The human CRC tissues and NM tissues were fixed in paraformaldehyde and embedded in paraffin. Immunohistochemical staining was carried out manually, as previously described [29,30]. Histological assessments were performed on 5 µm thick hematoxylin-eosin-stained sections. For immunohistochemistry, sections were deparaffinized, hydrated, and boiled in 10 mM of citric acid buffer (pH 6.0) in a retriever for 2.5 h. Tissue sections were incubated with blocking solutions (10% normal goat serum (NGS) with 3% bovine serum albumin (BSA) or only 3% BSA in PBS) for 1 h at room temperature to reduce non-specific background staining. Then, sections were incubated overnight at 4 °C with the primary antibodies for PGR (MA5-12658, Thermo Fisher Scientific Inc., Waltham, MA, USA; dilution 1:700), mPRα (ab75508, Abcam, Cambridge, UK; dilution 1:500), mPRβ (ab46534, Abcam; dilution 1:500), mPRγ (ab79517, Abcam; Cambridge, UK; dilution 1:500), PGRMC1 (PAB20135, Abnova Corporation, Taipei, Taiwan; dilution 1:300), PGRMC2 (ab125122, Abcam; Cambridge, UK; dilution 1:500), SERBP1 (ab28481, Abcam; Cambridge, UK; dilution 1:700), NENF (MAB6714, R&D Systems Europe Ltd. Abingdon, UK; dilution 25 µg), IgG (ab190475, Abcam; Cambridge, UK; dilution 1:700), and IgG2a (ab190463, Abcam; Cambridge, UK; dilution 1:500). After endogenous peroxidase blocking (0.5% H_2O_2 in PBS for 20 min in dark at room temperature), the primary antibodies were linked with Envision® anti-mouse or anti-rabbit polymer + HRP (Dako, Glostrup, Denmark) for 30 min at room temperature. The reaction product was visualized using 3'3-diaminobenzidine tetrahydrochloride (DAB, Dako, Glostrup, Denmark). Each step was followed by three washings using PBS with 0.05% Tween 20 (PBS-T). After staining with hematoxylin, the sections were dehydrated through ascending ethanol concentrations and cleared using xylene. They were then mounted with Pertex (Histolab Products AB, Spånga, Sweden).

2.5. ImageJ Analysis

The intensity of the staining was determined by measuring the optical density of the reaction product, which was analyzed using Fiji Software 1.8.0_172 (Fiji Is Just ImageJ). Six random areas from each section were quantified and the average optical density (OD) was calculated for each of these areas.

2.6. Immunocytochemical Staining

To minimize autofluorescence in dual staining, tissues were treated with 100 mM of NH4Cl for 10 min. The blocking solution, a mixture of 5% NGS and 1% BSA in PBST, was then applied for 1 h at room temperature. After blocking unspecific binding sites with 3% BSA in PBS with 0.05% Tween 20 for 30 min, the tissue slides were incubated with primary antibodies for PGRMC1 (PAB20135 from Abnova Corporation, Taipei, Taiwan; dilution 1:300) and NENF (MAB6714 from R&D Systems Europe Ltd., Abingdon, UK; dilution 25 µg) diluted in the blocking solution for 1 h. Following the previous step, the tissue slides were incubated in the dark with the secondary fluorescent antibodies Alexa Fluor 594 and 488 goat anti-mouse (ab150116 from Abcam, Cambridge, UK; dilution 1:500 and ab150113 from Abcam, Cambridge, UK; dilution 1:500, respectively) for 1 h. Cell nuclei were detected by incubating the tissue slides with DAPI.

2.7. RNA Isolation

The total RNA was isolated from the colorectal cancer and NM tissues, DLD-1, and HT-29 cell lines using the TRIzol-based extraction method (Invitrogen, Carlsbad, CA, USA; catalog #: 15596018). The quantity and quality of the extracted RNA were assessed by measuring its absorbance using the Synergy HTX Multi-Mode Reader (Agilent, Santa Clara, CA, USA). The integrity of the isolated RNA was confirmed by performing gel electrophoresis.

2.8. Real-Time RT-PCR

Before the reverse transcription (RT) reaction, 1 µg of the total RNA was treated with DNase I, Amplification Grade (Invitrogen, Carlsbad, CA, USA; catalog number 18068-015) following the manufacturer's instructions. The RT reaction was carried out with the SensiFAST cDNA Synthesis Kit (Bioline Reagents Ltd., London, UK; catalog #: BIO-65054), according to the manufacturer's protocol. The expressions of the target genes were quantified using the StepOnePlus™ Real-Time PCR System (Applied Biosystems™, Thermo Fisher Scientific, Life Sciences Solutions Group, Carlsbad, CA, USA) and Power SYBR™ Green PCR Master Mix (Applied Biosystems™, catalog #: 4368706, Thermo Fisher Scientific Baltics UAB, Vilnius, Lithuania).

The reaction conditions were as follows: an initial denaturation step at 95 °C for 10 min, followed by 40 cycles of amplification at 95 °C for 15 s, 56–60 °C for 45 s, and 70 °C for 45 s. A melting curve analysis was performed at the end of the PCR reaction to verify that only a single product was amplified. The amplification products were separated on 1.5% agarose gel and stained with ethidium bromide. The expression levels were normalized to the housekeeping gene peptidylprolyl isomerase A (*PPIA*). The sequences of the primers and the expected product sizes are listed in Supplementary Table S2. Each reaction product was verified using a sequencing analysis.

2.9. Cell Proliferation

The proliferation of the DLD-1 and HT-29 cell lines was assessed using two methods after being treated for 24, 48, and 72 h. The first method was the CellTiter 96® AQueous Non-Radioactive Cell Proliferation Assay (Promega, Madison, WI, USA, catalog #: G4000) and the second was the BrdU Cell Proliferation Assay Kit (Cell Signaling Technology, Danvers, MA, USA, catalog #: 6813). The medium containing the drugs was changed every 24 h, while the control groups were treated with a starvation medium (RPMI/McCoy's medium with 0.5% FBS and P/S solution). The metabolic activity of living cells was measured using the MTT (3-(4,5-dimethylthiazol-2-yl)-2,5-diphenyltetrazolium bromide) assay, which evaluates the conversion of a tetrazolium salt into a formazan product. The cells were subjected to the tetrazolium salt for 4 h and the measurement was performed using spectrophotometry. The BrdU assay evaluated the incorporation of 5-bromo-2'-deoxyuridine (BrdU) into the DNA of the cells that were exposed to 10 µM of the substance for 12 h. The cells were then fixed and treated with an anti-BrdU antibody, and the magnitude of the absorbance was used to assess the incorporation of BrdU into the DNA. The results were read using a plate reader Infinite M200 Pro (Tecan Trading AG, Männedorf, Switzerland) and are presented as a percentage of the control group, which was set at 100%. Each experiment was run three times with eight replicates.

2.10. Cell Invasion

The invasion intensity of the DLD-1 and HT-29 cells was determined using the CultreCoat® 96 Well Medium BME Cell Invasion Assay from R&D Systems (catalog #: 3482-096-K). In brief, 2.5×10^4 cells were placed in each well of a 96-well plate, with the top chamber coated in Medium Basement Membrane Extract (BME). The invasion of the cells in response to P4 (1 µM) and AG-205 (1 µM) was measured using Calcein AM after 24 h of treatment. Free Calcein fluoresces brightly, and was used to quantify the number of cells that invaded or migrated in comparison to a standard curve. The invasion intensity of the treated groups was expressed as a percentage of the control group, which was set at 100%. The results were obtained from three separate experiments, each consisting of eight replicates.

2.11. ELISA Evaluation

The levels of NENF and IL-8 were analyzed using sandwich enzyme-linked immunosorbent assay (ELISA) kits. No dilution was performed on the samples before analysis, and the experiments were performed according to the manufacturer's guidelines. The concentrations of NENF in the CRC patient serum were measured using the Human Neudesin ELISA Kit from EIAB Science Inc, Wuhan, China (catalog #: E13396h), with an intra-assay coefficient of variation (CV%) of ≤4.8% and an inter-assay CV% of ≤7.1%, according to the manufacturer. The levels of IL-8 in the cell lines' medium were measured using the ELISA Quantikine® Human IL-8/CXCL8 Immunoassay kit (catalog #: D8000C) from R&D Systems Europe Ltd., Abingdon, UK. The manufacturer reported an intra-assay CV% of 5.6% at an IL-8 mean concentration of 168 pg/mL.

2.12. Statistical Analysis

The obtained results were analyzed with the STATISTICA 13.0 PL software (StatSoft Inc., Tulsa, OK, USA) and the GraphPad Prism v.8.4.3 (GraphPad Software, Inc., San Diego, CA, USA). The results are expressed as mean ± SEM. The Mann–Whitney test was used to compare two independent samples. A receiver operator characteristic (ROC) curve was generated to calculate the area under the ROC curve (AUC). To indicate the optimal cut-off point (threshold value)m the Youden index was estimated. Differences were considered significant for a two-tailed $p < 0.05$ level and are denoted by an asterisk (* $p \leq 0.05$, ** $p \leq 0.01$, *** $p \leq 0.001$, and **** $p \leq 0.0001$).

3. Results

3.1. Nuclear and Membrane P4 Receptors Are Expressed in CRC Tissues and Cell Lines

We screened the CRC tissues and DLD-1 and HT-29 cells for the expression profiling of all PR types (Figures 1–3, Supplementary Figure S1). The expression of *PGR* was significantly down-regulated in the CRC compared to normal mucosa (NM) tissues (Figure 1a). The IHC analysis showed a weak nuclear PGR signal in the CRC tissues compared to an abundant expression in the glandular cells of the NM tissues (Figure 1b). Densitometric quantification and an optical density (OD) evaluation showed significantly decreased PGR protein expression in the CRC compared to NM tissues (Figure 1c). P4 or NENF treatment did not have any effect on the *PGR* expression level in both the DLD-1 and HT-29 cell lines (Figure 1d,e).

The expression of *mPRα* was unchanged in the CRC and NM tissues, whereas *mPRβ* and *mPRγ* were significantly down-regulated in the colorectal cancer compared to NM tissues (Figure 2a). The IHC analysis showed a weak mPRα signal in both the CRC and NM tissues, and weak mPRβ and mPRγ cytoplasmic expressions in the CRC tissues (Figure 2b). The OD evaluation revealed significantly decreased mPRβ and mPRγ protein expressions in the CRC tissues compared to the NM tissues (Figure 2c). P4 or NENF treatment did not affect the *mPRs* in the DLD-1 cells (Figure 2d), whereas P4 significantly up-regulated *mPRα* and *mPRγ* expressions in the HT-29 cells (Figure 2e).

The expression of *PGRMC1* and its potential cofactor SERPINE 1 mRNA binding protein 1 (*SERBP1*) was similar in the CRC and NM tissues, whereas the gene expression of *PGRMC2* was significantly down-regulated in the CRC tissues (Figure 3a). Immunolocalization studies detected PGRMC1, PGRMC2, and SERBP1 expression in the cytoplasm of both the CRC and NM tissues. IHC showed abundant PGRMC1 and SERPINE expressions in the CRC tissues (Figure 3b). The OD evaluation revealed a significantly decreased PGRMC2 protein expression in the CRC tissues compared to the NM tissues (Figure 3c). NENF treatment significantly down-regulated the expressions of *PGRMC1* and *SERBP1* in the DLD-1 cells (Figure 3d), while P4 or NENF treatment did not have any effect on *PGRMC's* expression in the HT-29 cells (Figure 3e).

3.2. NENF Level Is Upregulated in Colorectal Cancer

We assessed the *NENF* expression in colorectal cancer and its release in CRC tissues and DLD-1 and HT-29 cell lines. The mRNA of the *NENF* expression level was similar in the CRC and NM tissues (Figure 4a). IHC showed abundant NENF expression in the CRC tissues (Figure 4b). The OD evaluation indicated that the protein expression in the CRC tissues was comparable to that in the NM tissues (Figure 4c). However, the NENF concentration was significantly higher in the serum of the CRC patients compared to the healthy controls (Figure 4d). A receiver operator characteristic (ROC) curve analysis showed that the serum NENF score significantly differentiated the colorectal cancer patients from the healthy controls with a diagnostic sensitivity and positive predictive value of 83% and 81%, respectively (Supplementary Table S3, Supplementary Figure S2). P4 treatment did not affect the *NENF* in both the DLD-1 and HT-29 cells (Figure 4e,f), however, P4 treatment significantly increased the release of NENF in the DLD-1 cells (Figure 4g), but not in the HT-29 cells (Figure 4h). Immunocytochemical staining colocalized both PGRMC1 and NENF in the cytoplasm of the CRC and NM tissues (Figure 4i).

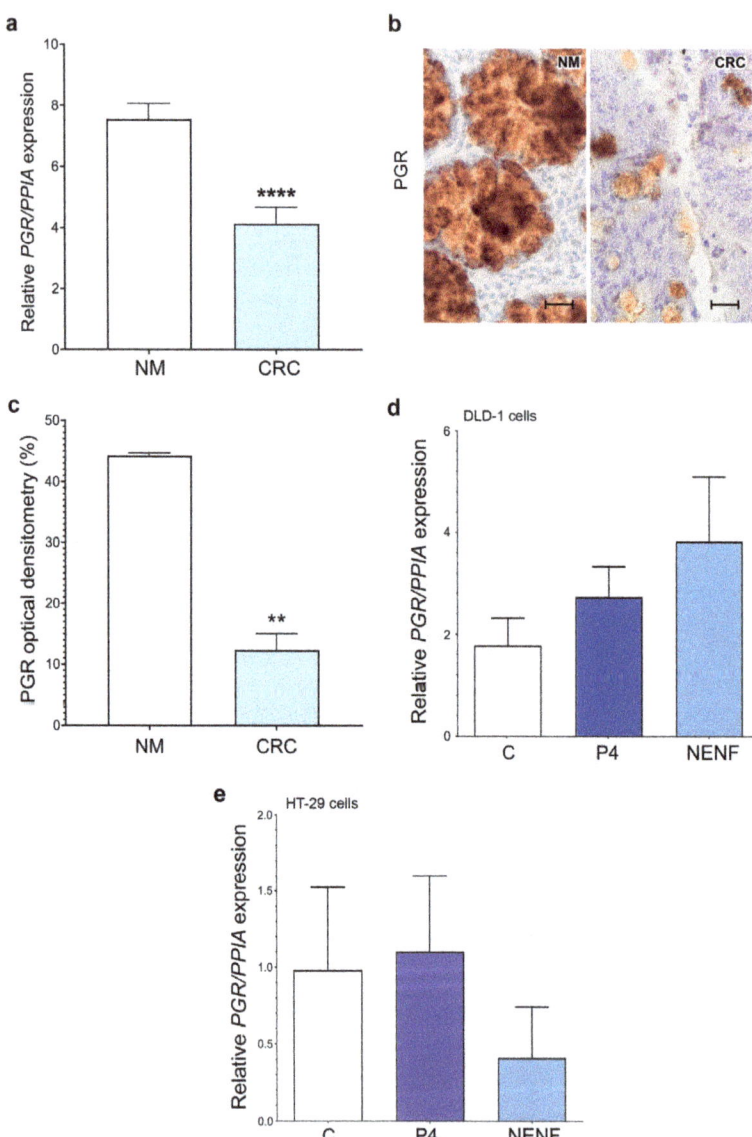

Figure 1. Characterization of *PGR* expression levels in colorectal cancer and DLD-1 and HT-29 cell lines. *PGR* expression at gene, (**a**) and protein (**b**,**c**) levels in NM ($n = 10$) and colorectal cancer (CRC) tissues ($n = 20$). Original magnification, 20×; scale bar, 20 μm. The columns represent the mean ± SEM relative to *PPIA*. The Mann–Whitney test was used to compare NM vs. CRC results. Statistical significance of NM vs. CRC: **** $p \leq 0.0001$, ** $p \leq 0.01$. *PGR* expression after treatment with 1 μM of P4 and 1 μg/mL of NENF in DLD-1 cell line (**d**) and HT-29 cell line (**e**) ($n = 3$ independent experiments). The Mann–Whitney test was used to compare C vs. P4 and C vs. NENF of DLD-1 and HT-29 cells. The differences are statistically non-significant. Abbreviations: C, control/non-treated group; CRC, colorectal cancer; NENF, neudesin; NM, normal mucosa; P4, progesterone; PGR, nuclear progesterone receptor; PPIA, peptidylprolyl isomerase A; SEM, standard error of the mean; and vs., versus.

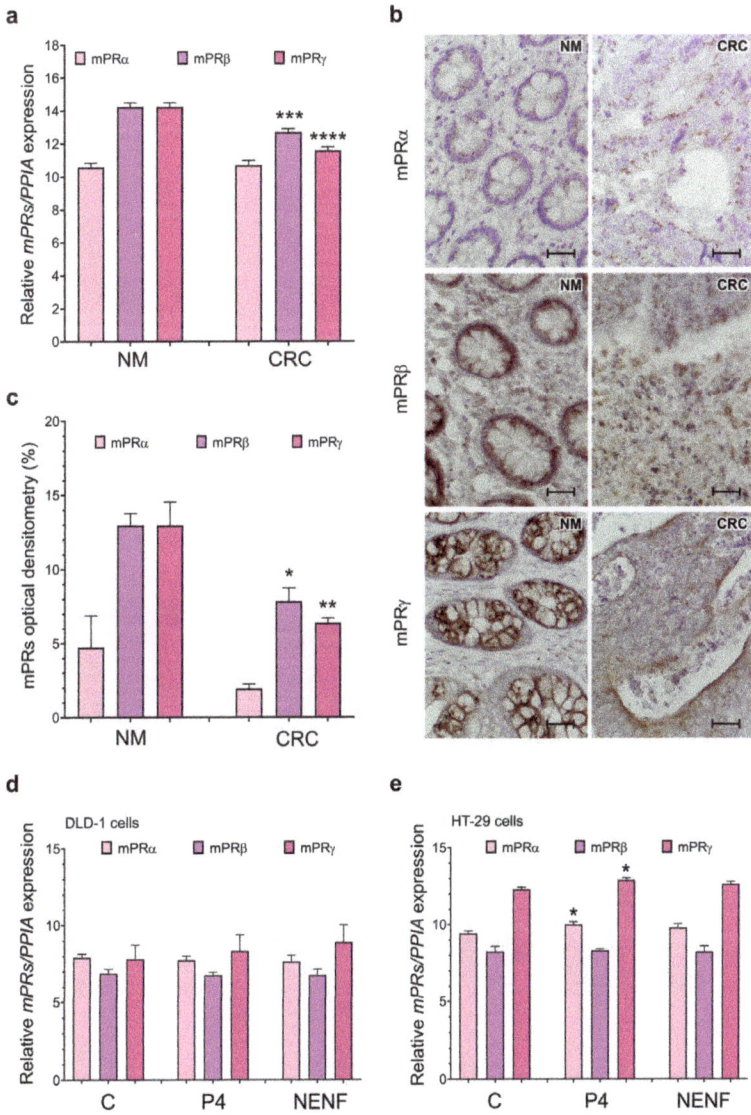

Figure 2. Characterization of *mPRα*, *mPRβ*, and *mPRγ* expression levels in colorectal cancer and DLD-1 and HT-29 cell lines. *mPRα*, *mPRβ*, and *mPRγ* expression at gene (**a**) and protein (**b,c**) levels in NM ($n = 10$) and colorectal cancer (CRC) tissues ($n = 20$). Original magnification, 20×; scale bar, 20 μm. The columns represent the mean ± SEM relative to *PPIA*. The Mann–Whitney test was used to compare NM vs. CRC results. Statistical significance of NM vs. CRC for *mPRβ*, *mPRγ*: * $p \leq 0.05$, ** $p \leq 0.01$, *** $p \leq 0.001$, and **** $p \leq 0.0001$. *mPRα*, *mPRβ*, and *mPRγ* expression after treatment with 1 μM of P4 and 1 μg/mL of NENF in DLD-1 cell line (**d**) and HT-29 cell line (**e**) ($n = 3$ independent experiments). The Mann–Whitney test was used to compare C vs. P4 and C vs. NENF. Statistical significance of C vs. P4 for *mPRα*, *mPRγ* of HT-29 cells: * $p \leq 0.05$. Other differences are statistically non-significant. Abbreviations: C, control/non-treated group; CRC, colorectal cancer; mPRα, membrane progesterone receptor alfa; mPRβ, membrane progesterone receptor beta; mPRγ, membrane progesterone receptor gamma; NENF, neudesin; NM, normal mucosa; P4, progesterone; PPIA, peptidylprolyl isomerase A; SEM, standard error of the mean; and vs., versus.

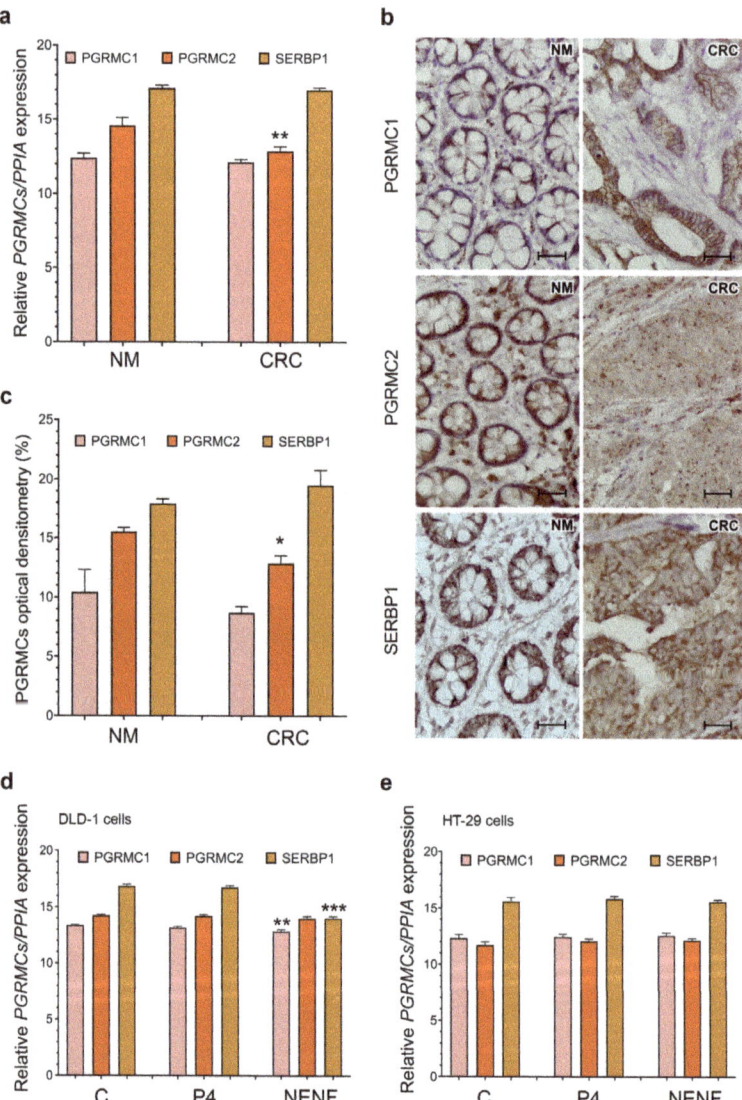

Figure 3. Characterization of *PGRMCs* expression levels in colorectal cancer and DLD-1 and HT-29 cell lines. *PGRMC1*, *PGRMC2*, and *SERBP1* expression at gene (**a**) and protein (**b**,**c**) levels in NM (n = 10) and colorectal cancer (CRC) tissues (n = 20). Original magnification, 20×; scale bar, 20 µm. The columns represent the mean ± SEM relative to *PPIA*. The Mann–Whitney test was used to compare NM vs. CRC results. Statistical significance of NM vs. CRC for *PGRMC2*: * $p \leq 0.05$, ** $p \leq 0.01$. Other differences are statistically non-significant. *PGRMC1*, *PGRMC2*, and *SERBP1* expression after treatment with 1 µM of P4 and 1 µg/mL of NENF in DLD-1 cell line (**d**) and HT-29 cell line (**e**) (n = 3 independent experiments. The Mann–Whitney test was used to compare C vs. P4 and C vs. NENF. Statistical significance of C vs. NENF for *PGRMC1* and *SERBP1* of DLD-1 cells: ** $p \leq 0.01$, *** $p \leq 0.001$. Other differences are statistically non-significant. Abbreviations: C, control/non-treated group; CRC, colorectal cancer; NENF, neudesin; NM, normal mucosa; P4, progesterone; PGRMC1, progesterone receptor membrane component 1; PGRMC2, progesterone receptor membrane component 2; PPIA, peptidylprolyl isomerase A; SEM, standard error of the mean; SERBP1, SERPINE1 mRNA binding protein; and vs., versus.

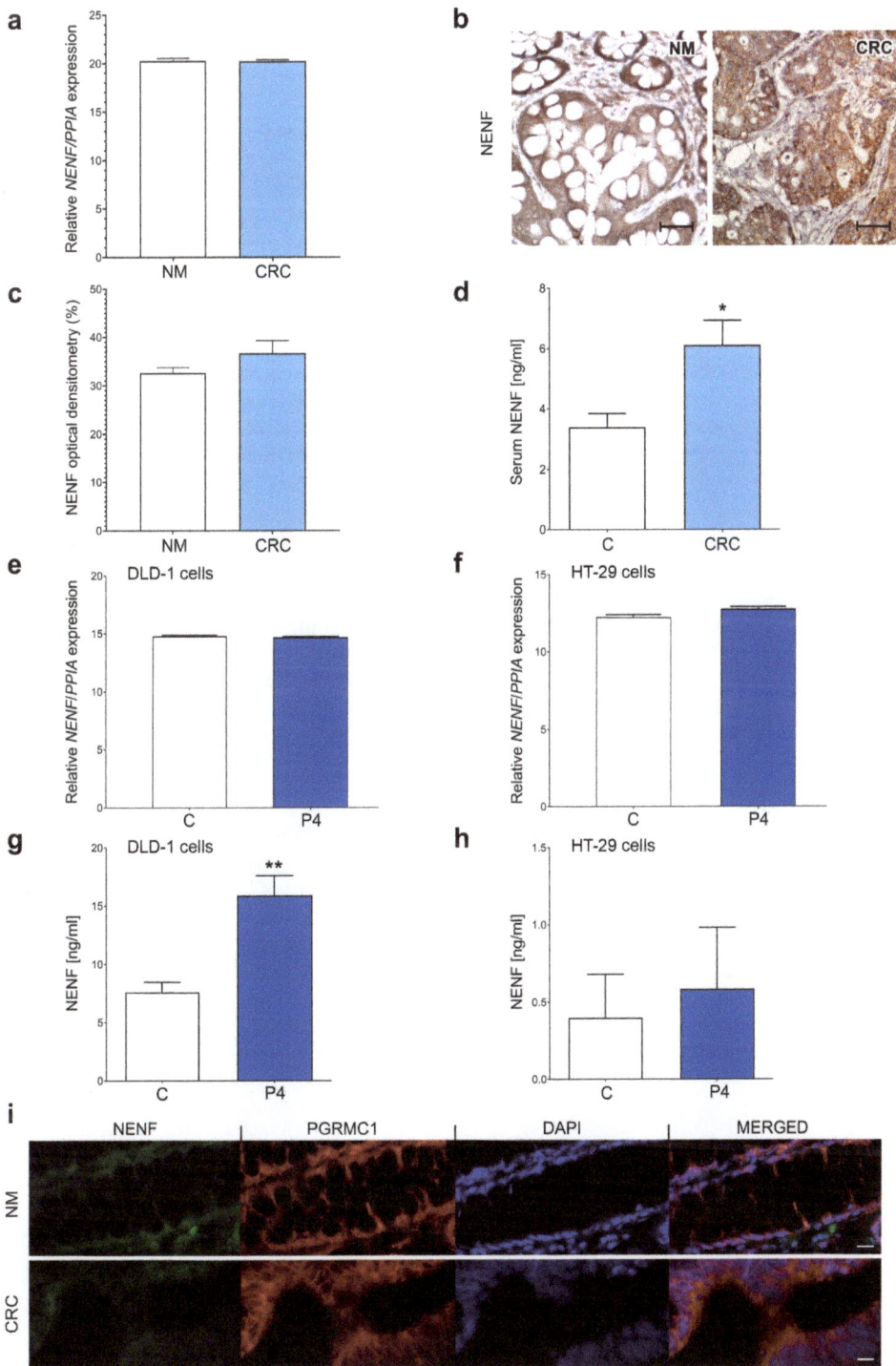

Figure 4. Characterization of *NENF* expression and release in colorectal cancer patients and DLD-1 and HT-29 cell lines. *NENF* expression at gene (**a**) and protein (**b**,**c**) levels in NM (*n* = 10) and colorectal

cancer (CRC) tissues ($n = 20$). Original magnification, 20×; scale bar, 20 µm. The columns represent the mean ± SEM relative to *PPIA*. The Mann–Whitney test was used to compare NM vs. CRC results. The differences are statistically non-significant. Serum NENF concentration in colorectal cancer patients ($n = 41$) compared to the control group ($n = 15$) (**d**). The Mann–Whitney test was used to compare serum NENF concentration in C vs. CRC results. Statistical significance: * $p \leq 0.05$. *NENF* expression after treatment with 1 µM of P4 in DLD-1 cell line (**e**) and HT-29 cell line (**f**) ($n = 3$ independent experiments). The Mann–Whitney test was used to compare *NENF* expression in C vs. CRC results in DLD-1 and HT-29 cells. The differences are statistically non-significant. NENF concentration in the medium after treatment with 1 µM of P4 in DLD-1 cell line (**g**) ($n = 6$) and HT-29 cell line (**h**) ($n = 6$). The Mann–Whitney test was used to compare NENF concentration in the medium C vs. P4 results of DLD-1 cells, statistical significance: ** $p \leq 0.01$. Double staining for PGRMC1 and NENF in NM tissues ($n = 10$) and colorectal cancer ($n = 20$) (**i**). PGRMC1-positive cells are in red, NENF-positive cells are in green, nucleus localization in cells is in blue, and PGRMC1 and NENF merged staining is in orange, scale bar, 20 µm. Abbreviations: C, control/non-treated group; CRC, colorectal cancer; NENF, neudesin; NM, normal mucosa; P4, progesterone; PGRMC1, progesterone receptor membrane component 1; PPIA, peptidylprolyl isomerase A; SD, standard deviation; SEM, standard error of the mean; and vs., versus.

3.3. P4 Treatment Affects the DLD-1 and HT-29 Cell Proliferation, but in Combination with NENF Also Promotes Cell Invasion

We examined the effects of P4 and NENF on the cell proliferation and invasion of DLD-1 and HT-29 colorectal cancer. P4 significantly stimulated cell proliferation after 48 h and 72 h in the DLD-1 cell line, and after 24 h, 48 h, and 72 h in the HT-29 cell lines (Figure 5a,b). PGRMC1 blockage with the PGRMC1 inhibitor AG-205 inhibited the P4 effect in both cell lines (Figure 5a,b). P4 or NENF treatment alone did not affect the cell invasion of the DLD-1 and HT-29 cells (Figure 5c,d). However, P4 and NENF co-treatment significantly increased the cell invasion of the DLD-1 and HT-29 cells, and this effect could be abolished by AG-205 (Figure 5c,d).

3.4. P4 and NENF Up-Regulate IL-8 Expression and Its Release in DLD-1 and HT-29 Cells

We assessed the expression of *IL-8* in the CRC tissues and checked the P4 and NENF treatment effect on the expression and release of IL-8 and its receptor CXCR1 in the DLD-1 and HT-29 cell lines. The expression of *IL-8* was significantly up-regulated in the CRC compared to the NM tissues (Figure 6a). NENF treatment significantly up-regulated *IL-8* in the DLD-1 and HT-29 cells (Figure 6b,c). P4 or NENF treatment significantly increased the IL-8 release in both the DLD-1 and HT-29 cells, whereas AG-205 significantly abolished this effect (Figure 6d). The expression of the IL-8 receptor *CXCR1* was unaffected in the CRC and NM tissues (Figure 6e). P4 or NENF treatment did not have *any effect on the *CXCR1* expression in both cell lines (Figure 6f,g).

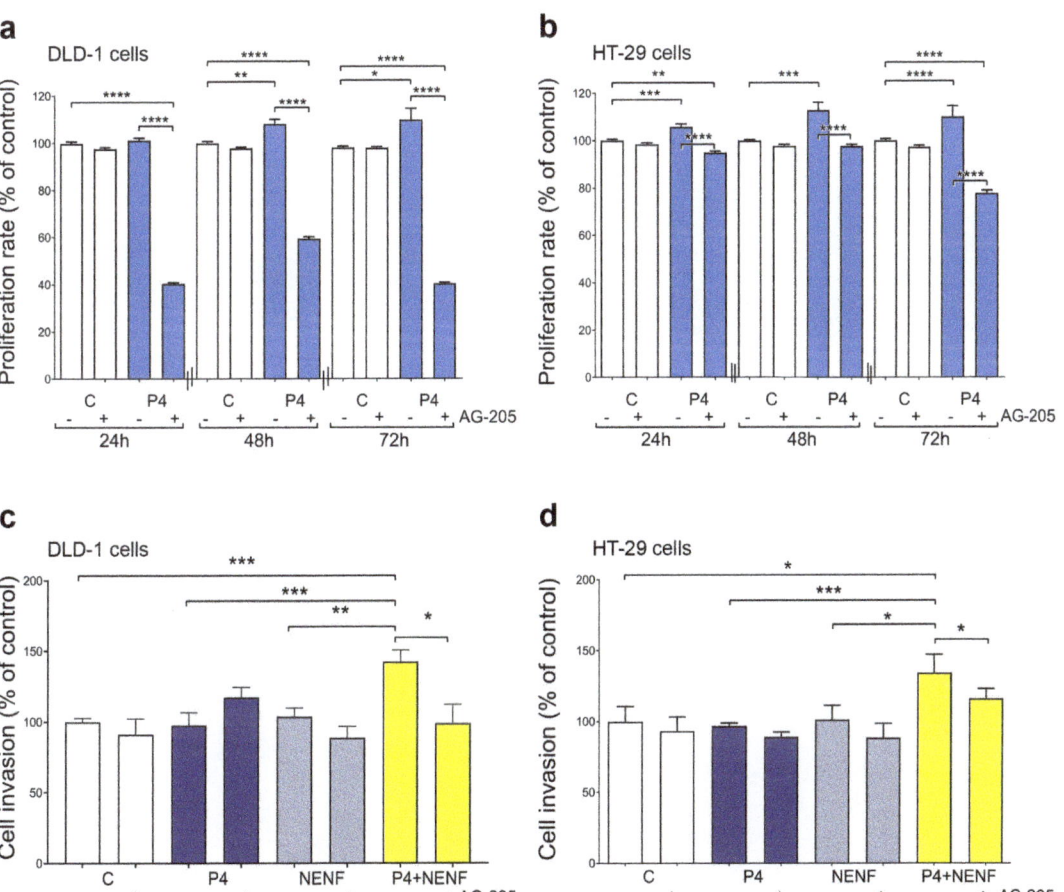

Figure 5. P4 and NENF treatment effect on cell proliferation and invasion in DLD-1 and HT-29 cell lines. Effects of 1 μM of P4 with or without 1 μM of PGRMC1 inhibitor (AG-205) on the proliferation rate of DLD-1 (**a**) and HT-29 (**b**) cell lines after 24 h, 48 h, and 72 h treatments (n = 3 independent experiments). The Mann–Whitney test was used to compare C vs. P4 results without or with 1 μM of PGRMC1 inhibitor (AG-205), and compare P4 without AG-205 vs. P4 with AG-205 of DLD-1 and HT-29 cells, statistical significance: * $p \leq 0.05$, ** $p \leq 0.01$, *** $p \leq 0.001$, and **** $p \leq 0.0001$. Other differences are statistically non-significant. Effects of 1 μM of P4 or/and 1 μg/mL of NENF treatments without or with 1 μM of AG-205 on cell migration of DLD-1 (**c**) and HT-29 (**d**) cell lines after 24 h treatment (n = 3 independent experiments). The Mann–Whitney test was HT-29 cells, statistical significance: * $p \leq 0.05$, ** $p \leq 0.01$, and *** $p \leq 0.001$. Other differences are statistically non-significant. Cell proliferation and cell invasion rates of the treated and non-treated (control) groups are presented as the percentage of the control, considered as 100%. Abbreviations: AG-205, PGRMC1 inhibitor; C, control/non-treated group; NENF, neudesin; P4, progesterone; PGRMC1, progesterone receptor membrane component 1; and vs., versus. Used to compare C vs. P4 + NENF results without or with 1 μM of PGRMC1 inhibitor (AG-205); compare P4 vs. P4 + NENF without or with AG-205; and compare NENF vs. P4 + NENF without or with AG-205 of DLD-1 and.

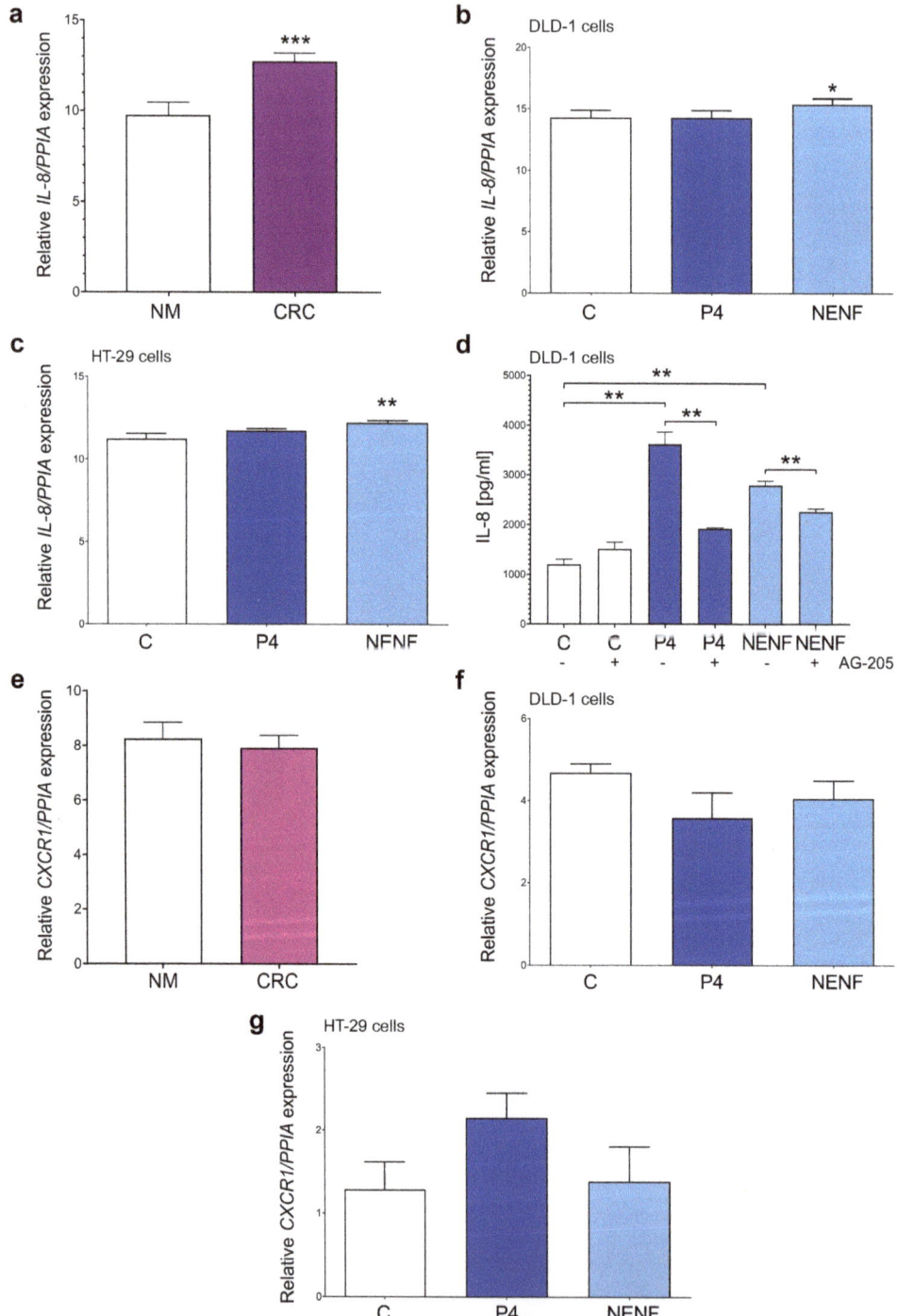

Figure 6. Characterization of *IL-8* and *CXCR1* expression and release in colorectal cancer patients and DLD-1 and HT-29 cell lines. *IL-8* expression in NM (*n* = 10) and colorectal cancer (*n* = 20) tissues (**a**). The columns represent the mean ± SEM relative to *PPIA*. The Mann–Whitney test was used to

compare NM vs. CRC results, statistical significance of NM vs. CRC for *IL-8*: *** $p \leq 0.001$. *IL-8* expression after treatment with 1 µM of P4 or with 1 µg/mL of NENF in DLD-1 (**b**) and HT-29 (**c**) cell lines (n = 3 independent experiments). The Mann–Whitney test was used to compare C vs. P4 and C vs. NENF results of DLD-1 and HT-29 cells, statistical significance of C vs. NENF: * $p \leq 0.05$, ** $p \leq 0.01$. IL-8 concentration in the medium of DLD-1 cell line after treatment with 1 µM of P4 or 1 ng/mL of NENF without or with 1 µM of AG-205 (**d**) (n = 6). The Mann–Whitney test was used to compare C vs. P4 without or with 1 µM of AG-205, C vs. NENF without or with 1 µM of AG-205, and also results of P4 without AG-205 vs. P4 with AG-205 of DLD-1 or NENF without AG-205 vs. NENF with AG-205 of DLD-1 cells, statistical significance: ** $p \leq 0.01$. *CXCR1* expression in NM (n = 10) and colorectal cancer (n = 20) tissues (**e**). The Mann–Whitney test was used to compare NM vs. CRC results. The differences are statistically non-significant. *CXCR1* expression after treatment with 1 µM of P4 or with 1 µg/mL of NENF in DLD-1 (**f**) and HT-29 (**g**) cell lines (n = 3 independent experiments). The Mann–Whitney test was used to compare C vs. P4 and C vs. NENF results of DLD-1 and HT-29 cells. The differences are statistically non-significant. Abbreviations: AG-205, PGRMC1 inhibitor; C, control/non-treated group; CRC, colorectal cancer; CXCR1, C-X-C Motif Chemokine Receptor 1; IL-8, interleukin 8; NENF, neudesin; NM, normal mucosa; P4, progesterone; PGRMC1, progesterone receptor membrane component 1; PPIA, peptidylprolyl isomerase A; SEM, standard error of the mean; and vs., versus.

4. Discussion

The available data in the medical literature on the risk of colorectal cancer (CRC) in the context of hormone therapy (HT) are still unconvincing. Recently, hormone therapy has been linked to a decreased risk of colorectal cancer (CRC) [31–33]. Lin et al. suggested that both estrogen–progestogen therapy (EPT) and estrogen therapy (ET), especially when used currently, are associated with a reduced risk of colorectal cancer in peri- or postmenopausal women. EPT demonstrates a more consistent association with the reduction in CRC risk, regardless of the duration of its use [34]. The use of hormone replacement therapy (HRT) has been connected to a significant reduction in the risk of both colorectal-cancer-specific mortality and all-cause mortality in women with colorectal cancer. The authors emphasized the hormone-dependent nature of CRC and its inverse association with tumor progression concerning estrogen receptor β (ERβ) expression [35]. However, the available literature also suggests that the use of HRT is not associated with an increased risk or even the possibility of CRC [36]. Clinical studies of the Women's Health Initiative (WHI) have revealed an increased risk of breast cancer in women using HRT [37]. In the context of breast cancer, progesterone plays a complex role by influencing cell growth and division, regulating autocrine mechanisms, and interacting with other growth factors. It can impact the development of both receptor-positive (estrogen/progesterone receptor-positive) and receptor-negative tumors, making its role in carcinogenesis multifaceted [38,39]. The studies describing the relationship between HT and ovarian cancer also are inconclusive [40]. They have suggested both an increased risk of ovarian cancer with long-term use of HRT [41,42], and no significant difference in ovarian cancer incidence between a HRT group and a placebo group [43].

The expression of all P4 receptors in colorectal cancer has not been well-characterized, and the existing data in the available literature have been conflicting and inconclusive (Supplementary Table S1) [2,3,23–28]. This suggests that further research is needed to fully understand the role of P4 receptors in colorectal cancer to determine whether they could serve as potential therapeutic targets. In our study, we found that the mRNA and protein levels of PGR, mPRβ, mPRγ, and PGRMC2 were significantly down-regulated in the CRC compared to the normal tissues. A low expression of PGR has been associated with a poor prognosis of CRC [28]. The abundant expression of PGRMC1 in advanced stages of CRC suggests its potential role in cancer progression. Most studies have, to date, focused on the expression and role of PGRMC1 in cancers other than colorectal cancer, such as ovarian [44], breast [45], endometrial [46], lung [47], and hepatocellular carcinoma [48]. These studies have shown that PGRMC1 plays a role in promoting tumor growth and cell proliferation [49], anchorage-independent growth, migration, invasion [50], resistance to chemotherapy [44], tumor angiogenesis regulation, and cancer cell apoptosis suppression [47]. Our present observations on the P4 receptor expression pattern in CRC are consistent with the recently reported marginal expression of PGR, with low expression levels of mPRβ, mPRγ, and PGRMC2, and abundant expression levels of PGRMC1 in high-grade human ovarian cancer [29]. A significantly up-regulated expression of PGRMC1 in advanced human ovarian cancers has been suggested to indicate its important role in disease progression [51]. Moreover, elevated PGRMC1 expression in breast cancer has been linked to more advanced stages and a poor prognosis [52]. Our results also suggested that PGRMC1 may play a pivotal role in CRC progression, however, the molecular mechanism of PGRMC1 action in cancers is still not fully understood.

P4 treatment has been shown to inhibit the proliferation of various colorectal cancer cell lines by stopping the G2/M phase of the cell cycle and inducing apoptosis [28]. However, the P4 doses used in this study were supraphysiological and most likely clinically irrelevant, due to the very rapid metabolism of P4 [28]. The inhibitory effect of P4 has also been shown in vivo, but information on the dose used in the mice treatment is lacking [28]. However, SW620 cells with a higher expression of the PGR were used for inoculation, suggesting that P4 may have an inhibitory effect on CRC with a high expression of PGR [28]. In our study, we chose cell lines with a very low/traceable PGR expression as a model to study advanced cancer stages. We showed that P4 treatment at clinically relevant doses stimulated cell proliferation in the DLD-1 and HT-29 cell lines, suggesting its potential role in the progression of colorectal cancer. DLD-1 and HT-29 cell lines have been widely used in various CRC studies [53,54]. The study by Tankiewicz-Kwedlo et al. demonstrated that the effects of Epo therapy on tumor growth dynamics were more pronounced in HT-29 cell xenografts compared to DLD-1 cell xenografts [55]. Sihong et al. suggested that HT-29 cells may have a low metastatic potential [56]. These variations in the behavior and response to treatment of DLD-1 and HT-29 cells may be attributed to their own distinct genetic profiles, microsatellite stability, potential mutations, gene expression patterns, and genomic alterations. This is of particular significance in cancer research, as it can aid in identifying potential drug targets and treatment strategies.

Similarly, P4 stimulation increased the proliferation of ovarian cancer cells in vitro and ovarian tumor growth in vivo through PGRMC1 [29]. In the present study, PGRMC1 inhibition abolished the effect of P4, suggesting that P4 in CRC may act primarily through PGRMC1. However, it is still uncertain whether PGRMC1 is an independent P4 receptor or requires additional P4-binding proteins for signaling. It has been suggested that PGRMC1 might act as a downstream mediator for other P4-binding proteins [10]. Possible binding partners for PGRMC1 include SERBP1 [10], microsomal cytochrome P450 (CYP) monooxygenase systems, and NENF [17,57]. The structural similarities between PGRMC1 and NENF have been demonstrated in previous studies [13]. Based on the fact that PGRMC1 is involved in the regulation of rapid non-genomic P4 actions, it has been hypothesized that NENF may also play a role in this type of P4 regulation [58]. Our results showed that PGRMC1 and NENF were co-expressed in the cytoplasm of the CRC cells, suggesting NENF as a potential cofactor for PGRMC1. Co-treatment with P4 and NENF increased the invasiveness of the CRC cells. This effect was abolished by a blockage of PGRMC1, indicating an important interaction between NENF and PGRMC1 for P4-mediated cell migration. Additionally, we showed that P4 increased the NENF secretion in the CRC cells, which suggests a direct effect of P4 on NENF production in colorectal cancer.

The role of NENF in cancer biology, progression, or metastasis has not been extensively investigated [16,17], but it has been proposed that it may play a significant role in the development of liver, bladder, and breast cancers [16]. The depletion of NENF has been shown to reduce cancer cell growth and invasiveness and impair the ability of liver cancer cells to form tumors in mice [16]. NENF also increases the tumorigenicity and invasiveness of MCF-7 breast cancer cells [15]. Recently, we found that the concentration of NENF in the cerebrospinal fluid of patients with astrocytic brain tumors was significantly higher compared to that of non-tumoral individuals [17]. We also observed a strong correlation between the serum NENF concentration and its levels in cerebrospinal fluid, and noted that these levels were strongly gender-dependent [17]. Our present study showed an abundant NENF expression in the CRC tissues and a significantly higher serum concentration of NENF in patients with CRC compared to healthy controls. A diagnostic analysis revealed the usefulness of serum NENF levels in identifying CRC patients from those without cancer, suggesting its potential role as a circulating biomarker for colorectal cancer.

Cancer cell proliferation and survival may be promoted by the activation of MAPK and PI3K/AKT signaling [13]. It has been shown that NENF and the pro-inflammatory cytokine IL-8 may also regulate these pathways [13,59]. However, the specific mechanism of IL-8 regulation in colorectal cancer requires further investigation. IL-8 is a versatile cytokine that has been shown to promote angiogenesis, attract immune cells, and stimulate tumor growth, invasion, and migration through both autocrine and paracrine effects [60–62]. Previous research has also shown that IL-8 can serve as a biomarker associated with a poor prognosis and chemoresistance for various types of cancer [63,64]. Additionally, IL-8 has been linked to adverse outcomes in brain tumors [65] and breast cancer [66]. In our study, we found that *IL-8* expression was significantly up-regulated in the CRC compared to the NM tissues. Moreover, NENF treatment increased the *IL-8* expression in both the DLD-1 and HT-29 cells, as well as the IL-8 release into the medium of the DLD-1 cells after P4 or NENF treatment without a PGRMC1 inhibitor (AG-205). This effect was abolished by PGRMC1 inhibition, suggesting that P4 and NENF require PGRMC1 to regulate IL-8 in colorectal cancer. Conversely, Emmanouil et al. suggested that HT-29 cells exhibit an increased metastatic potential and secrete angiogenic chemokines, notably IL-8 and VEGF, fostering neoangiogenesis and tumor advancement [67].

To perform *IL-8* knockdown, or even better, to knockout through CRISPR/CAS9 technology, experiments could elucidate the role of IL-8 in mediating the proliferation or invasion effects of P4 on CRC cells. This should be conducted in the future to enhance our knowledge on the mechanistic aspects of this. One limitation of our study was the absence of in vivo experiments, which could have provided additional support for our in vitro findings.

5. Conclusions

Taken together, our data provided novel insights into the actions of P4, PGRMC1, and NENF in colorectal cancer, emphasizing new potential actions that may regulate CRC biology (summarized in Figure 7). In this action, it seems that PGRMC1 and NENF may interact as possible cofactors in non-classical P4 signaling. Targeting the PGRMC1/NENF complex may open-up new therapeutic possibilities for patients with colorectal cancer. Therefore, future studies aimed at developing treatment strategies for CRC could consider not only PGRMC1 inhibition, but also a blockage of NENF production and secretion. Moreover, NENF, as a secreted protein, could become a promising circulating biomarker candidate to distinguish between colorectal cancer patients and healthy individuals.

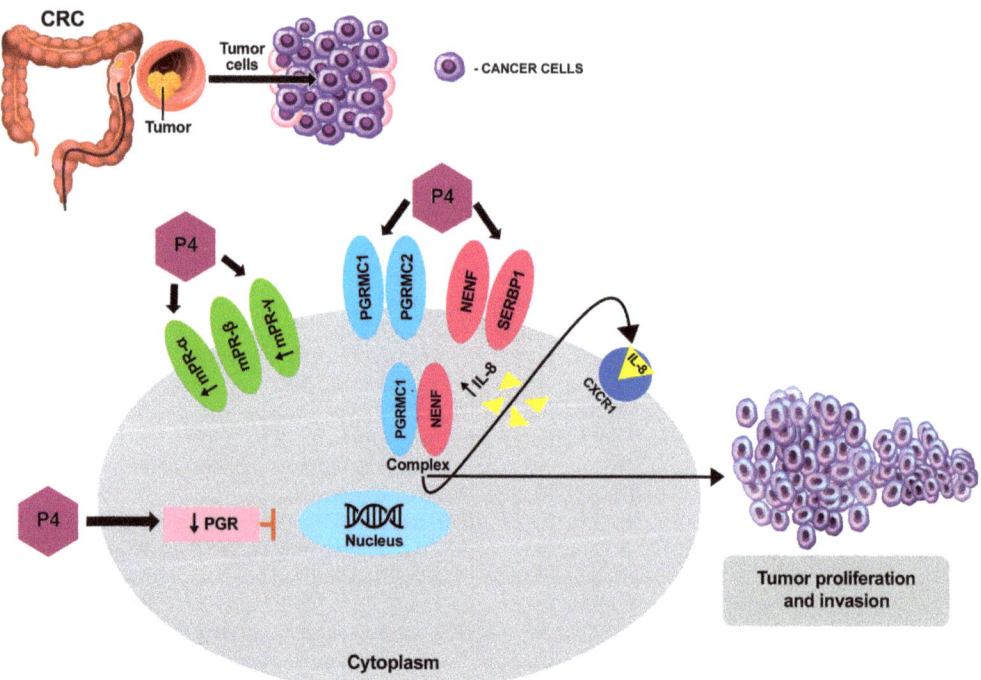

Figure 7. Schematic overview of the potential non-genomic P4 action in colorectal cancer. P4 may initiate rapid non-classical signaling through the complex of PGRMC1 and NENF, leading to increased proliferation and invasion of colorectal cancer cells. However, P4 cannot activate the classical genomic signaling pathway due to weak PGR expression in colorectal cancer cells (arrow: ↓*PGR*—weak PGR expression). P4 or NENF may significantly increase the release of IL-8 by colorectal cancer cells (arrow: ↑IL-8—increased release of IL-8). P4 significantly up-regulates *mPRα* and *mPRγ* expression in colorectal cancer cells (arrow: ↑*mPR-α*, ↑*mPR-γ*—increased expression of *mPR-α* and *mPR-γ*). NENF, neuron-derived neurotrophic factor; P4, progesterone; PGR, nuclear progesterone receptor; and PGRMC1, progesterone receptor membrane component 1.

Supplementary Materials: The following supporting information can be downloaded at: https://www.mdpi.com/article/10.3390/cancers15205074/s1. Table S1. Progesterone (P4) receptors characteristics regarding different clinicopathological parameters in colorectal cancer (CRC) patients. Table S2. Sequences of primers used in the q-RT-PCR. Table S3. Diagnostic usefulness of serum NENF evaluation in differentiating colorectal cancer patients from healthy individuals. Figure S1. Characterization of progesterone receptors and NENF expression profiles in DLD-1 and HT-29 cell lines. Figure S2. The area under the ROC curve (AUC) to distinguish between colorectal cancer patients and healthy individuals based on serum NENF levels.

Author Contributions: J.K., O.M.K.-L., D.P.-T. and V.D.-P.: conceptualization, methodology, investigation, data curation, formal analysis, visualization, software, writing—original draft, funding acquisition, writing—review and editing, supervision; W.L., E.P., M.S. and P.B.: investigation, methodology, writing—review and editing; J.D. and K.Z.: investigation, writing—review and editing; K.G.-U. and N.A.R.: Formal analysis, writing—review and editing; S.W.: formal analysis, data curation, funding acquisition, writing—review and editing, supervision. All authors have read and agreed to the published version of the manuscript.

Funding: This work was supported by the Medical University of Bialystok (research project no: SUB/1/DN/21/004/2209).

Institutional Review Board Statement: Studies involving human participants adhered to ethical standards as determined by the institutional and/or national research committees and were following the 1964 Helsinki Declaration and its subsequent amendments, or equivalent ethical standards. The study was approved by the Local Bioethics Committee of the Medical University of Bialystok (APK. 002.100.20.21; approved date: 25 February 2021). All subjects gave their informed consent for inclusion before they participated in the study.

Informed Consent Statement: All subjects gave their informed consent for inclusion before they participated in the study.

Data Availability Statement: The data generated in this study are available upon request from the corresponding author (V.D.-P., J.K.).

Acknowledgments: The authors are grateful to Piotr Abramowicz for his technical help in preparing the figure. We are grateful to Arkadiusz Surażyński for providing the fluorescence microscope, which enabled the imaging of the colocalization of PGRMC1 and NENF.

Conflicts of Interest: The authors declare that they have no competing interests.

References

1. Sasso, C.V.; Santiano, F.E.; Arboccó, F.C.V.; Zyla, L.E.; Semino, S.N.; Guerrero-Gimenez, M.E.; Creydt, V.P.; Fontana, C.M.L.; Carón, R.W. Estradiol and progesterone regulate proliferation and apoptosis in colon cancer. *Endocr. Connect.* **2019**, *8*, 217–229. [CrossRef] [PubMed]
2. Ye, S.B.; Cheng, Y.K.; Zhang, L.; Wang, X.P.; Wang, L.; Lan, P. Prognostic value of estrogen receptor-α and progesterone receptor in curatively resected colorectal cancer: A retrospective analysis with independent validations. *BMC Cancer* **2019**, *19*, 933. [CrossRef] [PubMed]
3. ElLateef, A.A.E.G.A.; El Sayed Mohamed, A.; Elhakeem, A.A.S.; Ahmed, S.F.M. Estrogen and progesterone expression in colorectal carcinoma: A clinicopathological study. *Asian Pac. J. Cancer Prev.* **2020**, *21*, 1155–1162. [CrossRef] [PubMed]
4. Koper-Lenkiewicz, O.M.; Dymicka-Piekarska, V.; Milewska, A.J.; Zińczuk, J.; Kamińska, J. The relationship between inflammation markers (Crp, il-6, scd40l) and colorectal cancer stage, grade, size and location. *Diagnostics* **2021**, *11*, 1382. [CrossRef]
5. Gadkar-Sable, S. Progesterone receptors: Various forms and functions in reproductive tissues. *Front. Biosci.* **2005**, *10*, 2118. [CrossRef]
6. Nagy, B.; Szekeres-Barthó, J.; Kovács, G.L.; Sulyok, E.; Farkas, B.; Várnagy, Á.; Vértes, V.; Kovács, K.; Bódis, J. Key to life: Physiological role and clinical implications of progesterone. *Int. J. Mol. Sci.* **2021**, *22*, 11039. [CrossRef]
7. Bramley, T. Non-genomic progesterone receptors in the mammalian ovary: Some unresolved issues. *Reproduction* **2003**, *125*, 3–15. [CrossRef]
8. Grimm, S.L.; Hartig, S.M.; Edwards, D.P. Progesterone Receptor Signaling Mechanisms. *J. Mol. Biol.* **2016**, *428*, 3831–3849. [CrossRef]
9. Mani, S.K.; Oyola, M.G. Progesterone signaling mechanisms in brain and behavior. *Front. Endocrinol.* **2012**, *3*, 7. [CrossRef]
10. Petersen, S.L.; Intlekofer, K.A.; Moura-Conlon, P.J.; Brewer, D.N.; Del Pino Sans, J.; Lopez, J.A. Novel progesterone receptors: Neural localization and possible functions. *Front. Neurosci.* **2013**, *7*, 164. [CrossRef]
11. Tokumoto, T.; Hossain, M.B.; Wang, J. Establishment of procedures for studying mPR-interacting agents and physiological roles of mPR. *Steroids* **2016**, *111*, 79–83. [CrossRef] [PubMed]
12. Wendler, A.; Wehling, M. Many or too many progesterone membrane receptors? Clinical implications. *Trends Endocrinol. Metab.* **2022**, *33*, 850–868. [CrossRef]
13. Kimura, I.; Yoshioka, M.; Konishi, M.; Miyake, A.; Itoh, N. Neudesin, a novel secreted protein with a unique primary structure and neurotrophic activity. *J. Neurosci. Res.* **2005**, *79*, 287–294. [CrossRef] [PubMed]
14. Kimura, I.; Konishi, M.; Asaki, T.; Furukawa, N.; Ukai, K.; Mori, M.; Hirasawa, A.; Tsujimoto, G.; Ohta, M.; Itoh, N.; et al. Neudesin, an extracellular heme-binding protein, suppresses adipogenesis in 3T3-L1 cells via the MAPK cascade. *Biochem. Biophys. Res. Commun.* **2009**, *381*, 75–80. [CrossRef] [PubMed]
15. Han, K.H.; Lee, S.H.; Ha, S.A.; Kim, H.K.; Lee, C.W.; Kim, D.H.; Gong, K.H.; Yoo, J.A.; Kim, S.; Kim, J.W. The functional and structural characterization of a novel oncogene GIG47 involved in the breast tumorigenesis. *BMC Cancer* **2012**, *12*, 274. [CrossRef]
16. Stefanska, B.; Cheishvili, D.; Suderman, M.; Arakelian, A.; Huang, J.; Hallett, M.; Han, Z.G.; Al-Mahtab, M.; Akbar, S.M.F.; Khan, W.A.; et al. Genome-wide study of hypomethylated and induced genes in patients with liver cancer unravels novel anticancer targets. *Clin. Cancer Res.* **2014**, *20*, 3118–3132. [CrossRef]
17. Koper-Lenkiewicz, O.M.; Kamińska, J.; Milewska, A.; Sawicki, K.; Jadeszko, M.; Mariak, Z.; Reszeć, J.; Dymicka-Piekarska, V.; Matowicka-Karna, J. Serum and cerebrospinal fluid Neudesin concentration and Neudesin Quotient as potential circulating biomarkers of a primary brain tumor. *BMC Cancer* **2019**, *19*, 319. [CrossRef]

18. Valadez-Cosmes, P.; Vázquez-Martínez, E.R.; Cerbón, M.; Camacho-Arroyo, I. Membrane progesterone receptors in reproduction and cancer. *Mol. Cell. Endocrinol.* **2016**, *434*, 166–175. [CrossRef]
19. Ganesh, K.; Stadler, Z.K.; Cercek, A.; Mendelsohn, R.B.; Shia, J.; Segal, N.H.; Diaz, L.A., Jr. Immunotherapy in colorectal cancer: Rationale, challenges and potential. *Nat. Rev. Gastroenterol. Hepatol.* **2019**, *16*, 361–375. [CrossRef]
20. Kabe, Y.; Nakane, T.; Koike, I.; Yamamoto, T.; Sugiura, Y.; Harada, E.; Sugase, K.; Shimamura, T.; Ohmura, M.; Muraoka, K.; et al. Haem-dependent dimerization of PGRMC1/Sigma-2 receptor facilitates cancer proliferation and chemoresistance. *Nat. Commun.* **2016**, *7*, 11030. [CrossRef]
21. Kabe, Y.; Handa, H.; Suematsu, M. Function and structural regulation of the carbon monoxide (CO)-responsive membrane protein PGRMC1. *J. Clin. Biochem. Nutr.* **2018**, *63*, 12–17. [CrossRef] [PubMed]
22. Kim, J.Y.; Kim, S.Y.; Choi, H.S.; An, S.; Ryu, C.J. Epitope mapping of anti-PGRMC1 antibodies reveals the non-conventional membrane topology of PGRMC1 on the cell surface. *Sci. Rep.* **2019**, *9*, 653. [CrossRef] [PubMed]
23. Kaklamanos, I.G.; Bathe, O.F.; Franceschi, D.; Lazaris, A.C.; Davaris, P.; Glinatsis, M.; Golematis, B.C. Expression of receptors for estrogen and progesterone in malignant colonic mucosa as a prognostic factor for patient survival. *J. Surg. Oncol.* **1999**, *72*, 225–229. [CrossRef]
24. Zavarhei, M.D.; Bidgoli, S.A.; Ziyarani, M.M.; Shariatpanahi, M.A.F. Progesterone Receptor Positive Colorectal Tumors Have Lower Thymidine Phosphorylase Expression: An Immunohistochemical Study. *Pak. J. Biol. Sci.* **2007**, *10*, 4485–4489. [CrossRef] [PubMed]
25. Qasim, B.J.; Ali, H.H.; Hussein, A.G. Immunohistochemical expression of estrogen and progesterone receptors in human colorectal adenoma and carcinoma using specified automated cellular image analysis system: A clinicopathological study. *Oman Med. J.* **2011**, *26*, 307–314. [CrossRef]
26. Liu, D. Gene signatures of estrogen and progesterone receptor pathways predict the prognosis of colorectal cancer. *FEBS J.* **2016**, *283*, 3115–3133. [CrossRef]
27. Silverstein, J.; Kidder, W.; Fisher, S.; Hope, T.A.; Maisel, S.; Ng, D.; Van Ziffle, J.; Atreya, C.E.; Van Loon, K. Hormone receptor expression of colorectal cancer diagnosed during the peri-partum period. *Endocr. Connect.* **2019**, *8*, 1149–1158. [CrossRef]
28. Zhang, Y.L.; Wen, X.D.; Guo, X.; Huang, S.Q.; Wang, T.T.; Zhou, P.T.; Li, W.; Zhou, L.F.; Hu, Y.H. Progesterone suppresses the progression of colonic carcinoma by increasing the activity of the GADD45α/JNK/c-Jun signalling pathway. *Oncol. Rep.* **2021**, *45*, 95. [CrossRef]
29. Ponikwicka-Tyszko, D.; Chrusciel, M.; Stelmaszewska, J.; Bernaczyk, P.; Chrusciel, P.; Sztachelska, M.; Scheinin, M.; Bidzinski, M.; Szamatowicz, J.; Huhtaniemi, I.T.; et al. Molecular mechanisms underlying mifepristone's agonistic action on ovarian cancer progression. *EBioMedicine* **2019**, *47*, 170–183. [CrossRef]
30. Ponikwicka-Tyszko, D.; Chrusciel, M.; Pulawska, K.; Bernaczyk, P.; Sztachelska, M.; Guo, P.; Li, X.; Toppari, J.; Huhtaniemi, I.T.; Wołczyński, S.; et al. Mifepristone Treatment Promotes Testicular Leydig Cell Tumor Progression in Transgenic Mice. *Cancers* **2020**, *12*, 3263. [CrossRef]
31. Genazzani, A.R.; Monteleone, P.; Giannini, A.; Simoncini, T. Hormone therapy in the postmenopausal years: Considering benefits and risks in clinical practice. *Hum. Reprod. Update* **2021**, *27*, 1115–1150. [CrossRef] [PubMed]
32. Deli, T.; Orosz, M.; Jakab, A. Hormone Replacement Therapy in Cancer Survivors—Review of the Literature. *Pathol. Oncol. Res.* **2020**, *26*, 63–78. [CrossRef] [PubMed]
33. Gambacciani, M.; Monteleone, P.; Sacco, A.; Genazzani, A.R. Hormone replacement therapy and endometrial, ovarian and colorectal cancer. *Best Pract. Res. Clin. Endocrinol. Metab.* **2003**, *17*, 139–147. [CrossRef]
34. Lin, K.J.; Cheung, W.Y.; Lai, J.Y.C.; Giovannucci, E.L. The effect of estrogen vs. combined estrogen-progestogen therapy on the risk of colorectal cancer. *Int. J. Cancer* **2012**, *130*, 419–430. [CrossRef]
35. Jang, Y.C.; Huang, H.L.; Leung, C.Y. Association of hormone replacement therapy with mortality in colorectal cancer survivor: A systematic review and meta-analysis. *BMC Cancer* **2019**, *19*, 1199. [CrossRef]
36. Dinger, J.C.; Heinemann, L.A.J.; Möhner, S.; Thai, D.M.; Assmann, A. Colon cancer risk and different HRT formulations: A case-control study. *BMC Cancer* **2007**, *7*, 76. [CrossRef]
37. Gurney, E.P.; Nachtigall, M.J.; Nachtigall, L.E.; Naftolin, F. The Women's Health Initiative trial and related studies: 10 years later: A clinician's view. *J. Steroid Biochem. Mol. Biol.* **2014**, *142*, 4–11. [CrossRef] [PubMed]
38. Li, Z.; Wei, H.; Li, S.; Wu, P.; Mao, X. The Role of Progesterone Receptors in Breast Cancer. *Drug Des. Devel. Ther.* **2022**, *16*, 305–314. [CrossRef]
39. Campagnoli, C.; Clavel-Chapelon, F.; Kaaks, R.; Peris, C.; Berrino, F. Progestins and progesterone in hormone replacement therapy and the risk of breast cancer. *J. Steroid Biochem. Mol. Biol.* **2005**, *96*, 95–108. [CrossRef]
40. Ho, S.-M. Estrogen, progesterone and epithelial ovarian cancer. *Reprod. Biol. Endocrinol.* **2003**, *1*, 73. [CrossRef]
41. Negri, E.; Tzonou, A.; Beral, V.; Lagiou, P.; Trichopoulos, D.; Parazzini, F.; Franceschi, S.; Booth, M.; La Vecchia, C. Hormonal therapy for menopause and ovarian cancer in a collaborative re-analysis of European studies. *Int. J. Cancer* **1999**, *80*, 848–851. [CrossRef]
42. Whittmore, A.S.; Harris, R.; Itnyre, J. Characteristics Relating to Ovarian Cancer Risk: Collaborative Analysis of 12 US Case-Control Studies. *Am. J. Epidemiol.* **1992**, *136*, 1184–1203. [CrossRef]

43. Hulley, S.; Furberg, C.; Barrett-Connor, E.; Cauley, J.; Grady, D.; Haskell, W.; Knopp, R.; Lowery, M.; Satterfield, S.; Schrott, H.; et al. Noncardiovascular Disease Outcomes During 6.8 Years of Hormone Therapy. *JAMA* **2002**, *288*, 58. [CrossRef]
44. Peluso, J.J.; Gawkowska, A.; Liu, X.; Shioda, T.; Pru, J.K. Progesterone receptor membrane component-1 regulates the development and cisplatin sensitivity of human ovarian tumors in athymic nude mice. *Endocrinology* **2009**, *150*, 4846–4854. [CrossRef]
45. Pedroza, D.A.; Subramani, R.; Tiula, K.; Do, A.; Rashiraj, N.; Galvez, A.; Chatterjee, A.; Bencomo, A.; Rivera, S.; Lakshmanaswamy, R. Crosstalk between progesterone receptor membrane component 1 and estrogen receptor α promotes breast cancer cell proliferation. *Lab. Investig.* **2021**, *101*, 733–744. [CrossRef] [PubMed]
46. Friel, A.M.; Zhang, L.; Pru, C.A.; Clark, N.C.; McCallum, M.L.; Blok, L.J.; Shioda, T.; Peluso, J.J.; Rueda, B.R.; Pru, J.K. Progesterone receptor membrane component 1 deficiency attenuates growth while promoting chemosensitivity of human endometrial xenograft tumors. *Cancer Lett.* **2015**, *356*, 434–442. [CrossRef] [PubMed]
47. Mir, S.U.R.; Jin, L.; Craven, R.J. Neutrophil gelatinase-associated lipocalin (NGAL) expression is dependent on the tumor-associated sigma-2 receptor S2RPgrmc1. *J. Biol. Chem.* **2012**, *287*, 14494–14501. [CrossRef]
48. Tsai, H.W.; Ho, C.L.; Cheng, S.W.; Lin, Y.J.; Chen, C.C.; Cheng, P.N.; Yen, C.J.; Chang, T.T.; Chiang, P.M.; Chan, S.H.; et al. Progesterone receptor membrane component 1 as a potential prognostic biomarker for hepatocellular carcinoma. *World J. Gastroenterol.* **2018**, *24*, 1152–1166. [CrossRef]
49. Peluso, J.J.; Pappalardo, A.; Losel, R.; Wehling, M. Progesterone membrane receptor component 1 expression in the immature rat ovary and its role in mediating progesterone's antiapoptotic action. *Endocrinology* **2006**, *147*, 3133–3140. [CrossRef] [PubMed]
50. Peluso, J.J.; Liu, X.; Gawkowska, A.; Lodde, V.; Wu, C.A. Progesterone inhibits apoptosis in part by PGRMC1-regulated gene expression. *Mol. Cell. Endocrinol.* **2010**, *320*, 153–161. [CrossRef] [PubMed]
51. Peluso, J.J.; Liu, X.; Saunders, M.M.; Claffey, K.P.; Phoenix, K. Regulation of ovarian cancer cell viability and sensitivity to cisplatin by progesterone receptor membrane component-1. *J. Clin. Endocrinol. Metab.* **2008**, *93*, 1592–1599. [CrossRef]
52. Neubauer, H.; Clare, S.E.; Wozny, W.; Schwall, G.P.; Poznanović, S.; Stegmann, W.; Vogel, U.; Sotlar, K.; Wallwiener, D.; Kurek, R.; et al. Breast cancer proteomics reveals correlation between estrogen receptor status and differential phosphorylation of PGRMC1. *Breast Cancer Res.* **2008**, *10*, R85. [CrossRef] [PubMed]
53. Tutino, V.; Defrancesco, M.L.; Tolomeo, M.; De Nunzio, V.; Lorusso, D.; Paleni, D.; Caruso, M.G.; Notarnicola, M.; Barile, M. The expression of riboflavin transporters in human colorectal cancer. *Anticancer Res.* **2018**, *38*, 2659–2667. [CrossRef]
54. Gornowicz, A.; Szymanowska, A.; Mojzych, M.; Bielawski, K.; Bielawska, A. The Effect of Novel 7-methyl-5-phenyl-pyrazolo[4,3-e]tetrazolo[4,5-b][1,2,4]triazine Sulfonamide Derivatives on Apoptosis and Autophagy in DLD-1 and HT-29 Colon Cancer Cells. *Int. J. Mol. Sci.* **2020**, *21*, 5221. [CrossRef] [PubMed]
55. Tankiewicz-Kwedlo, A.; Hermanowicz, J.; Surażyński, A.; Rożkiewicz, D.; Pryczynicz, A.; Domaniewski, T.; Pawlak, K.; Kemona, A.; Pawlak, D. Erythropoietin accelerates tumor growth through increase of erythropoietin receptor (EpoR) as well as by the stimulation of angiogenesis in DLD-1 and HT-29 xenografts. *Mol. Cell. Biochem.* **2016**, *421*, 1–18. [CrossRef] [PubMed]
56. You, S.; Zhou, J.; Chen, S.; Zhou, P.; Lv, J.; Han, X.; Sun, Y. PTCH1, a receptor of Hedgehog signaling pathway, is correlated with metastatic potential of colorectal cancer. *Upsala J. Med. Sci.* **2010**, *115*, 169–175. [CrossRef]
57. Ryu, C.S.; Klein, K.; Zanger, U.M. Membrane associated progesterone receptors: Promiscuous proteins with pleiotropic functions—Focus on interactions with cytochromes P450. *Front. Pharmacol.* **2017**, *8*, 159. [CrossRef]
58. Kimura, I.; Nakayama, Y.; Zhao, Y.; Konishi, M.; Itoh, N. Neurotrophic effects of neudesin in the central nervous system. *Front. Neurosci.* **2013**, *7*, 111. [CrossRef]
59. Zhao, Z.; Wang, S.; Lin, Y.; Miao, Y.; Zeng, Y.; Nie, Y.; Guo, P.; Jiang, G.; Wu, J. Epithelial-mesenchymal transition in cancer: Role of the IL-8/IL-8R axis. *Oncol. Lett.* **2017**, *13*, 4577–4584. [CrossRef]
60. Kamińska, J.; Lyson, T.; Chrzanowski, R.; Sawicki, K.; Milewska, A.J.; Tylicka, M.; Zińczuk, J.; Matowicka-Karna, J.; Dymicka-Piekarska, V.; Mariak, Z.; et al. Ratio of IL-8 in CSF Versus Serum Is Elevated in Patients with Unruptured Brain Aneurysm. *J. Clin. Med.* **2020**, *9*, 1761. [CrossRef]
61. Liu, Q.; Li, A.; Tian, Y.; Wu, J.D.; Liu, Y.; Li, T.; Chen, Y.; Han, X.; Wu, K. The CXCL8-CXCR1/2 pathways in cancer. *Cytokine Growth Factor Rev.* **2016**, *31*, 61–71. [CrossRef] [PubMed]
62. Koper, O.M.; Kamińska, J.; Sawicki, K.; Reszeć, J.; Rutkowski, R.; Jadeszko, M.; Mariak, Z.; Dymicka-Piekarska, V.; Kemona, H. Cerebrospinal fluid and serum IL-8, CCL2, and ICAM-1 concentrations in astrocytic brain tumor patients. *Irish J. Med. Sci.* **2018**, *187*, 767–775. [CrossRef] [PubMed]
63. Fousek, K.; Horn, L.A.; Palena, C. Interleukin-8: A chemokine at the intersection of cancer plasticity, angiogenesis, and immune suppression. *Pharmacol. Ther.* **2021**, *219*, 107692. [CrossRef]
64. Lin, L.; Li, L.; Ma, G.; Kang, Y.; Wang, X.; He, J. Overexpression of IL-8 and Wnt2 is associated with prognosis of gastric cancer. *Folia Histochem. Cytobiol.* **2022**, *60*, 66–73. [CrossRef]
65. Koper-Lenkiewicz, O.M.; Kamińska, J.; Reszeć, J.; Dymicka-Piekarska, V.; Ostrowska, H.; Karpińska, M.; Matowicka-Karna, J.; Tylicka, M. Elevated plasma 20S proteasome chymotrypsin-like activity is correlated with IL-8 levels and associated with an increased risk of death in glial brain tumor patients. *PLoS ONE* **2020**, *15*, e0238406. [CrossRef] [PubMed]

66. de Zuccari, D.A.P.C.; Leonel, C.; Castro, R.; Gelaleti, G.B.; Jardim, B.V.; Moscheta, M.G.; Regiani, V.R.; Ferreira, L.C.; Lopes, J.R.; Neto, D. de S.; et al. An immunohistochemical study of interleukin-8 (IL-8) in breast cancer. *Acta Histochem.* **2012**, *114*, 571–576. [CrossRef]
67. George, E.; Andrew, M.; Maria, T.; Argyro, V.; Elias, K. Angiodrastic Chemokines Production by Colonic Cancer Cell Lines. *Onco* **2022**, *2*, 69–84. [CrossRef]

Disclaimer/Publisher's Note: The statements, opinions and data contained in all publications are solely those of the individual author(s) and contributor(s) and not of MDPI and/or the editor(s). MDPI and/or the editor(s) disclaim responsibility for any injury to people or property resulting from any ideas, methods, instructions or products referred to in the content.

Review

Harnessing Ferroptosis to Overcome Drug Resistance in Colorectal Cancer: Promising Therapeutic Approaches

Xiaofei Cheng [1,*], Feng Zhao [2], Bingxin Ke [1], Dong Chen [1] and Fanlong Liu [1,*]

[1] Department of Colorectal Surgery, The First Affiliated Hospital, School of Medicine, Zhejiang University, Hangzhou 310003, China; 1514012@zju.edu.cn (B.K.); cdking@zju.edu.cn (D.C.)

[2] Department of Radiation Oncology, The First Affiliated Hospital, School of Medicine, Zhejiang University, Hangzhou 310030, China; zju_zhaofeng@zju.edu.cn

* Correspondence: xfcheng@zju.edu.cn (X.C.); fanlong_liu@zju.edu.cn (F.L.)

Simple Summary: In the realm of colorectal cancer treatment, drug resistance is a formidable obstacle. However, a ray of hope emerges in the form of ferroptosis, a unique iron-driven cell death mechanism. This review delves into the role of ferroptosis in colorectal cancer and its implications for drug resistance. Unlike conventional cell death pathways, such as apoptosis and necrosis, ferroptosis offers distinct advantages. We explore current research breakthroughs, including innovative strategies like ferroptosis inducers, iron metabolism and lipid peroxidation interventions, and combination therapies. Additionally, we investigate the potential of immunotherapy in modulating ferroptosis. While targeting ferroptosis presents notable strengths, it comes with challenges in specificity and drug development. Looking ahead, this review underscores the promise of ferroptosis-based therapies in colorectal cancer and emphasizes the need for ongoing research to unlock its full potential, offering renewed hope for colorectal cancer patients.

Abstract: Drug resistance remains a significant challenge in the treatment of colorectal cancer (CRC). In recent years, the emerging field of ferroptosis, a unique form of regulated cell death characterized by iron-dependent lipid peroxidation, has offered new insights and potential therapeutic strategies for overcoming drug resistance in CRC. This review examines the role of ferroptosis in CRC and its impact on drug resistance. It highlights the distinctive features and advantages of ferroptosis compared to other cell death pathways, such as apoptosis and necrosis. Furthermore, the review discusses current research advances in the field, including novel treatment approaches that target ferroptosis. These approaches involve the use of ferroptosis inducers, interventions in iron metabolism and lipid peroxidation, and combination therapies to enhance the efficacy of ferroptosis. The review also explores the potential of immunotherapy in modulating ferroptosis as a therapeutic strategy. Additionally, it evaluates the strengths and limitations of targeting ferroptosis, such as its selectivity, low side effects, and potential to overcome resistance, as well as challenges related to treatment specificity and drug development. Looking to the future, this review discusses the prospects of ferroptosis-based therapies in CRC, emphasizing the importance of further research to elucidate the interaction between ferroptosis and drug resistance. It proposes future directions for more effective treatment strategies, including the development of new therapeutic approaches, combination therapies, and integration with emerging fields such as precision medicine. In conclusion, harnessing ferroptosis represents a promising avenue for overcoming drug resistance in CRC. Continued research efforts in this field are crucial for optimizing therapeutic outcomes and providing hope for CRC patients.

Keywords: colorectal cancer (CRC); ferroptosis; drug resistance; treatment strategies; immunotherapy

1. Introduction

Drug resistance presents a pervasive challenge in colorectal cancer (CRC) treatment [1,2]. Despite significant advancements in chemotherapy and targeted therapies, the emergence

cancer cells, leading to therapeutic resistance [47]. Furthermore, the TME can elicit innate resistance to targeted therapies [50]. Epithelial–mesenchymal transitions (EMTs), a process associated with increased invasiveness and drug resistance, can be induced by changes in the TME. The TME can also modulate the immune response, which can impact therapeutic outcomes [51]. Additionally, interactions between cancer cells and their surrounding stromal cells, such as cancer-associated fibroblasts (CAFs) or tumor-associated macrophages (TAMs), create a supportive niche that enhances tumor cell survival and contributes to therapy resistance [52,53]. Understanding these influences can guide the development of strategies targeting the tumor microenvironment to overcome drug resistance and improve treatment outcomes in CRC patients.

Drug resistance stands out as a pivotal factor contributing to treatment failure in CRC. It is crucial to underscore that drug resistance significantly limits the effectiveness of therapies, resulting in tumor recurrence and progression during treatment. Despite advancements in therapeutic approaches, the emergence of drug-resistant cancer cells impedes the ability of drugs to effectively eradicate tumors. This phenomenon not only compromises the initial response to treatment but also poses long-term management challenges. The presence of drug resistance underscores the need for a comprehensive understanding of its underlying mechanisms and the development of innovative strategies to overcome this obstacle. By addressing drug resistance as the primary factor contributing to treatment failure in CRC, we can strive to improve patient outcomes and advance the success of therapeutic interventions.

3. Mechanisms of Ferroptosis

Ferroptosis is a unique form of regulated cell death characterized by the iron-dependent accumulation of lipid peroxides, leading to oxidative damage and cell membrane rupture [54]. It is distinct from other cell death modalities, such as apoptosis and necrosis, and is driven by dysregulation in pathways involved in iron metabolism, lipid peroxidation, and antioxidant defense. Key players in ferroptosis include glutathione peroxidase 4 (GPX4), which protects cells from lipid peroxidation, and system Xc-, a cystine/glutamate antiporter involved in maintaining intracellular redox balance [55]. Inclusion of a diagram depicting the iron death pathway can be found in Figure 1. Understanding the mechanisms of ferroptosis can provide insights into its role in various diseases and pave the way for the development of novel therapeutic strategies targeting this form of cell death.

3.1. Canonical GPX4-Regulated Pathway

The canonical GPX4-regulated pathway plays a pivotal role in the induction of ferroptosis, as revealed by the study conducted by Yang et al. [56]. Their findings demonstrated that the direct or indirect inhibition of GPX4 through glutathione (GSH) depletion is essential for triggering ferroptosis [56]. GPX4, functioning as a glutathione peroxidase, is responsible for catalyzing the conversion of phospholipid hydroperoxides (PLOOHs) into phospholipid alcohols, a critical process in maintaining cell membrane integrity. The availability of intracellular GSH, which is regulated by the cystine/glutamate antiporter (xCT) system and glutamate–cysteine ligase, influences GPX4 activity and determines the susceptibility to ferroptosis [57]. Inactivation of GPX4 leads to the accumulation of PLOOHs, resulting in cellular membrane damage and subsequent ferroptotic cell death [58]. Acetyl-CoA and the MEVALONATE pathway also play crucial roles in cells and influence their sensitivity to ferroptosis [59,60]. Isopentenyl pyrophosphate (IPP) serves as a pivotal intermediate in this process, participating in the biosynthesis of cell membranes, which is directly related to the occurrence of ferroptosis [60]. Additionally, selenium (Sec) functions as a cofactor for GPX4, a key protein in the ferroptosis pathway [61]. GPX4 maintains cell membrane integrity by reducing lipid peroxides, thus inhibiting the onset of ferroptosis. Targeting the canonical GPX4-regulated pathway shows great promise as a therapeutic strategy for selectively inducing ferroptosis in cancer cells. Further research efforts are warranted to deepen our understanding of this pathway and develop innovative interventions to exploit the therapeutic potential of ferroptosis in cancer treatment.

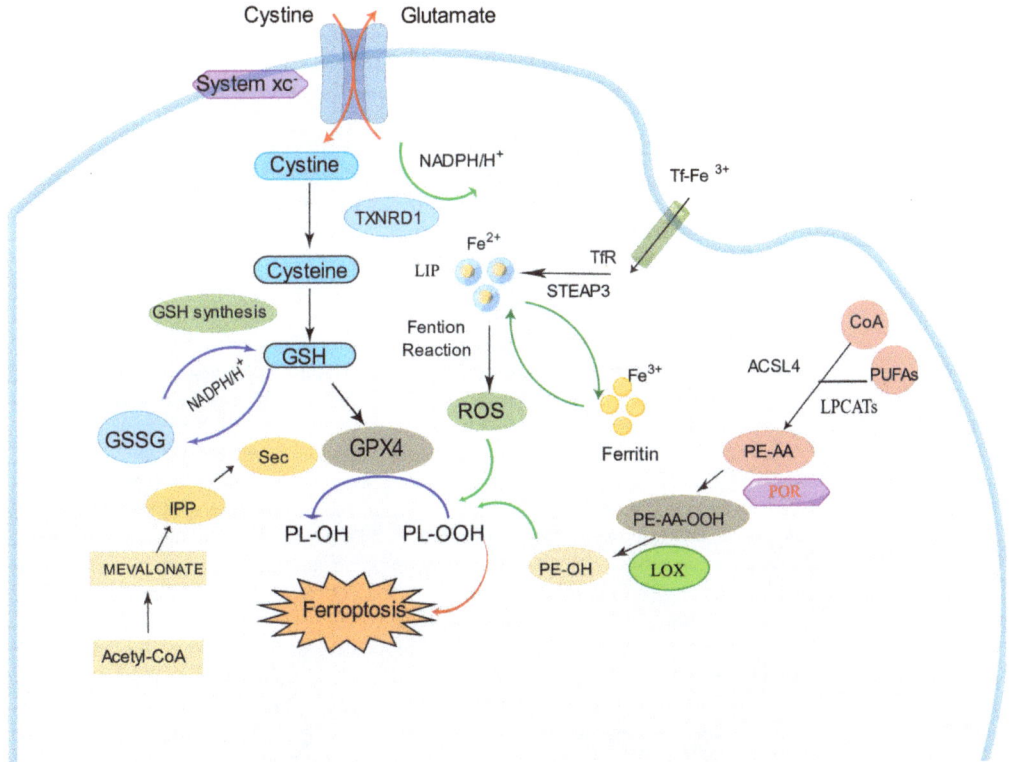

Figure 1. Classic signaling pathways of ferroptosis. Illustration depicting the fundamental signaling pathways associated with ferroptosis. The figure outlines the canonical GPX4-regulated pathway, focusing on the role of GPX4 in lipid peroxide reduction, the iron metabolism pathway elucidating the iron-dependent aspects of ferroptosis, and the lipid metabolism pathway highlighting lipid peroxidation and its contribution to cell membrane rupture. These interconnected pathways underscore the multifaceted nature of ferroptotic cell death. The figures were created using the Figdraw software (https://www.figdraw.com, accessed on 13 August 2023).

3.2. Iron Metabolism Pathway

Iron, particularly in its Fe^{2+} form, is a central player in driving ferroptosis through the generation of reactive oxygen species (ROS) via the Fenton reaction [62]. This reaction involves the interaction of Fe^{2+} with hydrogen peroxide (H_2O_2), leading to the production of highly reactive hydroxyl radicals (•OH). These hydroxyl radicals, in turn, initiate lipid peroxidation, a hallmark of ferroptosis, by attacking polyunsaturated fatty acids (PUFAs) in cell membranes [62]. Transferrin receptors (TFR) and ferritin, key components of the iron metabolism pathway, play significant roles in regulating intracellular iron levels. TFR facilitates the uptake of iron by binding to transferrin, which transports Fe^{3+} into the cell via receptor-mediated endocytosis [63]. Ferritin, on the other hand, acts as an iron storage protein, sequestering excess intracellular iron and preventing it from participating in the Fenton reaction [64]. This delicate balance between iron uptake and storage is crucial for determining a cell's susceptibility to ferroptosis.

Furthermore, the role of iron transporters, such as STEAP3 (six-transmembrane epithelial antigen of prostate 3), in facilitating iron uptake and releasing Fe^{2+} into the labile iron pool (LIP) cannot be overstated [65,66]. The LIP is a critical reservoir of Fe^{2+} that fuels the Fenton reaction and subsequently drives lipid peroxidation, a key event in ferroptosis [67]. Understanding the complex interactions between iron metabolism, ROS generation, and lipid peroxidation is critical for developing targeted interventions that exploit the vulnerabilities of cancer cells and enhance the effectiveness of ferroptosis-based therapies.

3.3. Lipid Metabolism Pathway

Ferroptosis is primarily driven by the peroxidation of membrane phospholipids, resulting in the formation of PLOOHs, which further decompose into 4-hydroxynonenal or malondialdehyde [68,69]. The accumulation of lipid peroxidation products induces membrane instability and permeabilization, ultimately leading to cell death [70].

In nonenzymatic lipid peroxidation, the conversion of polyunsaturated fatty acids (PUFAs) into acyl-CoA is facilitated by acyl-CoA synthetase long-chain family member 4 (ACSL4), which ligates PUFAs with coenzyme A (CoA) [59]. These acyl-CoA molecules can then be re-esterified into phospholipids by lysophosphatidylcholine acyltransferases (LPCATs), forming phospholipids. Therefore, the regulation of ACSL4 and LPCATs plays a critical role in determining the sensitivity to ferroptosis [69]. Enzymatic lipid peroxidation involves the activity of lipoxygenases (LOX) and cytochrome P450 oxidoreductase (POR) [71]. LOX enzymes, which contain nonheme iron, directly catalyze the deoxygenation of free and esterified PUFAs, resulting in the production of PLOOHs. Studies have demonstrated that overexpression of specific LOX isoforms increases cellular susceptibility to ferroptosis, while LOX inhibitors act as effective antioxidants, protecting cells from lipid peroxidation [72]. In 2020, Zou et al. identified POR as a crucial mediator of ferroptotic cell death in cancer cells through genome-wide CRISPR-Cas9 screens [73]. Previous research has suggested that P450 enzymes can accept electrons from POR and catalyze the peroxidation of PUFAs [74,75]. The pro-ferroptotic role of POR has been further supported by genetic depletion experiments across different cell types and lineages.

Understanding the intricate dynamics of the lipid metabolism pathway is essential for unraveling the mechanisms underlying ferroptosis. Further research in this field will enhance our knowledge of the regulatory mechanisms of lipid peroxidation and identify potential targets for therapeutic interventions aimed at promoting or inhibiting ferroptosis. These insights offer new avenues for the development of innovative strategies for cancer treatment.

4. Role of Ferroptosis in CRC and Its Impact on Drug Resistance

Ferroptosis has garnered significant attention in CRC research, offering novel insights and strategies for CRC treatment and prognosis. The relationships between certain genes and ferroptosis in CRC are illustrated in Figure 2. Elevated expression levels of several ferroptosis-related factors have been observed in CRC tissues, including TIGAR, AADAC, and a series of ferroptosis-related genes (FRGs) [76,77]. The dysregulated expression of these factors is closely associated with clinical outcomes in CRC patients, providing crucial prognostic and therapeutic predictive indicators [78,79]. Furthermore, researchers have discovered that by modulating the ferroptosis pathway, it is possible to enhance the sensitivity of CRC cells to specific drugs such as erastin and RSL3 [80,81]. These drugs induce ferroptosis in CRC cells, presenting a novel approach to CRC therapy. Some studies have identified gene signatures related to ferroptosis, including ACACA, GSS, and NFS1, among others, which have been validated as independent prognostic factors for CRC, outperforming the traditional TNM staging system in survival prediction [79]. The aberrant expression of these genes is closely correlated with CRC development and treatment response, offering diversified therapeutic options for patients.

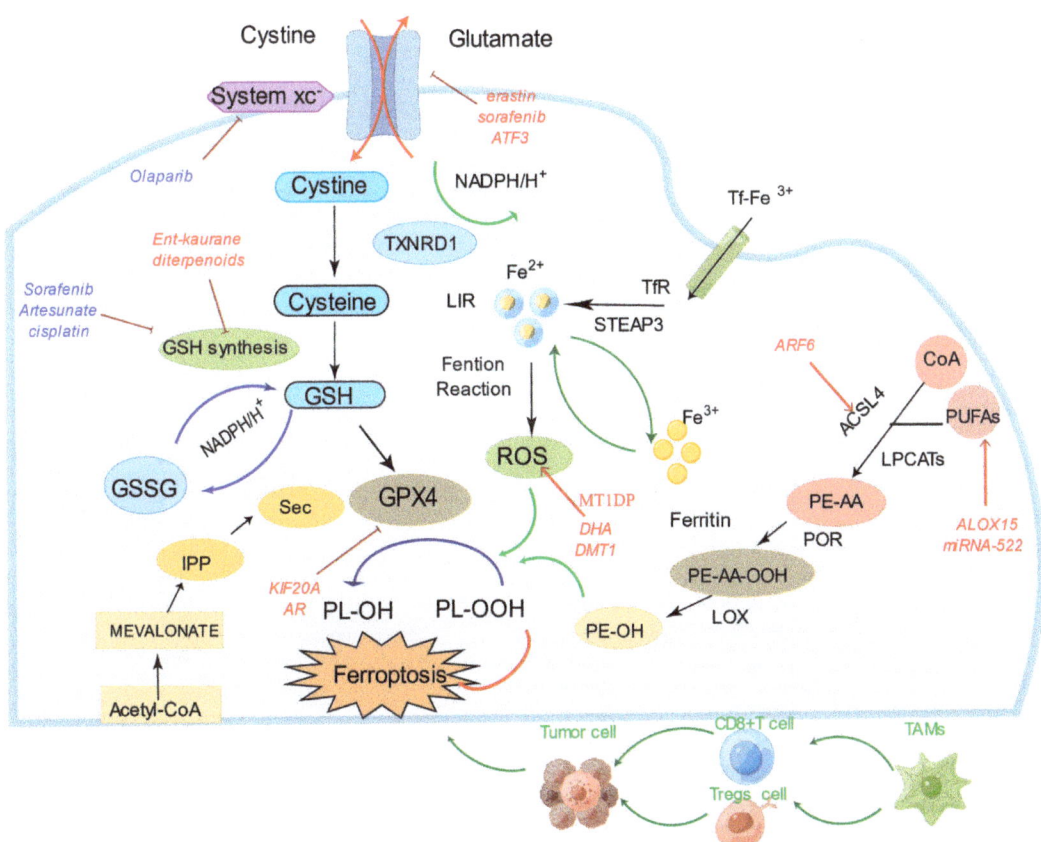

Figure 2. Mechanisms of drug resistance reversal through ferroptosis targeting. This figure illustrates diverse molecular interactions and sites of action for various agents aimed at reversing drug resistance through ferroptosis modulation. In the diagram, red text denotes sites of action for agents targeting ferroptosis reversal in chemotherapy, blue text represents sites of action for agents targeting ferroptosis reversal specifically, and green text indicates sites of action for agents targeting ferroptosis reversal in immunotherapy. The integrated approach to targeting these pathways highlights their potential synergistic effects in overcoming drug resistance. The figures were created using the Figdraw software (https://www.figdraw.com, accessed on 13 August 2023).

On a specific gene and protein level, certain studies emphasize the roles of particular factors in CRC, such as KRAS mutations, Nodal, MT1G, SFRS9, and p53 [82–86]. The dysregulated expression of these factors is intricately linked to CRC progression and drug resistance, rendering them potential targets for future therapeutic interventions. Additionally, researchers have explored novel approaches to modulate the ferroptosis pathway, including siRNA nanoparticles, GCH1 inhibitors, and the CUL9-HNRNPC-MATE1 negative feedback loop, with the potential to enhance the sensitivity of CRC cells to ferroptosis inducers and overcome drug resistance [87–89]. Furthermore, miRNAs and circular RNAs (circRNAs) play pivotal roles in CRC by influencing the development of the disease through the regulation of ferroptosis-related pathways [90–92]. Collectively, these studies reveal the diversity and complexity of ferroptosis in CRC, providing valuable clues and directions for future research and treatment. In conclusion, research on ferroptosis in CRC has made significant progress, spanning multiple levels, from molecular mechanisms to potential therapeutic strategies, all contributing to a deeper understanding of CRC's development and treatment. These studies offer substantial support for future personalized treatments and the overcoming of treatment resistance.

5. Potential Therapeutic Approaches Targeting Ferroptosis

Ferroptosis stands as a promising frontier in CRC therapy, offering a new perspective on overcoming the hurdles of treatment resistance. As we delve into the intricate mechanisms underlying these approaches, Figure 2 serves as a comprehensive visual guide, shedding light on the potential synergies between targeted drug interventions, immunotherapy, and the induction of ferroptosis. In the following sections, we will elaborate on the interplay between traditional chemotherapy drug resistance and ferroptosis, the intriguing relationship between targeted drug resistance and the ferroptosis process, and how the realm of immunotherapy intertwines with ferroptosis to shape the future of cancer treatment paradigms.

5.1. Targeted Drug Therapy for Ferroptosis

Chemotherapy has remained an essential approach in cancer treatment, but drug resistance remains a significant factor contributing to poor patient prognosis. Ferroptosis, with its molecular mechanisms, plays a crucial role in reducing chemotherapeutic drug resistance. Pathways involved in lipid metabolism, iron metabolism, and the classical GPX4 pathway are implicated in drug resistance in CRC and other malignancies. In lipid metabolism, ACSL4 participates in the lipid oxidation pathway by converting the AA and AdA in PUFAs into coenzyme derivatives, leading to the production of oxidized lipid molecules [93,94]. Another enzyme, LOX, mediates ferroptosis non-enzymatically by directly oxidizing PUFAs. ALOX15, a key player in gastric cancer, inhibits ferroptosis [95,96]. Decreasing miRNA-522 and increasing ALOX15 has emerged as a novel treatment strategy to reverse drug resistance, particularly resistance to cisplatin/paclitaxel [95].

Iron metabolic pathways are also involved in reversing drug resistance. Dihydroartemisinin (DHA), a safe and promising therapeutic agent, selectively induces ferroptosis in cancer cells. DHA intensifies the cytotoxicity of cisplatin by impairing mitochondrial homeostasis, increasing mitochondrial-derived ROS, and promoting ferroptosis through the accumulation of free iron and lipid peroxidation [97,98]. Blocking lysosomal iron translocation through the inhibition of DMT1 in CSC leads to iron accumulation in lysosomes, ROS production, and ferroptosis-mediated cell death [99,100]. Furthermore, indirectly triggering ferroptosis by blocking GSH synthesis or inhibiting system Xc- enhances the reversal of drug resistance. Ent-kaurane diterpenoids overcome cisplatin resistance by targeting peroxiredoxin I/II and consuming GSH to induce ferroptosis [101]. In head and neck cancer, inhibiting system Xc- and the Nrf2/Keap1/system Xc- signaling pathway can overcome cisplatin resistance [102,103].

In parallel, a series of studies has underscored the potential of targeted drug therapy for ferroptosis in CRC. For instance, adipose-derived exosomes upregulate the microsomal triglyceride transfer protein (MTTP), reducing ferroptosis susceptibility and contributing to oxaliplatin resistance [104]. Aspirin enhances the sensitivity of CRC cells with oncogenic PIK3CA activation to ferroptosis induction by inhibiting the AKT/mTOR pathway, suppressing SREBP-1 expression, and reducing SCD1-mediated lipogenesis [105]. Loss of the metabolic enzyme NFS1 in CRC cells heightens sensitivity to oxaliplatin and induces PANoptosis [106]. KIF20A, associated with oxaliplatin resistance, becomes a potential target for sensitizing CRC cells to this drug [107]. Moreover, Lipocalin 2 overexpression in colon cancer cells has conferred resistance to 5-fluorouracil through ferroptosis inhibition. FAM98A overexpression in CRC tissues has been linked to drug resistance, with metformin demonstrating potential in reversing FAM98A-mediated 5-FU resistance [108]. The cyclin-dependent kinase 1 (CDK1) has been identified as a key contributor to oxaliplatin resistance in CRC, highlighting the prospect of CDK1 inhibitors for treating oxaliplatin-resistant CRC patients [109]. Additionally, PYCR1, an oncogenic gene, has been associated with reducing the sensitivity of CRC cells to 5-fluorouracil and promoting tumor growth [110].

In summary, targeted drug therapy for ferroptosis holds immense potential in overcoming chemotherapy resistance in CRC. By addressing key molecular mechanisms and therapeutic targets, researchers aim to sensitize CRC cells to ferroptosis-inducing agents, ultimately improving treatment outcomes for CRC patients. These findings collectively provide a multifaceted approach to combatting drug resistance and enhancing the efficacy of CRC therapies.

5.2. Targeted Therapy for Ferroptosis

In contrast to chemotherapy, targeted therapy has emerged as an effective treatment with fewer side effects on normal cells. For example, the presence of RAS mutations in about half of metastatic CRC patients greatly limits the efficacy of cetuximab. β-elemene, a natural product derived from turmeric, combined with cetuximab, exhibits high cytotoxicity toward metastatic CRC cells with KRAS mutations [111]. This combination therapy induces ferroptosis and inhibits the epithelial–mesenchymal transition. Olaparib, a well-known poly (ADP-ribose) polymerase inhibitor, promotes ferroptosis by inhibiting SLC7A11-mediated GSH synthesis [112]. Combined with FINs, it sensitizes BRCA-activated ovarian cancer cells and xenografts [112]. In TNBC cells resistant to gefitinib, inhibiting and inducing ferroptosis enhances sensitivity to gefitinib [113]. Sorafenib, the first approved systemic medicine for advanced hepatocellular carcinoma, faces acquired resistance. Similarly, like erastin, cisplatin resistance can be overcome by depleting GSH accompanied by GPx inactivation. Combining erastin with cisplatin demonstrates enhanced antitumor activity compared to cisplatin alone, as their mechanisms of action differ [95,114].

Metallothionein (MT) is a multifunctional protein that plays a pivotal role in various aspects of ferroptosis regulation, drug resistance, ROS elimination, and Fe2+ metabolism. In the context of CRC, MT1G has emerged as a significant factor influencing ferroptosis susceptibility, contributing to drug resistance and disease progression. MT1G downregulation in clear cell renal cell carcinoma is associated with advanced stages and higher malignancy, potentially due to its negative regulatory role in ferroptosis and its influence on GSH metabolism [115]. Furthermore, in hepatocellular carcinoma, MT-1G is identified as a key player in the regulation of ferroptosis and is associated with sorafenib resistance. This highlights the importance of MTs in HCC treatment and potential strategies for overcoming resistance [116]. Additionally, in leukemia, MTs, particularly the MT2A and MT1M isoforms, are implicated in the modulation of ferroptosis induced by DHA [117]. Their role in regulating iron metabolism and cellular antioxidant responses underscores their impact on leukemic cell susceptibility to ferroptotic cell death. Moreover, metallothionein 1D pseudogene (MT1DP) is shown to be a crucial regulator of ferroptosis in non-small cell lung cancer (NSCLC), sensitizing cancer cells to erastin-induced ferroptosis by downregulating NRF2 and enhancing oxidative stress. This novel therapeutic strategy holds promise for the treatment of lung cancer [118]. Furthermore, MTs' interaction with ferritin, a protein involved in iron storage, is suggested to have a potential impact on iron release under oxidative conditions. In the context of Parkinson's disease (PD), where iron accumulation is linked to ferroptosis and disease pathogenesis, understanding the interplay between MTs and iron metabolism may offer insights into potential neuroprotective strategies [119].

Targeted therapy for ferroptosis offers innovative strategies for sensitizing cancer cells to ferroptosis-inducing agents, with a focus on KRAS-mutant CRC. Combining cetuximab with RSL3 enhances ferroptosis in KRAS-mutant CRC cells, providing a strategy to overcome drug resistance in this specific subgroup [120]. Additionally, apatinib has been identified as a ferroptosis promoter in CRC cells by targeting the ELOVL6/ACSL4 pathway, suggesting its potential as a valuable addition to CRC treatment strategies [121]. A novel ferroptosis inducer, talaroconvolutin A (TalaA), has demonstrated remarkable efficacy in suppressing CRC growth in mouse models, positioning it as a potent candidate for CRC therapy via ferroptosis induction [122]. Lastly, co-treatment with 3-Bromopyruvate (3-BP) and cetuximab has emerged as a promising strategy to overcome cetuximab resistance in CRC by inducing ferroptosis synergistically [123].

In summary, targeted therapy for ferroptosis offers an innovative and effective approach to combat drug resistance and enhance treatment outcomes in various cancer types, including CRC. These strategies hold promise for improving the prognosis and quality of life for cancer patients while minimizing the impact on healthy cells.

5.3. Immunotherapy Targeting Ferroptosis

Immunotherapy has emerged as a promising approach for cancer treatment, and recent research has shown that it can also regulate ferroptosis. Wang et al. found that CD8+ T cells activated by immunotherapy enhance ferroptosis-specific lipid peroxidation in tumor cells, leading to increased tumor cell death and improved anti-tumor efficacy [124]. This suggests that targeting ferroptosis could enhance the effectiveness of immunotherapy in cancer treatment. The mechanistic link between ferroptosis and cancer has been further explored in recent studies. One paper reviewed the regulatory mechanisms of mTORC1 and ferroptosis and proposed co-targeting mTOR and ferroptosis as a potential strategy for cancer treatment [125]. Another investigated ferroptosis as an autophagic cell death process and highlighted its relevance in cancer and cancer treatment [126]. One study reviewed the development of agents targeting molecules involved in ferroptosis, emphasizing the potential of ferroptosis as a therapeutic strategy for cancer [127]. A further study discussed the epigenetic regulators and metabolic changes associated with ferroptosis in cancer progression, suggesting that targeting ferroptosis-associated metabolism could improve the efficacy of cancer immunotherapy [128]. Others have demonstrated that immunotherapy sensitizes tumors to radiotherapy by promoting tumor cell ferroptosis, further supporting the potential synergy between immunotherapy and ferroptosis in cancer treatment [129].

The role of ferroptosis in cancer immunotherapy has also been recognized. One study highlighted ferroptosis as an effector pathway for cancer immunotherapy [130]. Another investigated the interaction between ferroptosis and immunotherapy in cancer cells and found that ferroptosis enhances the anti-tumor efficacy of immunotherapy through increased ferroptosis-specific lipid peroxidation and reduced cystine uptake induced by immunotherapy-activated CD8+ T cells [131]. Combining radiotherapy with PARP inhibitors, such as Niraparib, activates the cGAS signaling pathway and enhances ferroptosis, promoting an anti-tumor immune response [132]. Apolipoprotein L3 (APOL3) has been identified as a key modulator, positively affecting sensitivity to ferroptosis and improving CD8+ T cell-mediated anti-tumor responses in CRC [133]. Co-treatment with PR-619 and anti-PD1 inhibits CRC growth, induces ferroptosis, and enhances CD8+ T cell-mediated immunity [134]. Moreover, inhibiting CYP1B1, which contributes to ferroptosis resistance, enhances the sensitivity of CRC cells to anti-PD-1 antibody therapy [135]. These immunotherapy approaches can synergize with ferroptosis induction by enhancing the immune system's recognition and clearance of cancer cells undergoing ferroptotic cell death.

However, it is important to consider the limitations of immunotherapy targeting ferroptosis. Some tumors employ immune evasion mechanisms, such as downregulation of antigen presentation or upregulation of immunosuppressive factors, which can hinder the immune response and limit the effectiveness of immunotherapy. Additionally, tumor heterogeneity and individual variations in immune responses can impact the outcomes of immunotherapeutic interventions. Further research is needed to better understand these limitations and develop strategies to overcome them.

6. The Advantages and Limitations of Therapeutic Approaches Targeting Ferroptosis

Therapeutic strategies targeting ferroptosis offer several advantages compared to traditional treatment methods. One significant advantage is their high selectivity towards cancer cells, which can minimize damage to healthy tissues and reduce treatment-related side effects [136]. Additionally, ferroptosis induction may hold potential for overcoming drug resistance mechanisms, as it represents a distinct form of cell death that can bypass common resistance pathways. Moreover, targeting iron metabolism and lipid peroxidation, the core processes of ferroptosis, provides a unique opportunity to exploit vulnerabilities specific to cancer cells.

However, these therapeutic approaches targeting ferroptosis may also face certain challenges. One challenge is achieving treatment specificity and ensuring that the intervention selectively targets cancer cells without affecting normal cells. Another challenge lies in the development of effective and safe drugs that can modulate ferroptosis. The complex interplay of iron metabolism and lipid peroxidation pathways adds to the intricacy of drug development and requires thorough understanding to maximize therapeutic efficacy [136]. Furthermore, the potential irreversibility of ferroptotic cell death raises concerns regarding tissue damage and long-term effects.

7. Paving the Way for Future Developments in Therapeutic Approaches Targeting Ferroptosis

To further improve treatment outcomes and overcome the existing limitations, future research should focus on multiple fronts. Firstly, there is a need for continued exploration and development of novel therapeutic strategies that can specifically target ferroptosis. This includes the identification of new ferroptosis inducers and the refinement of existing ones to enhance their efficacy and safety profiles. Secondly, combining ferroptosis-based approaches with other treatment modalities, such as immunotherapy, targeted therapy, or precision medicine, holds great potential in optimizing treatment outcomes [137]. These synergistic combinations can exploit different pathways and vulnerabilities, leading to enhanced therapeutic efficacy.

Additionally, future developments should encompass the integration of ferroptosis-targeting strategies with emerging fields such as personalized medicine and precision oncology. Understanding the interplay between ferroptosis and specific molecular subtypes of cancer, as well as identifying predictive biomarkers, can enable the selection of appropriate patients and tailored treatment regimens for maximum effectiveness. The exploration of non-invasive imaging techniques to assess ferroptosis status in tumors and the development of strategies to modulate ferroptosis in a controlled manner are also promising directions for future research.

In conclusion, therapeutic approaches targeting ferroptosis offer distinct advantages, including selectivity, low side effects, and potential for overcoming drug resistance. However, challenges related to treatment specificity, drug development, and potential irreversibility exist. Future directions should focus on refining therapeutic strategies, exploring combination therapies, and integrating ferroptosis-based approaches with personalized medicine to optimize treatment outcomes and overcome current limitations.

8. Conclusions and Perspectives

The potential of targeting ferroptosis as a therapeutic strategy to overcome drug resistance in CRC is of significant importance. Ferroptosis, as a distinct form of cell death, offers unique advantages in selectively eliminating cancer cells and bypassing common resistance mechanisms. This detailed exploration of ferroptosis's role in CRC, its mechanisms, and its interplay with drug resistance mechanisms underscores the necessity for further research in this field. Promising advances have been made in identifying ferroptosis inducers, elucidating key regulators, and exploring combination therapies. However, there are challenges to address, including treatment specificity and the development of safe and effective drugs. Future directions should focus on refining therapeutic strategies, such as the development of novel ferroptosis inducers and the integration of ferroptosis-based approaches with other treatment modalities. Moreover, the incorporation of ferroptosis research into the realm of personalized medicine holds great potential. By understanding the complex interplay between ferroptosis and resistance mechanisms, identifying predictive biomarkers, and refining treatment selection, we can optimize therapeutic outcomes and provide hope for

CRC patients. Continued research efforts in this field will pave the way for more effective treatment strategies and better outcomes in CRC management.

Author Contributions: Conceptualization, X.C. and F.L.; investigation, F.Z. and B.K.; writing—original draft preparation, X.C. and writing—review and editing, X.C., D.C. and F.L. All authors have read and agreed to the published version of the manuscript.

Funding: This research received no external funding.

Conflicts of Interest: The authors declare no conflict of interest.

Abbreviations

PUFAs	polyunsaturated fatty acids	CRC	colorectal cancer
CoA	coenzyme A	5-FU	5-fluorouracil
LPCATs	Lys phosphatidylcholine acyltransferases	EGFR	epidermal growth factor receptor
LOX	lipoxygenases	VEGF	vascular endothelial growth factor
POR	P450 oxidoreductase	UGTs	UDP-glucuronosyltransferases
FRGs	ferroptosis-related genes	CYP	cytochrome P450
DHA	Dihydroartemisinin	ABC	ATP-binding cassette
MTTP	microsomal triglyceride transfer protein	MMR	mismatch repair
CDK1	cyclin-dependent kinase 1	HR	homologous recombination
MT	Metallothionein	MHC	major histocompatibility complex
MT1DP	metallothionein 1D pseudogene	CAFs	cancer-associated fibroblasts
TalaA	talaroconvolutin A	TAMs	tumor-associated macrophages
TME	tumor microenvironment	GPX4	glutathione peroxidase 4
TAMs	tumor-associated macrophages	GSH	glutathione
APOL3	Apolipoprotein L3	PLOOHs	phospholipid hydroperoxides
ABCB1	ATP-binding cassette subfamily B1	xCT	cystine/glutamate antiporter
CaSR	calcium-sensing receptor	ROS	reactive oxygen species
TME	tumor microenvironment	GLUT1	glucose transporter 1
EMT	Epithelial–mesenchymal transition	ACSL4	acyl-CoA synthetase long-chain family member 4

References

1. Luo, M.; Yang, X.; Chen, H.-N.; Nice, E.C.; Huang, C. Drug resistance in colorectal cancer: An epigenetic overview. *Biochim. Biophys. Acta Rev. Cancer* **2021**, *1876*, 188623. [CrossRef]
2. Friedmann Angeli, J.P.; Krysko, D.V.; Conrad, M. Ferroptosis at the crossroads of cancer-acquired drug resistance and immune evasion. *Nat. Rev. Cancer* **2019**, *19*, 405–414. [CrossRef]
3. Hutchinson, L. Colorectal cancer: A step closer to combating acquired resistance in CRC. *Nat. Rev. Gastroenterol. Hepatol.* **2012**, *9*, 427. [CrossRef]
4. Wang, Q.; Shen, X.; Chen, G.; Du, J. Drug Resistance in Colorectal Cancer: From Mechanism to Clinic. *Cancers* **2022**, *14*, 2928. [CrossRef]
5. Martini, G.; Ciardiello, D.; Vitiello, P.P.; Napolitano, S.; Cardone, C.; Cuomo, A.; Troiani, T.; Ciardiello, F.; Martinelli, E. Resistance to anti-epidermal growth factor receptor in metastatic colorectal cancer: What does still need to be addressed? *Cancer Treat. Rev.* **2020**, *86*, 102023. [CrossRef] [PubMed]
6. Van Cutsem, E.; Cervantes, A.; Adam, R.; Sobrero, A.; Van Krieken, J.H.; Aderka, D.; Aranda Aguilar, E.; Bardelli, A.; Benson, A.; Bodoky, G.; et al. ESMO consensus guidelines for the management of patients with metastatic colorectal cancer. *Ann. Oncol. Off. J. Eur. Soc. Med. Oncol.* **2016**, *27*, 1386–1422. [CrossRef] [PubMed]
7. Yaffee, P.; Osipov, A.; Tan, C.; Tuli, R.; Hendifar, A. Review of systemic therapies for locally advanced and metastatic rectal cancer. *J. Gastrointest. Oncol.* **2015**, *6*, 185–200. [CrossRef]
8. Tabernero, J.; Van Cutsem, E.; Díaz-Rubio, E.; Cervantes, A.; Humblet, Y.; André, T.; Van Laethem, J.-L.; Soulié, P.; Casado, E.; Verslype, C.; et al. Phase II trial of cetuximab in combination with fluorouracil, leucovorin, and oxaliplatin in the first-line treatment of metastatic colorectal cancer. *J. Clin. Oncol. Off. J. Am. Soc. Clin. Oncol.* **2007**, *25*, 5225–5232. [CrossRef] [PubMed]
9. Borner, M.; Koeberle, D.; Von Moos, R.; Saletti, P.; Rauch, D.; Hess, V.; Trojan, A.; Helbling, D.; Pestalozzi, B.; Caspar, C.; et al. Adding cetuximab to capecitabine plus oxaliplatin (XELOX) in first-line treatment of metastatic colorectal cancer: A randomized phase II trial of the Swiss Group for Clinical Cancer Research SAKK. *Ann. Oncol. Off. J. Eur. Soc. Med. Oncol.* **2008**, *19*, 1288–1292. [CrossRef]
10. Venook, A.P.; Niedzwiecki, D.; Lenz, H.-J.; Innocenti, F.; Fruth, B.; Meyerhardt, J.A.; Schrag, D.; Greene, C.; O'Neil, B.H.; Atkins, J.N.; et al. Effect of First-Line Chemotherapy Combined With Cetuximab or Bevacizumab on Overall Survival in Patients With KRAS Wild-Type Advanced or Metastatic Colorectal Cancer: A Randomized Clinical Trial. *JAMA* **2017**, *317*, 2392–2401. [CrossRef]

11. Hassannia, B.; Vandenabeele, P.; Vanden Berghe, T. Targeting Ferroptosis to Iron Out Cancer. *Cancer Cell* **2019**, *35*, 830–849. [CrossRef]
12. Jiang, X.; Stockwell, B.R.; Conrad, M. Ferroptosis: Mechanisms, biology and role in disease. *Nat. Rev. Mol. Cell Biol.* **2021**, *22*, 266–282. [CrossRef] [PubMed]
13. Chen, X.; Kang, R.; Kroemer, G.; Tang, D. Broadening horizons: The role of ferroptosis in cancer. *Nat. Rev. Clin. Oncol.* **2021**, *18*, 280–296. [CrossRef] [PubMed]
14. Zhang, C.; Liu, X.; Jin, S.; Chen, Y.; Guo, R. Ferroptosis in cancer therapy: A novel approach to reversing drug resistance. *Mol. Cancer* **2022**, *21*, 47. [CrossRef]
15. Misale, S.; Yaeger, R.; Hobor, S.; Scala, E.; Janakiraman, M.; Liska, D.; Valtorta, E.; Schiavo, R.; Buscarino, M.; Siravegna, G.; et al. Emergence of KRAS mutations and acquired resistance to anti-EGFR therapy in colorectal cancer. *Nature* **2012**, *486*, 532–536. [CrossRef]
16. Zhu, C.; Guan, X.; Zhang, X.; Luan, X.; Song, Z.; Cheng, X.; Zhang, W.; Qin, J.-J. Targeting KRAS mutant cancers: From druggable therapy to drug resistance. *Mol. Cancer* **2022**, *21*, 159. [CrossRef]
17. Douillard, J.-Y.; Oliner, K.S.; Siena, S.; Tabernero, J.; Burkes, R.; Barugel, M.; Humblet, Y.; Bodoky, G.; Cunningham, D.; Jassem, J.; et al. Panitumumab-FOLFOX4 treatment and RAS mutations in colorectal cancer. *N. Engl. J. Med.* **2013**, *369*, 1023–1034. [CrossRef] [PubMed]
18. Zahreddine, H.A.; Borden, K.L.B. Molecular Pathways: GLI1-Induced Drug Glucuronidation in Resistant Cancer Cells. *Clin. Cancer Res. Off. J. Am. Assoc. Cancer Res.* **2015**, *21*, 2207–2210. [CrossRef]
19. Mazerska, Z.; Mróz, A.; Pawłowska, M.; Augustin, E. The role of glucuronidation in drug resistance. *Pharmacol. Ther.* **2016**, *159*, 35–55. [CrossRef]
20. Zhao, M.; Ma, J.; Li, M.; Zhang, Y.; Jiang, B.; Zhao, X.; Huai, C.; Shen, L.; Zhang, N.; He, L.; et al. Cytochrome P450 Enzymes and Drug Metabolism in Humans. *Int. J. Mol. Sci.* **2021**, *22*, 12808. [CrossRef]
21. McFadyen, M.C.; McLeod, H.L.; Jackson, F.C.; Melvin, W.T.; Doehmer, J.; Murray, G.I. Cytochrome P450 CYP1B1 protein expression: A novel mechanism of anticancer drug resistance. *Biochem. Pharmacol.* **2001**, *62*, 207–212. [CrossRef]
22. Robey, R.W.; Pluchino, K.M.; Hall, M.D.; Fojo, A.T.; Bates, S.E.; Gottesman, M.M. Revisiting the role of ABC transporters in multidrug-resistant cancer. *Nat. Rev. Cancer* **2018**, *18*, 452–464. [CrossRef]
23. Hou, Y.-Q.; Wang, Y.-Y.; Wang, X.-C.; Liu, Y.; Zhang, C.-Z.; Chen, Z.-S.; Zhang, Z.; Wang, W.; Kong, D.-X. Multifaceted anti-colorectal tumor effect of digoxin on HCT8 and SW620 cells in vitro. *Gastroenterol. Rep.* **2020**, *8*, 465–475. [CrossRef] [PubMed]
24. Mashouri, L.; Yousefi, H.; Aref, A.R.; Ahadi, A.M.; Molaei, F.; Alahari, S.K. Exosomes: Composition, biogenesis, and mechanisms in cancer metastasis and drug resistance. *Mol. Cancer* **2019**, *18*, 75. [CrossRef] [PubMed]
25. Narayanankutty, A. PI3K/Akt/mTOR Pathway as a Therapeutic Target for Colorectal Cancer: A Review of Preclinical and Clinical Evidence. *Curr. Drug. Targets* **2019**, *20*, 1217–1226. [CrossRef]
26. Huang, J.; Chen, L.; Wu, J.; Ai, D.; Zhang, J.-Q.; Chen, T.-G.; Wang, L. Targeting the PI3K/AKT/mTOR Signaling Pathway in the Treatment of Human Diseases: Current Status, Trends, and Solutions. *J. Med. Chem.* **2022**, *65*, 16033–16061. [CrossRef] [PubMed]
27. Pashirzad, M.; Khorasanian, R.; Fard, M.M.; Arjmand, M.-H.; Langari, H.; Khazaei, M.; Soleimanpour, S.; Rezayi, M.; Ferns, G.A.; Hassanian, S.M.; et al. The Therapeutic Potential of MAPK/ERK Inhibitors in the Treatment of Colorectal Cancer. *Curr. Cancer Drug. Targets* **2021**, *21*, 932–943. [CrossRef] [PubMed]
28. Yang, G.; Huang, L.; Jia, H.; Aikemu, B.; Zhang, S.; Shao, Y.; Hong, H.; Yesseyeva, G.; Wang, C.; Li, S.; et al. NDRG1 enhances the sensitivity of cetuximab by modulating EGFR trafficking in colorectal cancer. *Oncogene* **2021**, *40*, 5993–6006. [CrossRef]
29. Koustas, E.; Karamouzis, M.V.; Mihailidou, C.; Schizas, D.; Papavassiliou, A.G. Co-targeting of EGFR and autophagy signaling is an emerging treatment strategy in metastatic colorectal cancer. *Cancer Lett.* **2017**, *396*, 94–102. [CrossRef]
30. Guièze, R.; Liu, V.M.; Rosebrock, D.; Jourdain, A.A.; Hernández-Sánchez, M.; Martinez Zurita, A.; Sun, J.; Ten Hacken, E.; Baranowski, K.; Thompson, P.A.; et al. Mitochondrial Reprogramming Underlies Resistance to BCL-2 Inhibition in Lymphoid Malignancies. *Cancer Cell* **2019**, *36*, 369–384.e13. [CrossRef]
31. Zhang, L.; Lu, Z.; Zhao, X. Targeting Bcl-2 for cancer therapy. *Biochim. Biophys. Acta Rev. Cancer* **2021**, *1876*, 188569. [CrossRef] [PubMed]
32. Nechiporuk, T.; Kurtz, S.E.; Nikolova, O.; Liu, T.; Jones, C.L.; D'Alessandro, A.; Culp-Hill, R.; d'Almeida, A.; Joshi, S.K.; Rosenberg, M.; et al. The TP53 Apoptotic Network Is a Primary Mediator of Resistance to BCL2 Inhibition in AML Cells. *Cancer Discov.* **2019**, *9*, 910–925. [CrossRef]
33. Hafezi, S.; Rahmani, M. Targeting BCL-2 in Cancer: Advances, Challenges, and Perspectives. *Cancers* **2021**, *13*, 1292. [CrossRef]
34. Russo, M.; Crisafulli, G.; Sogari, A.; Reilly, N.M.; Arena, S.; Lamba, S.; Bartolini, A.; Amodio, V.; Magrì, A.; Novara, L.; et al. Adaptive mutability of colorectal cancers in response to targeted therapies. *Science* **2019**, *366*, 1473–1480. [CrossRef] [PubMed]
35. Dong, L.; Jiang, H.; Kang, Z.; Guan, M. Biomarkers for chemotherapy and drug resistance in the mismatch repair pathway. *Clin. Chim. Acta* **2023**, *544*, 117338. [CrossRef]
36. Ray Chaudhuri, A.; Callen, E.; Ding, X.; Gogola, E.; Duarte, A.A.; Lee, J.-E.; Wong, N.; Lafarga, V.; Calvo, J.A.; Panzarino, N.J.; et al. Replication fork stability confers chemoresistance in BRCA-deficient cells. *Nature* **2016**, *535*, 382–387. [CrossRef]
37. Monteith, G.R.; Prevarskaya, N.; Roberts-Thomson, S.J. The calcium-cancer signalling nexus. *Nat. Rev. Cancer* **2017**, *17*, 367–380. [CrossRef]

38. Chen, Z.; Tang, C.; Zhu, Y.; Xie, M.; He, D.; Pan, Q.; Zhang, P.; Hua, D.; Wang, T.; Jin, L.; et al. TrpC5 regulates differentiation through the Ca2+/Wnt5a signalling pathway in colorectal cancer. *Clin. Sci.* **2017**, *131*, 227–237. [CrossRef] [PubMed]
39. Chen, Z.; Zhu, Y.; Dong, Y.; Zhang, P.; Han, X.; Jin, J.; Ma, X. Overexpression of TrpC5 promotes tumor metastasis via the HIF-1α-Twist signaling pathway in colon cancer. *Clin. Sci. (Lond.)* **2017**, *131*, 2439–2450. [CrossRef]
40. Wang, T.; Ning, K.; Lu, T.-X.; Hua, D. Elevated expression of TrpC5 and GLUT1 is associated with chemoresistance in colorectal cancer. *Oncol. Rep.* **2017**, *37*, 1059–1065. [CrossRef]
41. Wang, T.; Chen, Z.; Zhu, Y.; Pan, Q.; Liu, Y.; Qi, X.; Jin, L.; Jin, J.; Ma, X.; Hua, D. Inhibition of transient receptor potential channel 5 reverses 5-Fluorouracil resistance in human colorectal cancer cells. *J. Biol. Chem.* **2015**, *290*, 448–456. [CrossRef]
42. Landriscina, M.; Laudiero, G.; Maddalena, F.; Amoroso, M.R.; Piscazzi, A.; Cozzolino, F.; Monti, M.; Garbi, C.; Fersini, A.; Pucci, P.; et al. Mitochondrial chaperone Trap1 and the calcium binding protein Sorcin interact and protect cells against apoptosis induced by antiblastic agents. *Cancer Res.* **2010**, *70*, 6577–6586. [CrossRef]
43. Battista, T.; Fiorillo, A.; Chiarini, V.; Genovese, I.; Ilari, A.; Colotti, G. Roles of Sorcin in Drug Resistance in Cancer: One Protein, Many Mechanisms, for a Novel Potential Anticancer Drug Target. *Cancers* **2020**, *12*, 887. [CrossRef]
44. Maddalena, F.; Laudiero, G.; Piscazzi, A.; Secondo, A.; Scorziello, A.; Lombardi, V.; Matassa, D.S.; Fersini, A.; Neri, V.; Esposito, F.; et al. Sorcin induces a drug-resistant phenotype in human colorectal cancer by modulating Ca(2+) homeostasis. *Cancer Res.* **2011**, *71*, 7659–7669. [CrossRef]
45. Chakrabarty, S.; Radjendirane, V.; Appelman, H.; Varani, J. Extracellular calcium and calcium sensing receptor function in human colon carcinomas: Promotion of E-cadherin expression and suppression of beta-catenin/TCF activation. *Cancer Res.* **2003**, *63*, 67–71.
46. Quail, D.F.; Joyce, J.A. Microenvironmental regulation of tumor progression and metastasis. *Nat. Med.* **2013**, *19*, 1423–1437. [CrossRef] [PubMed]
47. Kadel, D.; Zhang, Y.; Sun, H.-R.; Zhao, Y.; Dong, Q.-Z.; Qin, L.-X. Current perspectives of cancer-associated fibroblast in therapeutic resistance: Potential mechanism and future strategy. *Cell Biol. Toxicol.* **2019**, *35*, 407–421. [CrossRef] [PubMed]
48. Sui, X.; Chen, R.; Wang, Z.; Huang, Z.; Kong, N.; Zhang, M.; Han, W.; Lou, F.; Yang, J.; Zhang, Q.; et al. Autophagy and chemotherapy resistance: A promising therapeutic target for cancer treatment. *Cell Death Dis.* **2013**, *4*, e838. [CrossRef] [PubMed]
49. Guo, C.; Liu, J.; Zhou, Q.; Song, J.; Zhang, Z.; Li, Z.; Wang, G.; Yuan, W.; Sun, Z. Exosomal Noncoding RNAs and Tumor Drug Resistance. *Cancer Res.* **2020**, *80*, 4307–4313. [CrossRef] [PubMed]
50. Straussman, R.; Morikawa, T.; Shee, K.; Barzily-Rokni, M.; Qian, Z.R.; Du, J.; Davis, A.; Mongare, M.M.; Gould, J.; Frederick, D.T.; et al. Tumour micro-environment elicits innate resistance to RAF inhibitors through HGF secretion. *Nature* **2012**, *487*, 500–504. [CrossRef]
51. Mantovani, A.; Allavena, P.; Marchesi, F.; Garlanda, C. Macrophages as tools and targets in cancer therapy. *Nat. Rev. Drug. Discov.* **2022**, *21*, 799–820. [CrossRef]
52. Kobayashi, H.; Enomoto, A.; Woods, S.L.; Burt, A.D.; Takahashi, M.; Worthley, D.L. Cancer-associated fibroblasts in gastrointestinal cancer. *Nat. Rev. Gastroenterol. Hepatol.* **2019**, *16*, 282–295. [CrossRef]
53. Erin, N.; Grahovac, J.; Brozovic, A.; Efferth, T. Tumor microenvironment and epithelial mesenchymal transition as targets to overcome tumor multidrug resistance. *Drug. Resist. Updat.* **2020**, *53*, 100715. [CrossRef]
54. Mou, Y.; Wang, J.; Wu, J.; He, D.; Zhang, C.; Duan, C.; Li, B. Ferroptosis, a new form of cell death: Opportunities and challenges in cancer. *J. Hematol. Oncol.* **2019**, *12*, 34. [CrossRef] [PubMed]
55. Du, Y.; Guo, Z. Recent progress in ferroptosis: Inducers and inhibitors. *Cell Death Discov.* **2022**, *8*, 501. [CrossRef] [PubMed]
56. Yang, W.S.; SriRamaratnam, R.; Welsch, M.E.; Shimada, K.; Skouta, R.; Viswanathan, V.S.; Cheah, J.H.; Clemons, P.A.; Shamji, A.F.; Clish, C.B.; et al. Regulation of ferroptotic cancer cell death by GPX4. *Cell* **2014**, *156*, 317–331. [CrossRef] [PubMed]
57. Maiorino, M.; Conrad, M.; Ursini, F. GPx4, Lipid Peroxidation, and Cell Death: Discoveries, Rediscoveries, and Open Issues. *Antioxid. Redox Signal.* **2018**, *29*, 61–74. [CrossRef]
58. Seiler, A.; Schneider, M.; Förster, H.; Roth, S.; Wirth, E.K.; Culmsee, C.; Plesnila, N.; Kremmer, E.; Rådmark, O.; Wurst, W.; et al. Glutathione peroxidase 4 senses and translates oxidative stress into 12/15-lipoxygenase dependent- and AIF-mediated cell death. *Cell Metab.* **2008**, *8*, 237–248. [CrossRef]
59. Doll, S.; Proneth, B.; Tyurina, Y.Y.; Panzilius, E.; Kobayashi, S.; Ingold, I.; Irmler, M.; Beckers, J.; Aichler, M.; Walch, A.; et al. ACSL4 dictates ferroptosis sensitivity by shaping cellular lipid composition. *Nat. Chem. Biol.* **2017**, *13*, 91–98. [CrossRef]
60. Yang, W.S.; Stockwell, B.R. Ferroptosis: Death by Lipid Peroxidation. *Trends Cell Biol.* **2016**, *26*, 165–176. [CrossRef]
61. Bersuker, K.; Hendricks, J.M.; Li, Z.; Magtanong, L.; Ford, B.; Tang, P.H.; Roberts, M.A.; Tong, B.; Maimone, T.J.; Zoncu, R.; et al. The CoQ oxidoreductase FSP1 acts parallel to GPX4 to inhibit ferroptosis. *Nature* **2019**, *575*, 688–692. [CrossRef] [PubMed]
62. Dixon, S.J.; Lemberg, K.M.; Lamprecht, M.R.; Skouta, R.; Zaitsev, E.M.; Gleason, C.E.; Patel, D.N.; Bauer, A.J.; Cantley, A.M.; Yang, W.S.; et al. Ferroptosis: An iron-dependent form of nonapoptotic cell death. *Cell* **2012**, *149*, 1060–1072. [CrossRef]
63. Hentze, M.W.; Muckenthaler, M.U.; Galy, B.; Camaschella, C. Two to tango: Regulation of Mammalian iron metabolism. *Cell* **2010**, *142*, 24–38. [CrossRef] [PubMed]
64. Torti, S.V.; Torti, F.M. Iron and cancer: More ore to be mined. *Nat. Rev. Cancer* **2013**, *13*, 342–355. [CrossRef] [PubMed]
65. Wang, D.; Wu, H.; Yang, J.; Li, M.; Ling, C.; Gao, Z.; Lu, H.; Shen, H.; Tang, Y. Loss of SLC46A1 decreases tumor iron content in hepatocellular carcinoma. *Hepatol. Commun.* **2022**, *6*, 2914–2924. [CrossRef] [PubMed]

66. Ohgami, R.S.; Campagna, D.R.; Greer, E.L.; Antiochos, B.; McDonald, A.; Chen, J.; Sharp, J.J.; Fujiwara, Y.; Barker, J.E.; Fleming, M.D. Identification of a ferrireductase required for efficient transferrin-dependent iron uptake in erythroid cells. *Nat. Genet.* **2005**, *37*, 1264–1269. [CrossRef] [PubMed]
67. Stockwell, B.R.; Friedmann Angeli, J.P.; Bayir, H.; Bush, A.I.; Conrad, M.; Dixon, S.J.; Fulda, S.; Gascón, S.; Hatzios, S.K.; Kagan, V.E.; et al. Ferroptosis: A Regulated Cell Death Nexus Linking Metabolism, Redox Biology, and Disease. *Cell* **2017**, *171*, 273–285. [CrossRef]
68. Lee, J.; You, J.H.; Shin, D.; Roh, J.-L. Inhibition of Glutaredoxin 5 predisposes Cisplatin-resistant Head and Neck Cancer Cells to Ferroptosis. *Theranostics* **2020**, *10*, 7775–7786. [CrossRef]
69. Dixon, S.J.; Winter, G.E.; Musavi, L.S.; Lee, E.D.; Snijder, B.; Rebsamen, M.; Superti-Furga, G.; Stockwell, B.R. Human Haploid Cell Genetics Reveals Roles for Lipid Metabolism Genes in Nonapoptotic Cell Death. *ACS Chem. Biol.* **2015**, *10*, 1604–1609. [CrossRef]
70. Rashba-Step, J.; Tatoyan, A.; Duncan, R.; Ann, D.; Pushpa-Rekha, T.R.; Sevanian, A. Phospholipid peroxidation induces cytosolic phospholipase A2 activity: Membrane effects versus enzyme phosphorylation. *Arch. Biochem. Biophys.* **1997**, *343*, 44–54. [CrossRef] [PubMed]
71. Wenzel, S.E.; Tyurina, Y.Y.; Zhao, J.; St Croix, C.M.; Dar, H.H.; Mao, G.; Tyurin, V.A.; Anthonymuthu, T.S.; Kapralov, A.A.; Amoscato, A.A.; et al. PEBP1 Wardens Ferroptosis by Enabling Lipoxygenase Generation of Lipid Death Signals. *Cell* **2017**, *171*, 628–641.e26. [CrossRef]
72. Ye, L.F.; Stockwell, B.R. Transforming Lipoxygenases: PE-Specific Enzymes in Disguise. *Cell* **2017**, *171*, 501–502. [CrossRef]
73. Zou, Y.; Li, H.; Graham, E.T.; Deik, A.A.; Eaton, J.K.; Wang, W.; Sandoval-Gomez, G.; Clish, C.B.; Doench, J.G.; Schreiber, S.L. Cytochrome P450 oxidoreductase contributes to phospholipid peroxidation in ferroptosis. *Nat. Chem. Biol.* **2020**, *16*, 302–309. [CrossRef]
74. Shen, F.; Jiang, G.; Philips, S.; Gardner, L.; Xue, G.; Cantor, E.; Ly, R.C.; Osei, W.; Wu, X.; Dang, C.; et al. Cytochrome P450 Oxidoreductase (POR) Associated with Severe Paclitaxel-Induced Peripheral Neuropathy in Patients of European Ancestry from ECOG-ACRIN E5103. *Clin. Cancer Res. Off. J. Am. Assoc. Cancer Res.* **2023**, *29*, 2494–2500. [CrossRef] [PubMed]
75. Huang, N.; Agrawal, V.; Giacomini, K.M.; Miller, W.L. Genetics of P450 oxidoreductase: Sequence variation in 842 individuals of four ethnicities and activities of 15 missense mutations. *Proc. Natl. Acad. Sci. USA* **2008**, *105*, 1733–1738. [CrossRef]
76. Liu, M.-Y.; Li, H.-M.; Wang, X.-Y.; Xia, R.; Li, X.; Ma, Y.-J.; Wang, M.; Zhang, H.-S. TIGAR drives colorectal cancer ferroptosis resistance through ROS/AMPK/SCD1 pathway. *Free. Radic. Biol. Med.* **2022**, *182*, 219–231. [CrossRef]
77. Sun, R.; Lin, Z.; Wang, X.; Liu, L.; Huo, M.; Zhang, R.; Lin, J.; Xiao, C.; Li, Y.; Zhu, W.; et al. AADAC protects colorectal cancer liver colonization from ferroptosis through SLC7A11-dependent inhibition of lipid peroxidation. *J. Exp. Clin. Cancer Res.* **2022**, *41*, 284. [CrossRef] [PubMed]
78. Shao, Y.; Jia, H.; Huang, L.; Li, S.; Wang, C.; Aikemu, B.; Yang, G.; Hong, H.; Yang, X.; Zhang, S.; et al. An Original Ferroptosis-Related Gene Signature Effectively Predicts the Prognosis and Clinical Status for Colorectal Cancer Patients. *Front. Oncol.* **2021**, *11*, 711776. [CrossRef]
79. Du, S.; Zeng, F.; Sun, H.; Liu, Y.; Han, P.; Zhang, B.; Xue, W.; Deng, G.; Yin, M.; Cui, B. Prognostic and therapeutic significance of a novel ferroptosis related signature in colorectal cancer patients. *Bioengineered* **2022**, *13*, 2498–2512. [CrossRef]
80. Sui, X.; Zhang, R.; Liu, S.; Duan, T.; Zhai, L.; Zhang, M.; Han, X.; Xiang, Y.; Huang, X.; Lin, H.; et al. RSL3 Drives Ferroptosis Through GPX4 Inactivation and ROS Production in Colorectal Cancer. *Front. Pharm.* **2018**, *9*, 1371. [CrossRef] [PubMed]
81. Han, Y.; Gao, X.; Wu, N.; Jin, Y.; Zhou, H.; Wang, W.; Liu, H.; Chu, Y.; Cao, J.; Jiang, M.; et al. Long noncoding RNA LINC00239 inhibits ferroptosis in colorectal cancer by binding to Keap1 to stabilize Nrf2. *Cell Death Dis.* **2022**, *13*, 742. [CrossRef]
82. Yan, H.; Talty, R.; Jain, A.; Cai, Y.; Zheng, J.; Shen, X.; Muca, E.; Paty, P.B.; Bosenberg, M.W.; Khan, S.A.; et al. Discovery of decreased ferroptosis in male colorectal cancer patients with KRAS mutations. *Redox Biol.* **2023**, *62*, 102699. [CrossRef]
83. Wu, T.; Wan, J.; Qu, X.; Xia, K.; Wang, F.; Zhang, Z.; Yang, M.; Wu, X.; Gao, R.; Yuan, X.; et al. Nodal promotes colorectal cancer survival and metastasis through regulating SCD1-mediated ferroptosis resistance. *Cell Death Dis.* **2023**, *14*, 229. [CrossRef]
84. Peng, B.; Peng, J.; Kang, F.; Zhang, W.; Peng, E.; He, Q. Ferroptosis-Related Gene MT1G as a Novel Biomarker Correlated With Prognosis and Immune Infiltration in Colorectal Cancer. *Front. Cell Dev. Biol.* **2022**, *10*, 881447. [CrossRef]
85. Wang, R.; Xing, R.; Su, Q.; Yin, H.; Wu, D.; Lv, C.; Yan, Z. Knockdown of SFRS9 Inhibits Progression of Colorectal Cancer Through Triggering Ferroptosis Mediated by GPX4 Reduction. *Front. Oncol.* **2021**, *11*, 683589. [CrossRef] [PubMed]
86. Xie, Y.; Zhu, S.; Song, X.; Sun, X.; Fan, Y.; Liu, J.; Zhong, M.; Yuan, H.; Zhang, L.; Billiar, T.R.; et al. The Tumor Suppressor p53 Limits Ferroptosis by Blocking DPP4 Activity. *Cell Rep.* **2017**, *20*, 1692–1704. [CrossRef]
87. Yu, Z.; Tong, S.; Wang, C.; Wu, Z.; Ye, Y.; Wang, S.; Jiang, K. PPy@Fe$_3$O$_4$ nanoparticles inhibit the proliferation and metastasis of CRC via suppressing the NF-κB signaling pathway and promoting ferroptosis. *Front. Bioeng. Biotechnol.* **2022**, *10*, 1001994. [CrossRef] [PubMed]
88. Hu, Q.; Wei, W.; Wu, D.; Huang, F.; Li, M.; Li, W.; Yin, J.; Peng, Y.; Lu, Y.; Zhao, Q.; et al. Blockade of GCH1/BH4 Axis Activates Ferritinophagy to Mitigate the Resistance of Colorectal Cancer to Erastin-Induced Ferroptosis. *Front. Cell Dev. Biol.* **2022**, *10*, 810327. [CrossRef] [PubMed]
89. Yang, L.; WenTao, T.; ZhiYuan, Z.; Qi, L.; YuXiang, L.; Peng, Z.; Ke, L.; XiaoNa, J.; YuZhi, P.; MeiLing, J.; et al. Cullin-9/p53 mediates HNRNPC degradation to inhibit erastin-induced ferroptosis and is blocked by MDM2 inhibition in colorectal cancer. *Oncogene* **2022**, *41*, 3210–3221. [CrossRef] [PubMed]

90. Fan, H.; Ai, R.; Mu, S.; Niu, X.; Guo, Z.; Liu, L. MiR-19a suppresses ferroptosis of colorectal cancer cells by targeting IREB2. *Bioengineered* **2022**, *13*, 12021–12029. [CrossRef]
91. Li, Q.; Li, K.; Guo, Q.; Yang, T. CircRNA circSTIL inhibits ferroptosis in colorectal cancer via miR-431/SLC7A11 axis. *Environ. Toxicol.* **2023**, *38*, 981–989. [CrossRef] [PubMed]
92. Yang, Y.; Lin, Z.; Han, Z.; Wu, Z.; Hua, J.; Zhong, R.; Zhao, R.; Ran, H.; Qu, K.; Huang, H.; et al. miR-539 activates the SAPK/JNK signaling pathway to promote ferropotosis in colorectal cancer by directly targeting TIPE. *Cell Death Discov.* **2021**, *7*, 272. [CrossRef]
93. Wei, Z.; Hang, S.; Wiredu Ocansey, D.K.; Zhang, Z.; Wang, B.; Zhang, X.; Mao, F. Human umbilical cord mesenchymal stem cells derived exosome shuttling mir-129-5p attenuates inflammatory bowel disease by inhibiting ferroptosis. *J. Nanobiotechnol.* **2023**, *21*, 188. [CrossRef]
94. Tang, F.; Zhou, L.-Y.; Li, P.; Jiao, L.-L.; Chen, K.; Guo, Y.-J.; Ding, X.-L.; He, S.-Y.; Dong, B.; Xu, R.-X.; et al. Inhibition of ACSL4 Alleviates Parkinsonism Phenotypes by Reduction of Lipid Reactive Oxygen Species. *Neurotherapeutics* **2023**, *20*, 1154–1166. [CrossRef] [PubMed]
95. Zhang, H.; Deng, T.; Liu, R.; Ning, T.; Yang, H.; Liu, D.; Zhang, Q.; Lin, D.; Ge, S.; Bai, M.; et al. CAF secreted miR-522 suppresses ferroptosis and promotes acquired chemo-resistance in gastric cancer. *Mol. Cancer* **2020**, *19*, 43. [CrossRef] [PubMed]
96. Prevete, N.; Liotti, F.; Illiano, A.; Amoresano, A.; Pucci, P.; de Paulis, A.; Melillo, R.M. Formyl peptide receptor 1 suppresses gastric cancer angiogenesis and growth by exploiting inflammation resolution pathways. *Oncoimmunology* **2017**, *6*, e1293213. [CrossRef]
97. Du, J.; Wang, X.; Li, Y.; Ren, X.; Zhou, Y.; Hu, W.; Zhou, C.; Jing, Q.; Yang, C.; Wang, L.; et al. DHA exhibits synergistic therapeutic efficacy with cisplatin to induce ferroptosis in pancreatic ductal adenocarcinoma via modulation of iron metabolism. *Cell Death Dis.* **2021**, *12*, 705. [CrossRef]
98. Chen, Y.; Yang, Z.; Wang, S.; Ma, Q.; Li, L.; Wu, X.; Guo, Q.; Tao, L.; Shen, X. Boosting ROS-Mediated Lysosomal Membrane Permeabilization for Cancer Ferroptosis Therapy. *Adv. Health Mater.* **2023**, *12*, e2202150. [CrossRef]
99. Yu, H.; Yang, C.; Jian, L.; Guo, S.; Chen, R.; Li, K.; Qu, F.; Tao, K.; Fu, Y.; Luo, F.; et al. Sulfasalazine-induced ferroptosis in breast cancer cells is reduced by the inhibitory effect of estrogen receptor on the transferrin receptor. *Oncol. Rep.* **2019**, *42*, 826–838. [CrossRef]
100. Hamad, M.; Mohammed, A.K.; Hachim, M.Y.; Mukhopadhy, D.; Khalique, A.; Laham, A.; Dhaiban, S.; Bajbouj, K.; Taneera, J. Heme Oxygenase-1 (HMOX 1) and inhibitor of differentiation proteins (ID1, ID3) are key response mechanisms against iron-overload in pancreatic β-cells. *Mol. Cell. Endocrinol.* **2021**, *538*, 111462. [CrossRef]
101. Sun, Y.; Qiao, Y.; Liu, Y.; Zhou, J.; Wang, X.; Zheng, H.; Xu, Z.; Zhang, J.; Zhou, Y.; Qian, L.; et al. ent-Kaurane diterpenoids induce apoptosis and ferroptosis through targeting redox resetting to overcome cisplatin resistance. *Redox Biol.* **2021**, *43*, 101977. [CrossRef]
102. Lin, L.; Shi, Q.; Su, C.-Y.; Shih, C.C.Y.; Lee, K.-H. Antitumor agents 247. New 4-ethoxycarbonylethyl curcumin analogs as potential antiandrogenic agents. *Bioorganic Med. Chem.* **2006**, *14*, 2527–2534. [CrossRef]
103. Tu, H.; Tang, L.J.; Luo, X.J.; Ai, K.L.; Peng, J. Insights into the novel function of system Xc- in regulated cell death. *Eur. Rev. Med. Pharm. Sci.* **2021**, *25*, 1650–1662. [CrossRef]
104. Zhang, Q.; Deng, T.; Zhang, H.; Zuo, D.; Zhu, Q.; Bai, M.; Liu, R.; Ning, T.; Zhang, L.; Yu, Z.; et al. Adipocyte-Derived Exosomal MTTP Suppresses Ferroptosis and Promotes Chemoresistance in Colorectal Cancer. *Adv. Sci.* **2022**, *9*, e2203357. [CrossRef] [PubMed]
105. Chen, H.; Qi, Q.; Wu, N.; Wang, Y.; Feng, Q.; Jin, R.; Jiang, L. Aspirin promotes RSL3-induced ferroptosis by suppressing mTOR/SREBP-1/SCD1-mediated lipogenesis in PIK3CA-mutatnt colorectal cancer. *Redox Biol.* **2022**, *55*, 102426. [CrossRef] [PubMed]
106. Lin, J.-F.; Hu, P.-S.; Wang, Y.-Y.; Tan, Y.-T.; Yu, K.; Liao, K.; Wu, Q.-N.; Li, T.; Meng, Q.; Lin, J.-Z.; et al. Phosphorylated NFS1 weakens oxaliplatin-based chemosensitivity of colorectal cancer by preventing PANoptosis. *Signal. Transduct. Target. Ther.* **2022**, *7*, 54. [CrossRef]
107. Yang, C.; Zhang, Y.; Lin, S.; Liu, Y.; Li, W. Suppressing the KIF20A/NUAK1/Nrf2/GPX4 signaling pathway induces ferroptosis and enhances the sensitivity of colorectal cancer to oxaliplatin. *Aging* **2021**, *13*, 13515–13534. [CrossRef] [PubMed]
108. Chaudhary, N.; Choudhary, B.S.; Shah, S.G.; Khapare, N.; Dwivedi, N.; Gaikwad, A.; Joshi, N.; Raichanna, J.; Basu, S.; Gurjar, M.; et al. Lipocalin 2 expression promotes tumor progression and therapy resistance by inhibiting ferroptosis in colorectal cancer. *Int. J. Cancer* **2021**, *149*, 1495–1511. [CrossRef]
109. Zeng, K.; Li, W.; Wang, Y.; Zhang, Z.; Zhang, L.; Zhang, W.; Xing, Y.; Zhou, C. Inhibition of CDK1 Overcomes Oxaliplatin Resistance by Regulating ACSL4-mediated Ferroptosis in Colorectal Cancer. *Adv. Sci.* **2023**, *10*, e2301088. [CrossRef]
110. Zhou, B.; Mai, Z.; Ye, Y.; Song, Y.; Zhang, M.; Yang, X.; Xia, W.; Qiu, X. The role of PYCR1 in inhibiting 5-fluorouracil-induced ferroptosis and apoptosis through SLC25A10 in colorectal cancer. *Hum. Cell* **2022**, *35*, 1900–1911. [CrossRef]
111. Chen, P.; Li, X.; Zhang, R.; Liu, S.; Xiang, Y.; Zhang, M.; Chen, X.; Pan, T.; Yan, L.; Feng, J.; et al. Combinative treatment of β-elemene and cetuximab is sensitive to KRAS mutant colorectal cancer cells by inducing ferroptosis and inhibiting epithelial-mesenchymal transformation. *Theranostics* **2020**, *10*, 5107–5119. [CrossRef] [PubMed]

112. Hong, T.; Lei, G.; Chen, X.; Li, H.; Zhang, X.; Wu, N.; Zhao, Y.; Zhang, Y.; Wang, J. PARP inhibition promotes ferroptosis via repressing SLC7A11 and synergizes with ferroptosis inducers in BRCA-proficient ovarian cancer. *Redox Biol.* **2021**, *42*, 101928. [CrossRef]
113. Song, X.; Wang, X.; Liu, Z.; Yu, Z. Role of GPX4-Mediated Ferroptosis in the Sensitivity of Triple Negative Breast Cancer Cells to Gefitinib. *Front. Oncol.* **2020**, *10*, 597434. [CrossRef] [PubMed]
114. Fu, D.; Wang, C.; Yu, L.; Yu, R. Induction of ferroptosis by ATF3 elevation alleviates cisplatin resistance in gastric cancer by restraining Nrf2/Keap1/xCT signaling. *Cell. Mol. Biol. Lett.* **2021**, *26*, 26. [CrossRef] [PubMed]
115. Zhang, W.; Luo, M.; Xiong, B.; Liu, X. Upregulation of Metallothionein 1G (MT1G) Negatively Regulates Ferroptosis in Clear Cell Renal Cell Carcinoma by Reducing Glutathione Consumption. *J. Oncol.* **2022**, *2022*, 4000617. [CrossRef] [PubMed]
116. Sun, X.; Niu, X.; Chen, R.; He, W.; Chen, D.; Kang, R.; Tang, D. Metallothionein-1G facilitates sorafenib resistance through inhibition of ferroptosis. *Hepatology* **2016**, *64*, 488–500. [CrossRef]
117. Grignano, E.; Cantero-Aguilar, L.; Tuerdi, Z.; Chabane, T.; Vazquez, R.; Johnson, N.; Zerbit, J.; Decroocq, J.; Birsen, R.; Fontenay, M.; et al. Dihydroartemisinin-induced ferroptosis in acute myeloid leukemia: Links to iron metabolism and metallothionein. *Cell Death Discov.* **2023**, *9*, 97. [CrossRef]
118. Gai, C.; Liu, C.; Wu, X.; Yu, M.; Zheng, J.; Zhang, W.; Lv, S.; Li, W. MT1DP loaded by folate-modified liposomes sensitizes erastin-induced ferroptosis via regulating miR-365a-3p/NRF2 axis in non-small cell lung cancer cells. *Cell Death Dis.* **2020**, *11*, 751. [CrossRef]
119. Miyazaki, I.; Asanuma, M. Multifunctional Metallothioneins as a Target for Neuroprotection in Parkinson's Disease. *Antioxidants* **2023**, *12*, 894. [CrossRef]
120. Yang, J.; Mo, J.; Dai, J.; Ye, C.; Cen, W.; Zheng, X.; Jiang, L.; Ye, L. Cetuximab promotes RSL3-induced ferroptosis by suppressing the Nrf2/HO-1 signalling pathway in KRAS mutant colorectal cancer. *Cell Death Dis.* **2021**, *12*, 1079. [CrossRef]
121. Tian, X.; Li, S.; Ge, G. Apatinib Promotes Ferroptosis in Colorectal Cancer Cells by Targeting ELOVL6/ACSL4 Signaling. *Cancer Manag. Res.* **2021**, *13*, 1333–1342. [CrossRef] [PubMed]
122. Xia, Y.; Liu, S.; Li, C.; Ai, Z.; Shen, W.; Ren, W.; Yang, X. Discovery of a novel ferroptosis inducer-talaroconvolutin A-killing colorectal cancer cells in vitro and in vivo. *Cell Death Dis.* **2020**, *11*, 988. [CrossRef] [PubMed]
123. Mu, M.; Zhang, Q.; Zhao, C.; Li, X.; Chen, Z.; Sun, X.; Yu, J. 3-Bromopyruvate overcomes cetuximab resistance in human colorectal cancer cells by inducing autophagy-dependent ferroptosis. *Cancer Gene Ther.* **2023**, *30*, 1414–1425. [CrossRef] [PubMed]
124. Wang, W.; Green, M.; Choi, J.E.; Gijón, M.; Kennedy, P.D.; Johnson, J.K.; Liao, P.; Lang, X.; Kryczek, I.; Sell, A.; et al. CD8+ T cells regulate tumour ferroptosis during cancer immunotherapy. *Nature* **2019**, *569*, 270–274. [CrossRef]
125. Lei, G.; Zhuang, L.; Gan, B. mTORC1 and ferroptosis: Regulatory mechanisms and therapeutic potential. *Bioessays* **2021**, *43*, e2100093. [CrossRef]
126. Gao, M.; Monian, P.; Pan, Q.; Zhang, W.; Xiang, J.; Jiang, X. Ferroptosis is an autophagic cell death process. *Cell Res.* **2016**, *26*, 1021–1032. [CrossRef]
127. Li, Z.; Chen, L.; Chen, C.; Zhou, Y.; Hu, D.; Yang, J.; Chen, Y.; Zhuo, W.; Mao, M.; Zhang, X.; et al. Targeting ferroptosis in breast cancer. *Biomark. Res.* **2020**, *8*, 58. [CrossRef]
128. Wu, Y.; Zhang, S.; Gong, X.; Tam, S.; Xiao, D.; Liu, S.; Tao, Y. The epigenetic regulators and metabolic changes in ferroptosis-associated cancer progression. *Mol. Cancer* **2020**, *19*, 39. [CrossRef]
129. Lang, X.; Green, M.D.; Wang, W.; Yu, J.; Choi, J.E.; Jiang, L.; Liao, P.; Zhou, J.; Zhang, Q.; Dow, A.; et al. Radiotherapy and Immunotherapy Promote Tumoral Lipid Oxidation and Ferroptosis via Synergistic Repression of SLC7A11. *Cancer Discov.* **2019**, *9*, 1673–1685. [CrossRef]
130. Zou, Y.; Schreiber, S.L. Progress in Understanding Ferroptosis and Challenges in Its Targeting for Therapeutic Benefit. *Cell Chem. Biol.* **2020**, *27*, 463–471. [CrossRef]
131. Gu, X.; Liu, Y.e.; Dai, X.; Yang, Y.-G.; Zhang, X. Deciphering the potential roles of ferroptosis in regulating tumor immunity and tumor immunotherapy. *Front. Immunol.* **2023**, *14*, 1137107. [CrossRef] [PubMed]
132. Shen, D.; Luo, J.; Chen, L.; Ma, W.; Mao, X.; Zhang, Y.; Zheng, J.; Wang, Y.; Wan, J.; Wang, S.; et al. PARPi treatment enhances radiotherapy-induced ferroptosis and antitumor immune responses via the cGAS signaling pathway in colorectal cancer. *Cancer Lett.* **2022**, *550*, 215919. [CrossRef] [PubMed]
133. Lv, Y.; Tang, W.; Xu, Y.; Chang, W.; Zhang, Z.; Lin, Q.; Ji, M.; Feng, Q.; He, G.; Xu, J. Apolipoprotein L3 enhances CD8+ T cell antitumor immunity of colorectal cancer by promoting LDHA-mediated ferroptosis. *Int. J. Biol. Sci.* **2023**, *19*, 1284–1298. [CrossRef]
134. Wu, J.; Liu, C.; Wang, T.; Liu, H.; Wei, B. Deubiquitinase inhibitor PR-619 potentiates colon cancer immunotherapy by inducing ferroptosis. *Immunology* **2023**, *170*, 439–451. [CrossRef] [PubMed]
135. Chen, C.; Yang, Y.; Guo, Y.; He, J.; Chen, Z.; Qiu, S.; Zhang, Y.; Ding, H.; Pan, J.; Pan, Y. CYP1B1 inhibits ferroptosis and induces anti-PD-1 resistance by degrading ACSL4 in colorectal cancer. *Cell Death Dis.* **2023**, *14*, 271. [CrossRef] [PubMed]

136. Shen, Z.; Song, J.; Yung, B.C.; Zhou, Z.; Wu, A.; Chen, X. Emerging Strategies of Cancer Therapy Based on Ferroptosis. *Adv. Mater.* **2018**, *30*, e1704007. [CrossRef]
137. Qiao, C.; Wang, H.; Guan, Q.; Wei, M.; Li, Z. Ferroptosis-based nano delivery systems targeted therapy for colorectal cancer: Insights and future perspectives. *Asian J. Pharm. Sci.* **2022**, *17*, 613–629. [CrossRef]

Disclaimer/Publisher's Note: The statements, opinions and data contained in all publications are solely those of the individual author(s) and contributor(s) and not of MDPI and/or the editor(s). MDPI and/or the editor(s) disclaim responsibility for any injury to people or property resulting from any ideas, methods, instructions or products referred to in the content.

Article

Synaptotagmin 1 Suppresses Colorectal Cancer Metastasis by Inhibiting ERK/MAPK Signaling-Mediated Tumor Cell Pseudopodial Formation and Migration

Jianyun Shi [1,†], Wenjing Li [1,†], Zhenhua Jia [1], Ying Peng [1], Jiayi Hou [2], Ning Li [3], Ruijuan Meng [4], Wei Fu [4], Yanlin Feng [1], Lifei Wu [1], Lan Zhou [1], Deping Wang [1], Jing Shen [1], Jiasong Chang [1], Yanqiang Wang [5,*] and Jimin Cao [1,*]

1. Key Laboratory of Cellular Physiology at Shanxi Medical University, Ministry of Education, and the Department of Physiology, Shanxi Medical University, Taiyuan 030606, China
2. Department of Clinical Laboratory, Shanxi Provincial Academy of Traditional Chinese Medicine, Taiyuan 030071, China
3. Department of Gastrointestinal and Pancreatic Surgery & Hernia and Abdominal Surgery, Shanxi Provincial People's Hospital, Taiyuan 030045, China
4. Department of Radiology, The First Hospital of Shanxi Medical University, Shanxi Medical University, Taiyuan 030606, China
5. Translational Medicine Research Center, Shanxi Medical University, Taiyuan 030606, China
* Correspondence: yqwang15@sxmu.edu.cn (Y.W.); caojimin@sxmu.edu.cn (J.C.)
† These authors contributed equally to this work.

Simple Summary: Colorectal cancer (CRC) is one major cause of cancer mortality worldwide. Emerging evidence shows that synaptotagmin 1 (SYT1) takes roles in a variety of cancers. However, the role of SYT1 in colorectal cancer remains an enigma. Here, we first assess SYT1 expression levels and discover that its expression is downregulated in CRC tissues and CRC cell lines. We further confirm that SYT1 overexpression suppresses CRC metastasis both in vivo and in vitro using mouse CRC xenograft metastasis model and colon cancer cells. The inhibitory effect of SYT1 overexpression on CRC metastasis is associated with reductions of CRC cell pseudopodial formation, migration, and invasion. Mechanistically, SYT1 overexpression inhibits EMT via negatively regulating the ERK/MAPK signaling, thereby resulting in suppression of CRC cell migration and invasion. Our findings provide new insights into CRC development and indicate the potential of SYT1 as a bio marker and potential therapeutic target for CRC.

Abstract: Although synaptotagmin 1 (SYT1) has been identified participating in a variety of cancers, its role in colorectal cancer (CRC) remains an enigma. This study aimed to demonstrate the effect of SYT1 on CRC metastasis and the underlying mechanism. We first found that SYT1 expressions in CRC tissues were lower than in normal colorectal tissues from the CRC database and collected CRC patients. In addition to this, SYT1 expression was also lower in CRC cell lines than in the normal colorectal cell line. SYT1 expression was downregulated by TGF-β (an EMT mediator) in CRC cell lines. In vitro, SYT1 overexpression repressed pseudopodial formation and reduced cell migration and invasion of CRC cells. SYT1 overexpression also suppressed CRC metastasis in tumor-bearing nude mice in vivo. Moreover, SYT1 overexpression promoted the dephosphorylation of ERK1/2 and downregulated the expressions of Slug and Vimentin, two proteins tightly associated with EMT in tumor metastasis. In conclusion, SYT1 expression is downregulated in CRC. Overexpression of SYT1 suppresses CRC cell migration, invasion, and metastasis by inhibiting ERK/MAPK signaling-mediated CRC cell pseudopodial formation. The study suggests that SYT1 is a suppressor of CRC and may have the potential to be a therapeutic target for CRC.

Keywords: synaptotagmin 1; pseudopodial formation; cell migration; metastasis; ERK/MAPK signaling

Citation: Shi, J.; Li, W.; Jia, Z.; Peng, Y.; Hou, J.; Li, N.; Meng, R.; Fu, W.; Feng, Y.; Wu, L.; et al. Synaptotagmin 1 Suppresses Colorectal Cancer Metastasis by Inhibiting ERK/MAPK Signaling-Mediated Tumor Cell Pseudopodial Formation and Migration. *Cancers* 2023, *15*, 5282. https://doi.org/10.3390/cancers15215282

Academic Editor: Shihori Tanabe

Received: 18 September 2023
Revised: 21 October 2023
Accepted: 1 November 2023
Published: 3 November 2023

Copyright: © 2023 by the authors. Licensee MDPI, Basel, Switzerland. This article is an open access article distributed under the terms and conditions of the Creative Commons Attribution (CC BY) license (https://creativecommons.org/licenses/by/4.0/).

1. Introduction

Colorectal cancer (CRC) represents a significant health problem as the world's third most commonly diagnosed and second leading cause of malignancy-associated mortality worldwide [1]. Approximately 9.4% of cancer-related deaths were due to CRC in 2020 and the prevalence of CRC is increasing in recent years [2]. Statistically, 20% of patients diagnosed with CRC have metastasis [3]. Recent studies reported that many factors may contribute to the incidence of CRC, including genetics, diet habits, colon polyp, environment, etc. [3,4]. The carcinogenesis of CRC is a multi-step process which involves a quantity of genomic alterations [5]. If the tumor suppressor gene mutates, the transition from non-invasive to invasive disease may take place. Therefore, it is urgent to explore the roles of these suppressor genes/oncogenes in colorectal cancer for cancer prevention, early diagnosis, and therapeutic development.

Synaptotagmins (SYTs) are a family of structurally related proteins which are highly conserved from invertebrates to human, and expressed in almost all tissues [6]. SYT1 was initially found to work as a Ca^{2+} senser in neurotransmitter release [7]. SYT1 has also been found expressed in non-neuronal cells and plays multiple functions in these cells. For example, SYT1 is required for spindle stability and metaphase-to-anaphase transition in mouse oocytes [6]; SYT1 is expressed in both intestinal epithelial cells and Caco-2BBe cell lines, and is required in cAMP-mediated endocytosis of intestinal epithelial NHE3 cells [8,9], suggesting that SYT1 plays a pivotal role in the physiological activity of the intestinal system. In addition, SYT1 plays important roles in many malignancies. For example, Nord et al. [10] reported that SYT1 is a new oncogene in glioblastoma. Liu et al. [11] demonstrated that dysregulated SYT1 is associated with the survival of head and neck squamous cell carcinoma. Yang et al. [12] found that SYT1 is significantly downregulated in adamantinomatous craniopharyngioma. These studies suggest that SYT1 takes roles in a variety of cancers. However, the potential role of SYT1 in CRC remains unclear.

Here, we comprehensively investigated the expression of SYT1 and its functional roles in the progression of CRC both in vitro and in vivo. Interestingly, we found that SYT1 was downregulated both in CRC tissues and CRC cell lines, and overexpression of SYT1 could suppress CRC cell metastasis in tumor-bearing mice in vivo and migration in vitro. We also demonstrated that SYT1 exerted the above effects through suppressing ERK/MAPK signaling-mediated tumor cell pseudopodia formation. The study may help in understanding the role of SYT1 in the development of CRC.

2. Materials and Methods

2.1. Public Dataset Acquisition

The gene expression and clinical information data in each tumor and normal sample were obtained from the TCGA database (https://portal.gdc.cancer.gov/, accessed on 17 September 2023). The publicly available TCGA database included 538 tumor tissues, excluding cases with insufficient or missing data on local invasion, lymph node metastasis, distant metastasis, age, overall survival, and TNM staging.

2.2. Collection of Human CRC Tissues

CRC tissues and paired adjacent normal tissues were obtained from 15 patients during tumor resection surgeries at the Shanxi Provincial Academy of Traditional Chinese Medicine (Taiyuan, China), and were used to check the expression levels of SYT1. The study was approved by the Medical Research Ethics Committee of the Shanxi Provincial Academy of Traditional Chinese Medicine, Taiyuan, China (approval no.: 2019-06KY005).

2.3. Mouse Model of Xenograft CRC Metastasis

The CRC metastatic animal model was constructed by tail vein injection of 1.5×10^6 HCT116 cells in 5-week-old female BALB/C nude mice (Gempharmatech Co., Ltd., Nanjing, China). Mice were kept in a sterile environment during the whole experiment. Eight weeks after tumor injection, the potential tumor metastasis to lung and liver of each mouse was scanned by a microPET-CT (Pingseng Healthcare Co., Ltd., Kunshan, China) after injection of 200 µCi ^{18}F-FDG. Maximum standardized uptake (SUVmax) of regions of interest (ROI) were analyzed after manual definition. Mice were then sacrificed and lung and liver tissues were harvested and fixed with 4% paraformaldehyde, embedded in paraffin, and stained with hematoxylin and eosin (H&E). The expressions of Slug and Vimentin were also stained using immunohistochemistry. The animal use protocol was approved by the Animal Ethics Committee of the Shanxi Medical University (Taiyuan, China) (approval no.: SYDL2023-085) and abided by the standards of the National Institute of Health Guide for the Care and Use of Laboratory Animals.

2.4. Immunohistochemical and Immunofluorescent Staining

Five-µm tissue sections attached to slides were deparaffinized, blocked, and incubated with primary antibody overnight at 4 °C. After washing with PBS, tissue sections were incubated with HRP-conjugated secondary antibody for 1 h, and, then, the peroxidase activity was visualized by reacting with 3,3-diaminobenzidine (DAB). Tissue sections were washed in water, counterstained with hematoxylin, and cover-slipped. Positive signals were observed and photographed under a microscope. To perform immunofluorescent staining, tissue sections were stained with Alexa Fluor 488 and 594 goat anti-mouse or anti-rabbit IgG (H + L) for 1 h at room temperature. Nuclei were stained with DAPI. The sample was incubated with the primary antibodies SYT1 (1:100), Ki67 (1:400), PCNA (1:400), Slug (1:200), Vimentin (1:200), and p-ERK (1:200). Antibody information is shown in Table S2.

2.5. Cell Lines and Cell Culture

Human colonic carcinoma cell lines SW620, SW480, and HCT116, and the normal colorectal cell line NCM460, were purchased from the Shanghai Institutes for Biological Sciences, at the Chinese Academy of Sciences. SW620 and SW480 cells were maintained in L15 medium (Hyclone, Logan, Utah) supplemented with 10% FBS (Gibco, Grand Island, NY, USA). HCT116 and NCM460 cells were cultured in DMEM medium (BOSTER) containing 10% FBS. All cells were cultured at 37 °C with 5% CO_2.

2.6. Western Blotting

Western blotting was carried out as we previously described [13]. In brief, cultured cells were collected and washed with PBS, and then were lysed with RIPA protein lysate buffer. The total protein concentration was determined by BCA protein assay kit (Beyotime, Haimen, China). Equal amounts of proteins were separated by SDS-PAGE and transferred onto a PVDF membrane (Bio-Rad, Hercules, CA, USA). Proteins were probed with specific primary antibodies followed by detection with secondary antibodies conjugated with HRP. Signals were detected on a gel imaging system by using ECL (Thermo Scientific, Waltham, MA, USA). The primary antibodies include Slug (1:500), Vimentin (1:500), and GAPDH antibody (1:1000).

2.7. RT-qPCR

Total RNA was extracted from SW480 cells and HCT116 cells by TRIZOL reagent (Invitrogen, Carlsbad, CA, USA). RNA extraction and cDNA synthesis were performed according to the manufacturer's instructions. Then, cDNA amplification was performed by qPCR using Premix Ex Taq™ kit (Takara, Osaka, Japan). qPCR procedures were run at 95 °C for 30 s, followed by 40 cycles at 95 °C for 5 s and, then, 60 °C for 30 s. The primers used in qPCR were designed with Primer 5.0 software and were synthesized with the help of Sangon Biotech (Shanghai, China). Primer sequences are shown in Table 1.

Table 1. Primer sequences.

Gene	Primer Sequence (5′-3′)	
SYT1	Forward	5′-AAAGTCCACCGAAAAACCCTT-3′
	Reverse	5′-CCACCCAATTCCGAGTATGGT-3′
GAPDH	Forward	5′-GGAGCGAGATCCCTCCAAAAT-3′
	Reverse	5′-GGCTGTTGTCATACTTCTCATGG-3′

2.8. Plasmids and Transfection

The complementary DNA (cDNA) of SYT1 was amplified by PCR and cloned into pcDNA3.1(+) plasmid. The pcDNA3.1(+)-SYT1 (SYT1) and pcDNA3.1(+) (control) plasmids were respectively amplified and purified, and, then, were transfected into SW480 and HCT116 cells using LipoFit 3.0 (Hanbio, Shanghai, China) as per the manufacturer's protocol. After 48 h transfection, the cells were gathered and employed for further analyses.

2.9. Transwell Migration Assay

The transwell system with 24-well polycarbonate membranes and 8 μm pores (Corning Costar, Corning, NY, USA) was used to perform migration assays. The volume of the upper chamber was 200 μL serum-free medium containing 1×10^5 cells. And 600 μL medium supplemented with 20% FBS was added to the lower chamber. The cells were cultured at 37 °C with 5% CO_2 for 48 h. Then, the wells were removed and gently washed twice with PBS. Subsequently, the wells were fixed with 4% paraformaldehyde for 30 min, and stained with 0.1% crystal violet for 10 min, and, then, images were captured under a microscope.

2.10. Wound Healing Assay

Wound healing assay was performed to observe cell migration ability. SW480 and HCT116 cells, which transiently overexpressed SYT1, were seeded in 6-well plates at 5×10^5/well. A single scratch was made across the center of the cell monolayer using a micropipette tip. Then, the cells were washed with PBS to remove cell debris. Images were captured under a microscope at 24 h, 48 h, and 72 h post wounding.

2.11. Statistical Analysis

GraphPad Prism 5.0 statistical software was used to perform statistical analysis. Data were presented as mean ± standard deviation (SD) of at least three independent experiments. The two-tailed *t*-test was used for comparison of two groups and one-way ANOVA for multi-group comparison, and $p < 0.05$ was considered statistically significant.

3. Results

3.1. SYT1 Expression Is Downregulated in Human CRC Tissues and Cell Lines

We first searched the SYT1 expression levels in human CRC tissues and normal colorectal tissues from the TCGA database. Results showed that the mRNA level of SYT1 in CRC tissues were significantly lower than that in the normal colorectal tissues (Figure 1A,B). Results of a Human Protein Atlas (HPA) search showed that the protein expression level of SYT1 in CRC tissues was also lower than in the normal colorectal tissues (Figure 1C).

We then validated the above database search results of SYT1 expression in our collected human CRC tissues and adjacent normal colorectal tissues by immunofluorescent staining. Results showed that SYT1 protein expression was lower in the CRC tissues than in the adjacent normal colorectal tissues (FigureS 1D and S1). This result was in concordance with the results of database search.

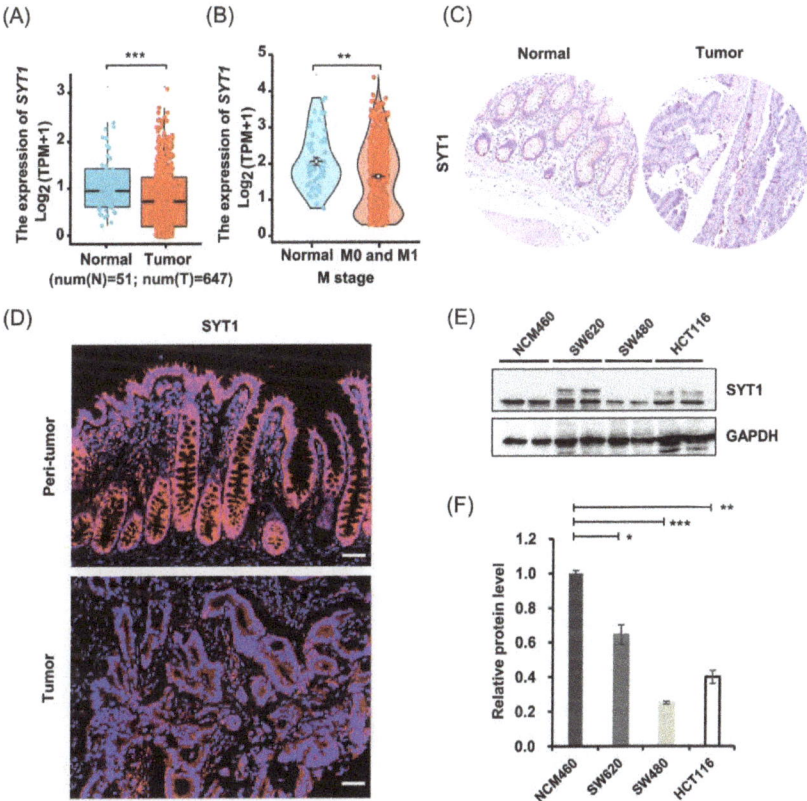

Figure 1. SYT1 was downregulated in human CRC issues and cell lines. (**A**) Relative mRNA levels of SYT1 in CRC tissues and normal colorectal tissues derived from TCGA database; $n = 51$ for normal colorectal tissues and $n = 647$ for CRC tissues. (**B**) Colonic SYT1 mRNA level in CRC patients with distant metastasis (M); $n = 51$ for normal colorectal tissues and $n = 564$ for CRC tissues. N: normal (blue); T: tumor (red). (**C**) Immunohistochemical stains of SYT1 protein in CRC tissues and normal colorectal tissues obtained from HPA database. (**D**) Representative immunofluorescent stains of SYT1 protein in CRC tissues and paired adjacent normal colorectal tissues of the collected CRC cases. Scale bar, 50 µm. (**E**) Western blots of SYT1 proteins in normal colonic cell line NCM460 and CRC cell lines HCT116, SW480, and SW620. (**F**) Grey value statistics of western blots for E; * $p < 0.05$, ** $p < 0.01$, and *** $p < 0.001$. The uncropped bolts are shown in Supplementary Materials Figure S2.

We further validated the above results of SYT1 expression in three human CRC cell lines (HCT116, SW620, and SW480) and a normal colorectal cell line (NCM460) by Western blotting. Results showed that SYT1 expression was significantly lower in CRC cells than in normal colorectal cells (Figure 1E,F). This cell result was also in agreement with the tissue result, indicating that SYT1 expression is really downregulated in CRC tissues and cells.

Univariate analysis using Cox regression revealed that several factors, including tumor status, distant metastasis, lymph node status, gender, age, and SYT1 expression, were significantly associated with the overall survival of CRC patients (Table S1).

Taken together, above results suggest that downregulation of SYT1 is closely associated with the malignant progression of CRC; therefore, SYT1 may act as a suppressor gene in CRC.

3.2. SYT1 Overexpression Represses Pseudopodial Formation of CRC Cells

Actin cytoskeleton reorganization regulates cell morphological changes, namely, pseudopodium formation, and results in the directional migration and invasion of cancer cells [14,15]. Emerging evidence indicates that epithelial to mesenchymal transition (EMT) is a key driver of CRC progression, and transforming growth factor β (TGF-β) is one of the main mediators of EMT [16,17]. We have shown in Figure 1 that CRC tissues and CRC cells showed a downregulated SYT1 expression compared with respective normal colorectal tissues and cells. We further found that SYT1 expression was downregulated in HCT116 cells after TGF-β stimulation (Figure 2A–C), suggesting that SYT1 may have some effect on EMT. To further explore the role of SYT1 in CRC, we chose SW480 cells and HCT116 cells which had lower SYT1 expression to perform further study. SYT1-overexpressing plasmid was designed and synthesized, and plasmid transfection efficacy was checked in CRC cell lines SW480 and HCT116. Transfection results showed that both the mRNA levels (Figure 2D,E) and the protein levels (Figure 2F,G) of SYT1 in SW480 and HCT116 cells were significantly elevated upon SYT1 overexpression, indicating that the transfection is successful and has high efficacy.

We then checked the effect of SYT1 overexpression on the pseudopodium formation of CRC cells using the plasmid and actin tracker green microfilament green fluorescent probe. The results showed that SYT1 overexpression significantly inhibited the formation of pseudopodia in CRC cells (Figure 2H,I). Because pseudopodial formation is an early sign of tumor cell movement, we speculated that SYT1 overexpression may inhibit CRC cell migration, invasion, and metastasis. The following cellular and animal experiments were performed to validate this speculation.

3.3. SYT1 Overexpression Suppresses CRC Cell Migration and Invasion In Vitro

CRC cell migration and invasion abilities were, respectively, examined by transwell assay and wound healing assay. Overexpression of SYT1 substantially reduced the cell motility of HCT116 cells and SW480 cells compared to the control as shown in the transwell assay (Figure 3A,B). Overexpression of SYT1 also inhibited the migration of HCT116 cells and SW480 cells as indicated by the wound closure rates after scratching in the wound healing assay (Figure 3C–E). These results indicate that the expression levels of SYT1 might be negatively correlated with the migration and invasion abilities of colon cancer cells.

3.4. SYT1 Overexpression Represses Metastasis of CRC Cells in Mice In Vivo

Above cellular experiments showed that SYT1 overexpression suppressed CRC cell migration and invasion, suggesting that SYT1 may inhibit CRC metastasis in vivo. We thus established a CRC xenograft metastasis nude mice model to examine whether SYT1 overexpression could suppress CRC metastasis at the integrative level. Eight weeks after CRC cell injection into the nude mice, the general information of the mice (including survival) was recorded, and microPET-CT images were taken after tail vein injection of ^{18}F-FDG. Results showed that the metastasis-free survival rate of the mice was higher in SYT1-overexpressing mice than in the control mice without SYT1 overexpression (Figure 4A). Consistently, high-metabolism lesions (concentration of ^{18}F-FDG) indicative of metastatic tumors were found in the lungs and livers of the control mice, whereas no obvious concentration of ^{18}F-FDG was found in the SYT1-overexpressing mice (Figure 4B,C). Histological examination showed a large amount of white clump-like dense tissues (tumors) in the lungs of the control mice, while no obvious clumps were observed in the SYT1-overexpressing mice. In addition, the surface of livers was rough and the elasticity was worse in the control mice compared with the SYT1-overexpressing mice (Figure 4D), which suggest tumor metastasis to the liver. H&E stains showed that the metastatic lesion area of livers and lungs was, obviously, smaller in SYT1-overexpressing mice than in control mice (Figure 4E,F). Immunohistochemical stains of the metastatic lesions in livers and lungs showed that Ki67 and PCNA protein levels were notably lower in the SYT1-overexpressing group than in the control mice (Figure 4G). These findings suggest that SYT1 overexpression weakens CRC metastasis in mice in vivo.

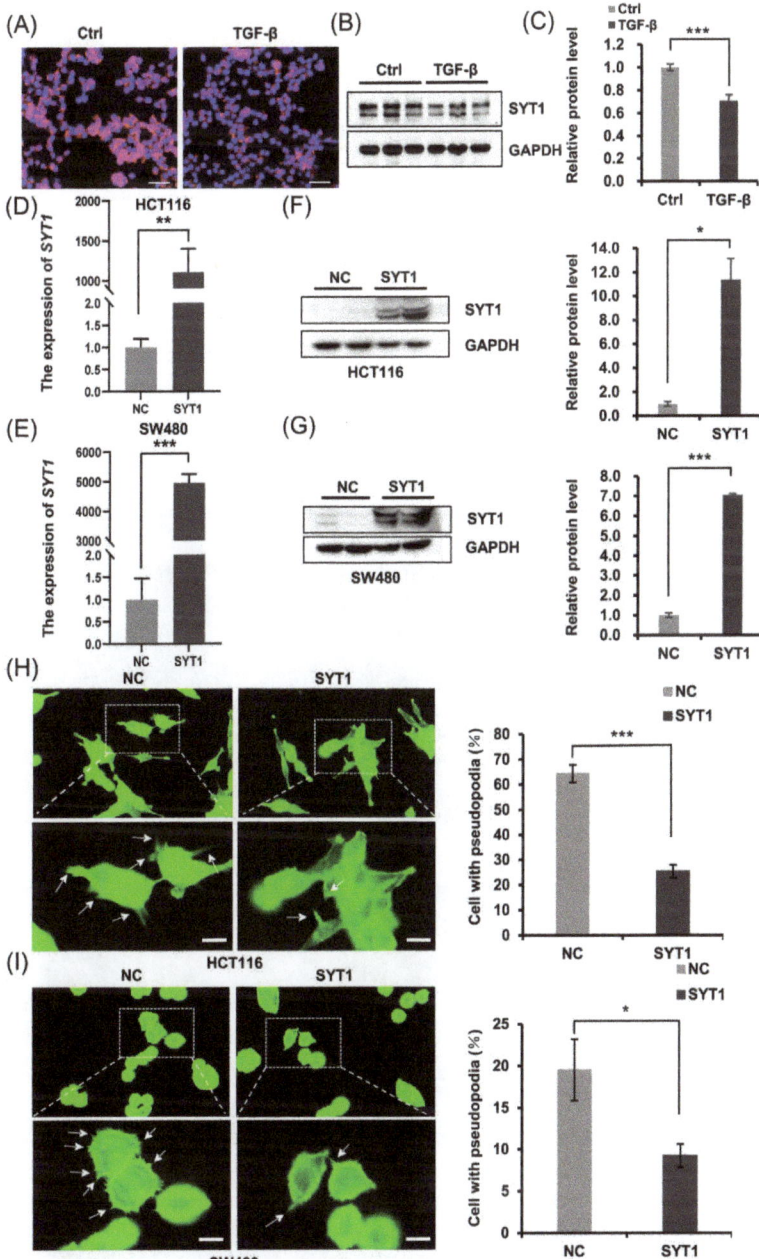

Figure 2. SYT1 overexpression suppressed pseudopodium formation in CRC cells. (**A**) Immunofluorescent stains of SYT1 in HCT116 cells with and without TGF-β treatment. (**B**,**C**) Western blots of SYT1 protein and respective quantitative gray values in HCT116 cells with or without TGF-β treatment. (**D**,**E**) RT-qPCR showing the mRNA levels of SYT1 in HCT116 cells (**D**) and SW480 cells

(**E**) with or without pcDNA3.1-SYT1 transfection. (**F**,**G**) Western blots of SYT1 protein in HCT116 cells (**F**) and SW480 cells (**G**) with or without pcDNA3.1-SYT1 transfection. (**H**,**I**) Confocal images of phalloidin (green) showing the pseudopodial formation in HCT116 cells (**H**) and SW480 cells (**I**) with or without SYT1 overexpression. The framed areas were enlarged to better present the pseudopodial protrusions. The white arrow indicates the cell pseudopodium. Scale bar, 25 μm. Quantifications of the pseudopodial protrusions are represented on the right side. Mean ± SD; $n > 50$ cells from three biological repeats; * $p < 0.05$, ** $p < 0.01$, and *** $p < 0.001$. The uncropped bolts are shown in Supplementary Materials Figure S2.

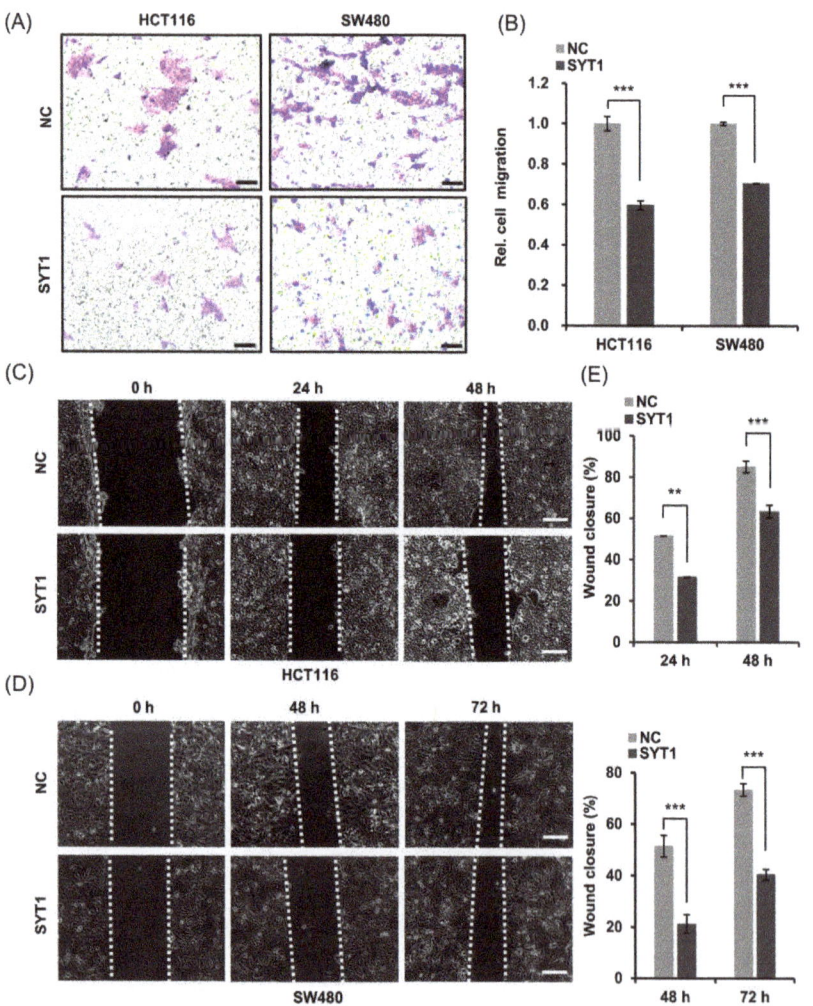

Figure 3. SYT1 overexpression inhibited the migration and invasion abilities of CRC cells in vitro. (**A**) Transwell migration assays of HCT116 cells and SW480 cells with or without SYT1 overexpression. The assays were performed after culture for 48 h. Scale bar, 50 μm. (**B**) Quantitative analysis of migrated cells across the transwell membrane. (**C**,**D**) Wound healing assays showing the invasion abilities of HCT116 cells (**C**) and SW480 cells (**D**) with or without SYT1 overexpression. Scale bar, 50 μm. (**E**) Quantitative analysis of the wound healing rates; ** $p < 0.01$, and *** $p < 0.001$.

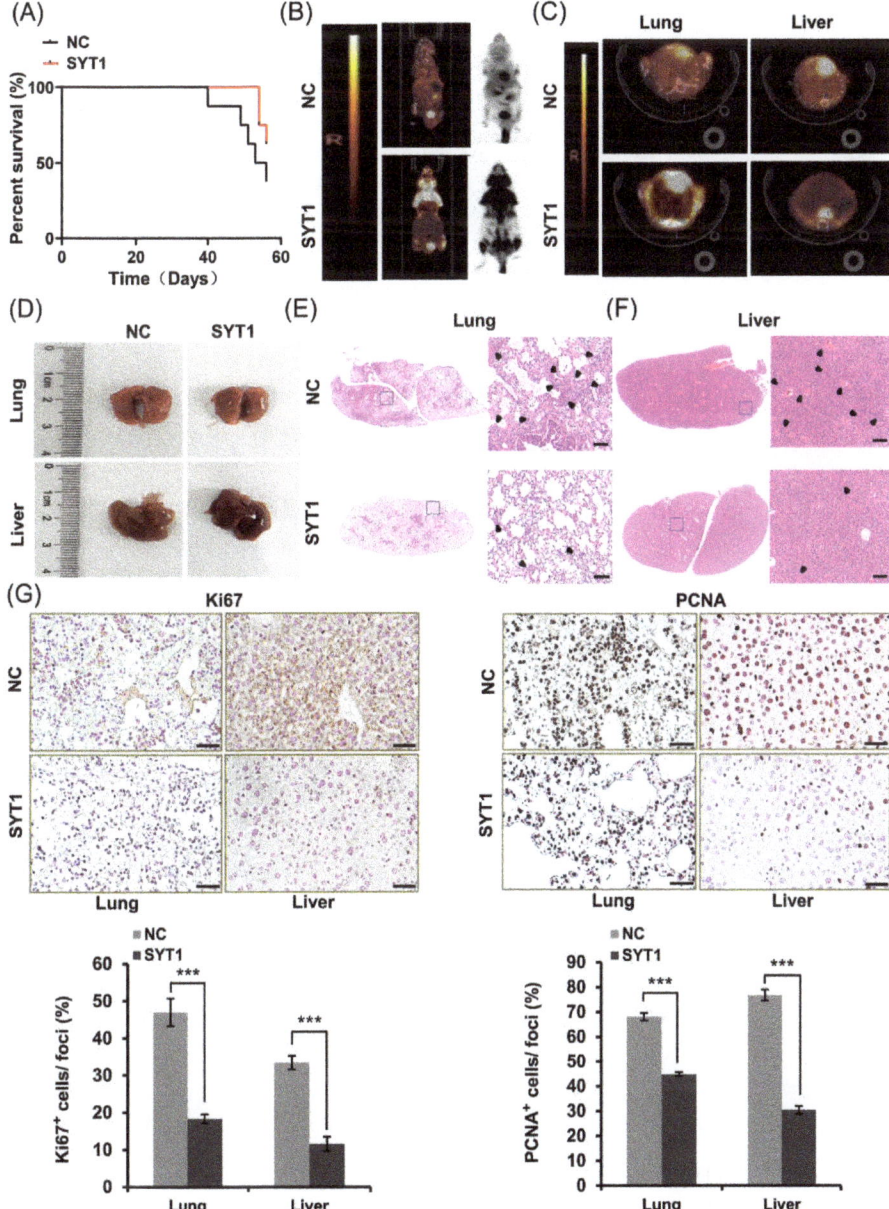

Figure 4. SYT1 overexpression suppressed CRC metastasis in mice in vivo. (**A**) Statistical results of metastasis-free survival rates in control and SYT1-overexpressing nude mice 8 weeks after CRC cell injection. (**B**) PET-CT images of CRC metastasis nude mice model after tail vein injection of ^{18}F-FDG. (**C**) PET-CT images of liver and lung of the CRC metastasis nude mice model. (**D**) Gross anatomy of representative lung or liver tissues. (**E**,**F**) H&E stains of lung and liver tissue sections. Framed areas were enlarged to better show the metastases. The black arrowhead indicates the lesion site. Scale bar, 100 μm. (**G**) Immunohistochemical stains of Ki67 and PCNA proteins in the metastatic lesion and precancerous tissue in lungs and livers. Scale bar, 50 μm. n = 8 biological replicates for each group; *** $p < 0.001$.

3.5. SYT1 Overexpression Downregulates EMT-Associated Slug and Vimentin

The spread of tumor cells is one of the typical behavioral characteristics of malignant tumors. These invasive cells undergo a transformation from an epithelial to a mesenchymal phenotype (epithelial–mesenchymal transition, EMT) [18,19], which plays a crucial role in the initial stage of metastasis. EMT make cells losing their epithelial characteristics such as motility limitation and strong cell-cell junction while obtaining mesenchymal characteristics associated with motility enhancement, cell–cell junction weakening and polarized actin cytoskeleton assembly, resulting in formation of protrusive and invasive pseudopodial structures, so that cells can acquire migration ability and undergo actin cytoskeletal reorganization [20]. Shankar et al. [15] reported that actin-dependent pseudopodial protrusion and tumor cell migration are determinants of EMT. We showed in Figure 2 that SYT1 was negatively associated with the formation of pseudopodia in HCT116 cells and SW480 cells. This phenomenon suggests that SYT1 is potentially associated with the EMT process. We, thus, detected the protein levels of Slug and Vimentin (these two EMT-related proteins) and also SYT1 in the lung and liver tissues of mice. Immunohistochemical stains of lung and liver showed that Slug and Vimentin were lower in SYT1-overexpressing mice than in the control mice, while SYT1 expression level showed the opposite (Figure 5A,B). Western blots of Slug and Vimentin (Figure 5C,D) showed the same trend as the immunohistochemical stains. In line with this, Slug and Vimentin was also downregulated in tumors in situ of SYT1 overexpression group mice (Figure 5E–G), suggesting that SYT1 suppressed the colon cancer cells metastasis.

In addition to this, SYT1 overexpression induced EMT-like cellular morphology in HCT116 and SW480 cells (Figure 6A,B). Furthermore, we detected a decrease in Slug and Vimentin protein levels in HCT116 and SW480 cells of SYT1 overexpression, compared with the control group (Figure 6C–F). Taken together, our results indicate SYT1 participates in inhibiting invasion and metastasis via regulating the EMT process of CRC cell.

3.6. SYT1 Overexpression Inhibits EMT via Negatively Regulating the ERK/MAPK Signaling

Although the above results demonstrated that SYT1 overexpression inhibited CRC metastasis both in vitro and in vivo, the underlying mechanism by which SYT1 overexpression suppressed CRC metastasis remained unclear. It is known that ERK/MAPK signaling pathway plays a vital role in the occurrence and development of various malignant tumors, and participates in the regulation of EMT process related to tumor migration [21–23]. Slug and Vimentin are two downstream components of the ERK/MAPK signaling pathway [24–27]. U0126 is a highly selective inhibitor of ERK/MAPK signaling [28,29]. In the present study, we first tested the effect of SYT1 overexpression on ERK1/2 phosphorylation. The results showed that SYT1 overexpression had no effect on the total protein expression of ERK1/2, but significantly decreased the p-ERK1/2 level in HCT116 and SW480 cells (Figure 7A,B). Consistently, SYT1 overexpression decreased the p-ERK1/2 protein levels both in the xenograft metastasis model and in the orthotopic transplantation tumor model (Figure 7C–F). Given the above, we speculated that SYT1 might act as a tumor suppressor via regulating the ERK/MAPK signaling. To test this speculation, we treated HCT116 and SW480 cells with SYT1 overexpression plasmid and the control plasmid plus U0126. Results showed that the inhibitory effect of SYT1 overexpression on cell migration and pseudopodium formation was further enhanced upon U0126 treatment, and this was further supported by wound healing and immunostaining assays (Figure 8A–D). We further confirmed that Slug, Vimentin, and p-ERK1/2 were downregulated in CRC cells under SYT1 overexpression plus U0126 compared to SYT1 overexpression alone (Figure 8E–H). Collectively, our results demonstrate that SYT1 suppresses EMT and CRC cell migration and invasion by negative regulating the ERK/MAPK signaling.

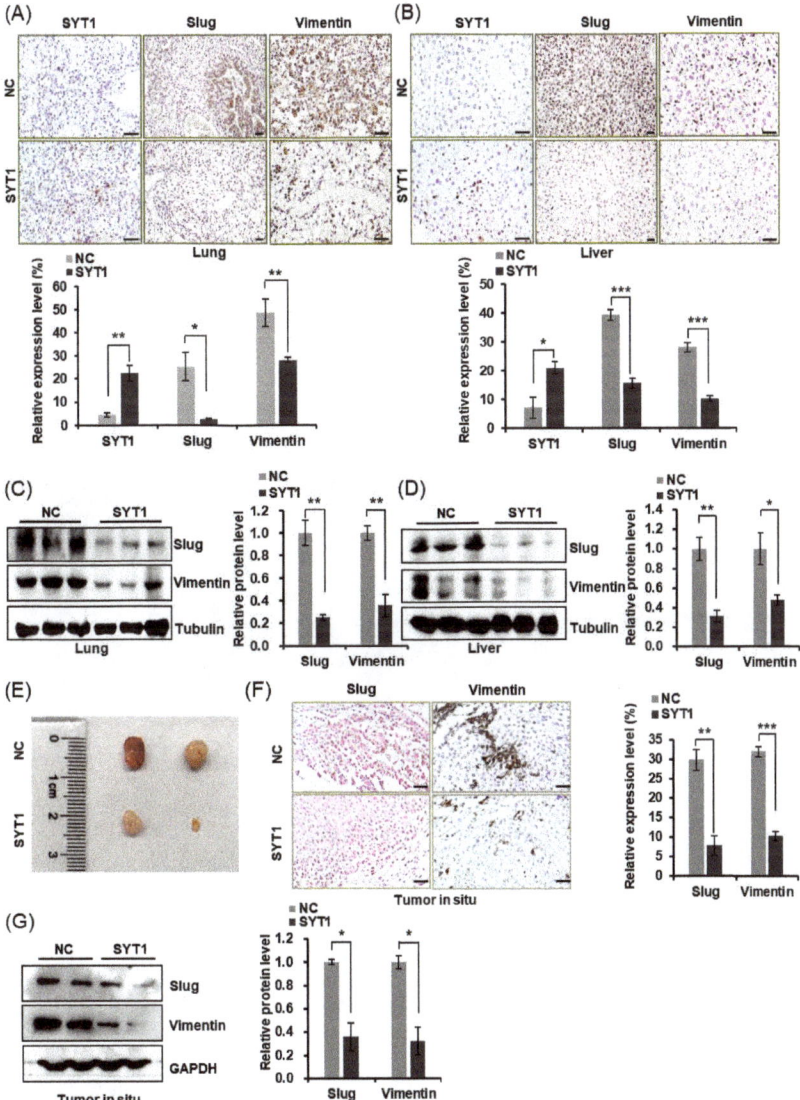

Figure 5. SYT1 overexpression downregulated the expressions of EMT-associated Slug and Vimentin in CRC xenograft metastasis nude mice model in vivo. (**A**,**B**) Immunostaining for SYT1, Slug, and Vimentin in lung (**A**) and liver (**B**) tissues of the control and SYT1 overexpression groups mice metastasis model. Scale bar, 50 μm. (**C**,**D**) The protein levels of Slug and Vimentin in liver and lung tissues of the control and SYT1 overexpression groups mice metastasis model detected by Western blot assays. (**E**) Gross images of tumors in situ of the control and SYT1 overexpression groups mice. (**F**) Immunostaining for Slug and Vimentin in tumors in situ of the control and SYT1 overexpression groups mice. Scale bar, 50 μm. (**G**) The protein levels of Slug and Vimentin in tumors in situ of SYT1 overexpression and the control groups detected by western blot assays. Tubulin or GAPDH was used as the loading control; * $p < 0.05$, ** $p < 0.01$, and *** $p < 0.001$. The uncropped bolts are shown in Supplementary Materials Figure S2.

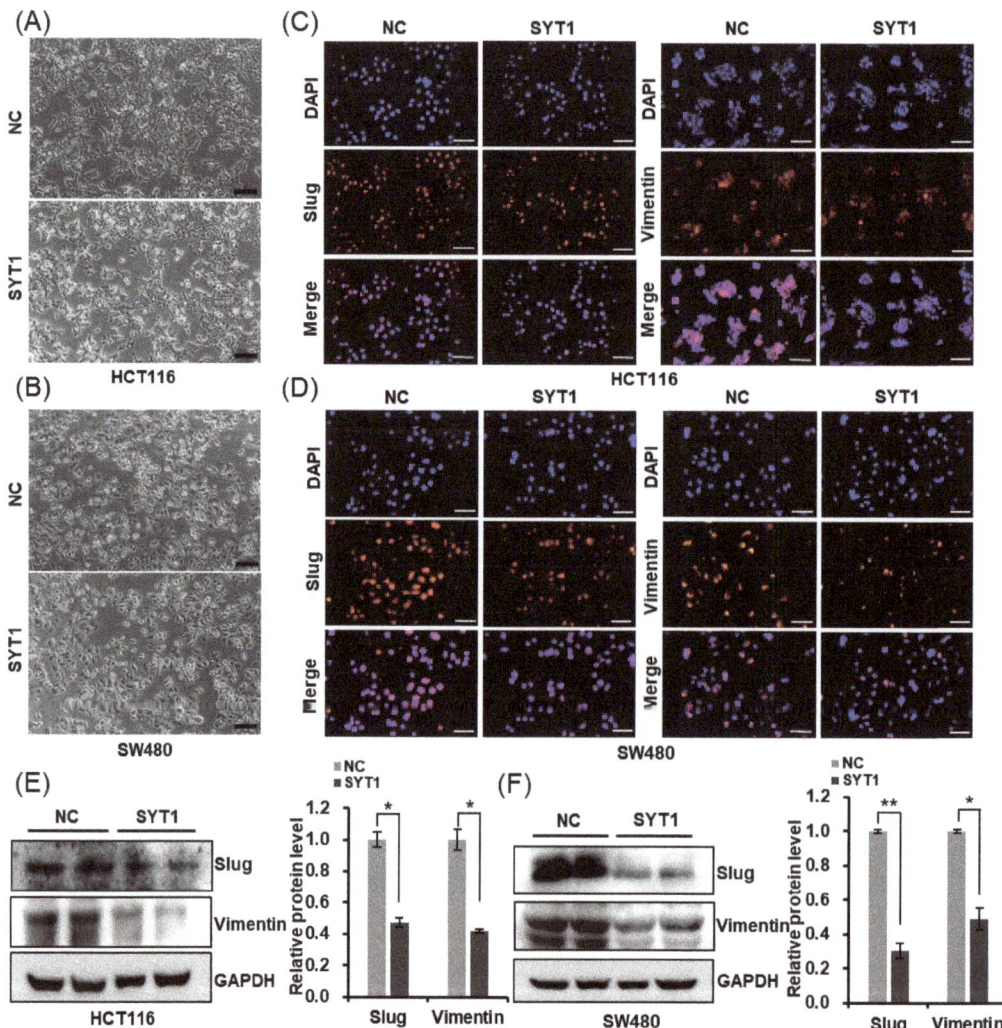

Figure 6. SYT1 suppresses the epithelial–mesenchymal transition (EMT) process of CRC cells in vitro. (**A**,**B**) Morphological changes of HCT116 (**A**) and SW480 (**B**) cells in the control and SYT1 overexpression groups. Scale bar, 50 μm. (**C**,**D**) Immunofluorescent stains of Slug and Vimentin in HCT116 (**C**) and SW480 (**D**) cells at control or SYT1 overexpression. Scale bar, 100 μm. (**E**,**F**) Western blots of EMT markers in HCT116 (**E**) and SW480 (**F**) cells at control or SYT1 overexpression. GAPDH was the loading control; * $p < 0.05$, ** $p < 0.01$. The uncropped bolts are shown in Supplementary Materials Figure S2.

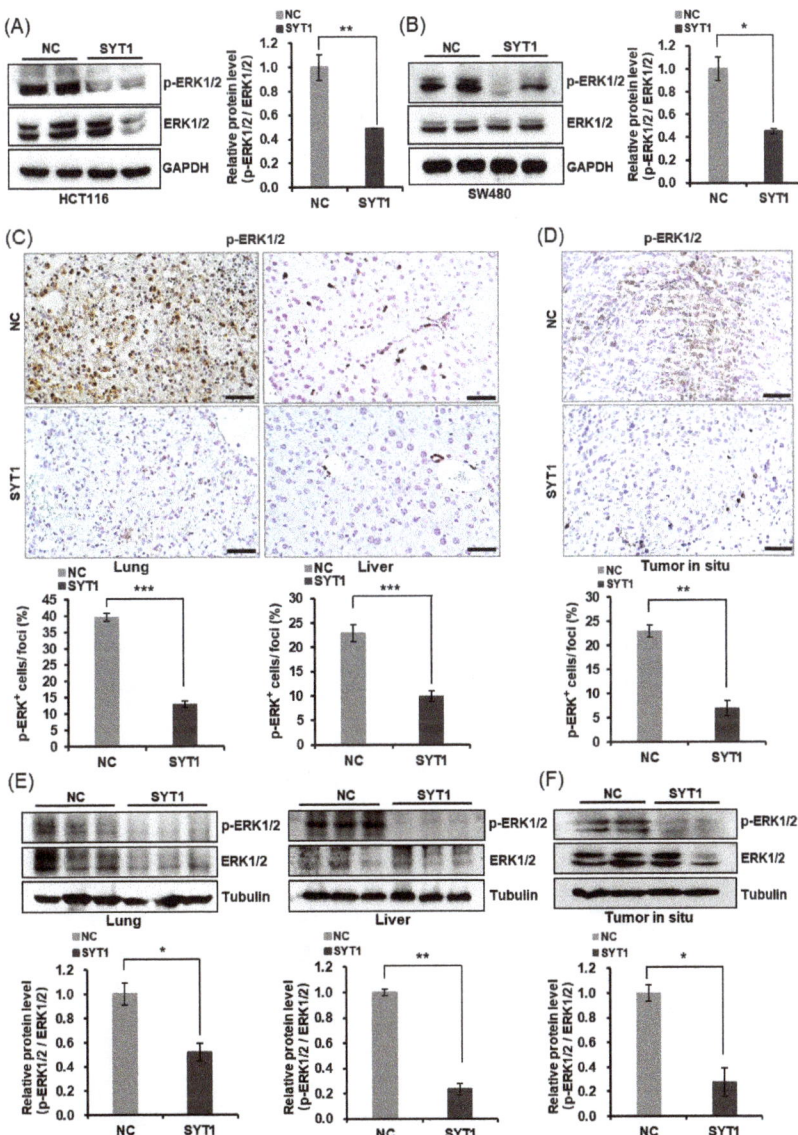

Figure 7. SYT1 inhibited ERK/MAPK signaling. (**A,B**) Western blots of MAPK42/44 (ERK1/2) and phosphorylated MAPK42/44 (p-ERK1/2) in HCT116 (**A**) and SW480 (**B**) cells at control or SYT1 overexpression. (**C**) Immunohistochemical stains of p-ERK1/2 in liver and lung tissues from control and SYT1-overexpressing CRC metastasis mice model. Scale bar, 50 μm. (**D**) Immunohistochemical stains of p-ERK1/2 in in situ tumor of control and SYT1-overexpressing mice. Scale bar, 50 μm. (**E**) Western blots of p-ERK1/2 in liver and lung tissues of control and SYT1-overexpressing CRC metastasis mice model. (**F**) Western blots of p-ERK1/2 in orthotopic transplantation tumor of control and SYT1-overexpressing mice. Tubulin or GAPDH was used as the loading control; * $p < 0.05$, ** $p < 0.01$, and *** $p < 0.001$. The uncropped bolts are shown in Supplementary Materials Figure S2.

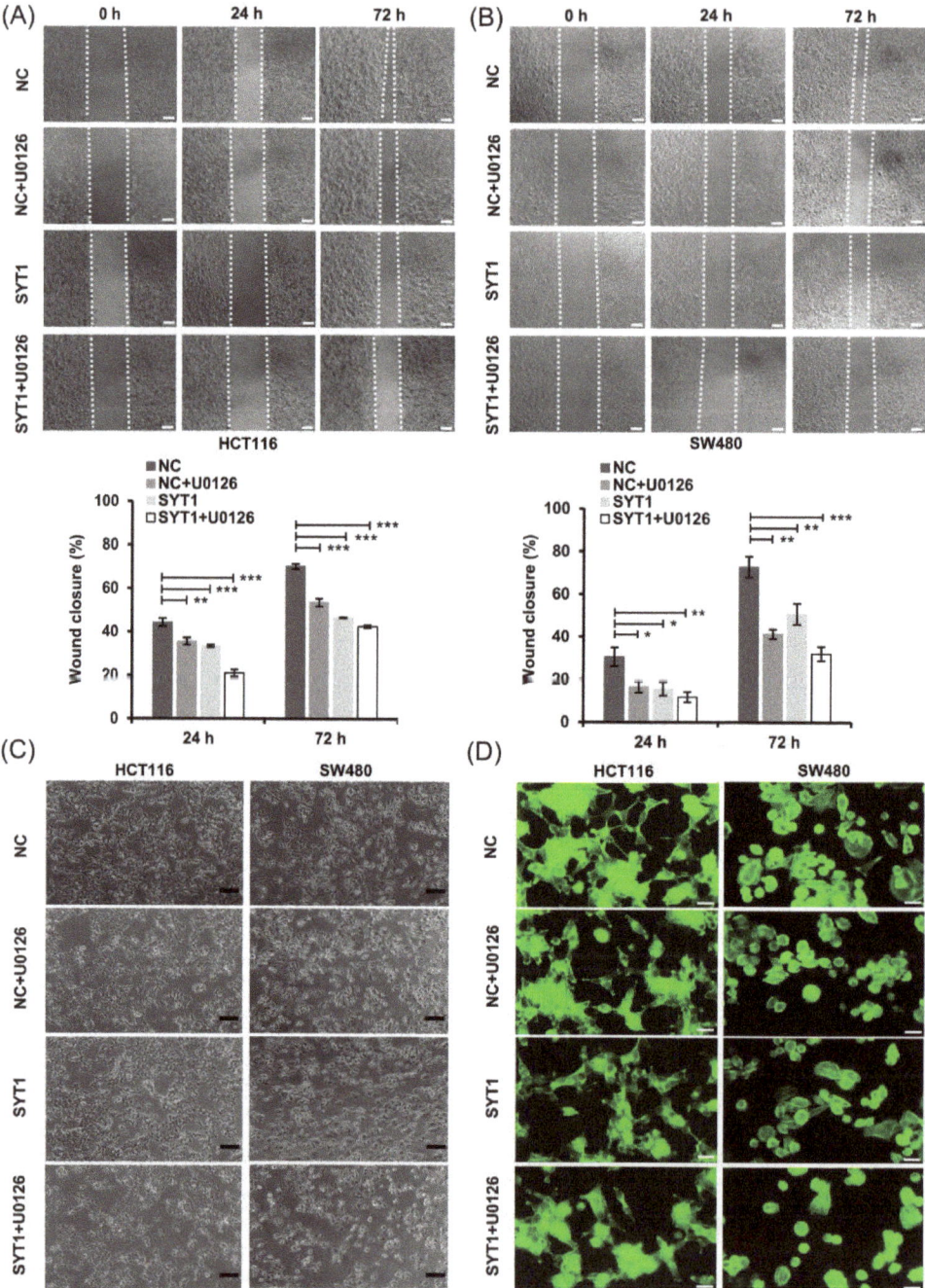

Figure 8. *Cont.*

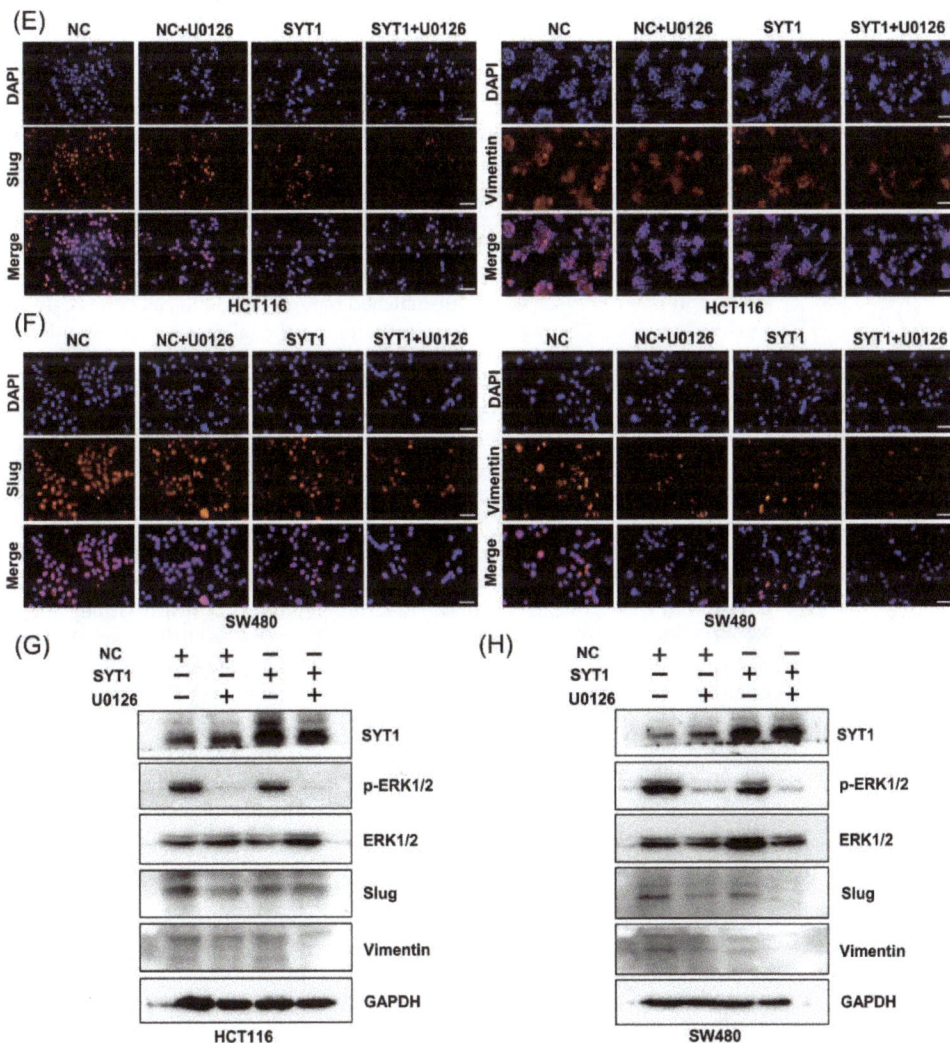

Figure 8. U0126 enhances the inhibitory effect of SYT1 overexpression on EMT and cell migration and invasion. (**A,B**) Wound healing assays showing the migration abilities of HCT116 (**A**) and SW480 (**B**) cells in groups of control, SYT1 overexpression, control + U0126, and SYT1 overexpression + U0126. Scale bar, 100 μm. (**C**) Morphological changes of HCT116 and SW480 cells in groups of control, SYT1 overexpression, control + U0126, and SYT1 overexpression + U0126. Scale bar, 100 μm. (**D**) Confocal images of phalloidin (green) in HCT116 and SW480 cells in groups of control, SYT1 overexpression, control + U0126 and SYT1 overexpression + U0126. Scale bar, 50 μm. (**E,F**) Immunofluorescent stains of Slug and Vimentin in HCT116 (**E**) and SW480 (**F**) cells of control, SYT1 overexpression, control + U0126 and SYT1 overexpression + U0126 groups. Scale bar, 100 μm. (**G,H**) Western blots of SYT1, Slug, Vimentin, and p-ERK1/2 in HCT116 (**G**) and SW480 (**H**) cells of the control, SYT1 overexpression, control + U0126, and SYT1 overexpression + U0126 groups. GAPDH was used as the loading control; * $p < 0.05$, ** $p < 0.01$, and *** $p < 0.001$. The uncropped bolts are shown in Supplementary Materials Figure S2.

4. Discussion

CRC has become a familiar malignant tumor with a high incidence worldwide [30]. Many patients are diagnosed with advanced stage, and recurrence or metastasis may occur after surgery [31]. About 20% of CRC patients suffer from metastatic disease at diagnosis [3]. In spite of recent advances in the management of CRC, metastatic disease remains challenging, and patients are rarely cured. Metastasis is the main cause of death in CRC [32]. Although CRC treatments have been improved, the prognosis remains pessimistic. Therefore, it is of great significance to further investigate the molecular mechanisms of CRC occurrence and metastasis.

CRC is a polygenic disease and is caused by genetic and epigenetic changes in oncogenes, tumor suppressor genes, mismatched repair genes, and cell cycle regulation genes in colonic mucosal cells [33]. Gene mutation is a vital factor in the incidence and development of CRC. For example, high expression and mutation of APC may provide valuable prognostic information for the clinical outcomes of CRC [34]. Similarly, vascular endothelial growth factor plays an important role in CRC and may be used as a prognostic indicator [35]. Furthermore, various types of suppressor genes and oncogenes have been identified related with the diagnosis and prognosis of CRC. Synaptotagmins (SYTs) are a family of synaptic vesicle transport proteins that largely serve as Ca^{2+} sensors in vesicular trafficking and exocytosis [36,37]. SYTs are highly conserved from invertebrates to human and are present in almost all tissues [6]. For instance, SYT1-6 and 9-13 have been detected in brain tissue, while SYT5, SYT9 and SYT13 are also found expressed in β-cells [33]. In addition to this, SYT7, SYT8, and SYT15 are expressed in heart, kidney, and pancreas. Nevertheless, the SYTs family has been gradually uncovered to be connected with human diseases including cancers. SYT7 was demonstrated to play an essential role in non-small cell lung carcinoma, head and neck squamous cell carcinoma (HNSCC), gastric cancer, and CRC [38–41]. Fu et al. [42] reported that SYT8 promoted pancreatic cancer progression via the TNNI2/ERRα/SIRT1 signaling pathway. Zhang et al. [43] indicated that SYT13 promoted the malignant phenotypes of breast cancer cells by activating the FAK/AKT signaling pathway. These studies suggest a tight association of SYTs with cancers.

Although the roles of SYT1 have been demonstrated in a variety of malignancies, such as glioblastoma and HNSCC [11], the potential role of SYT1 in CRC remains unclear. Here, we comprehensively explored the expression of SYT1 and its functional role in the progression of CRC both in vitro and in vivo. We demonstrated that the expression of SYT1 was significantly downregulated in human CRC tissues compared with the adjacent normal colorectal tissues. We further found that SYT1 expression level was negatively correlated with advanced tumor stage, cervical lymph node metastasis, and advanced clinical stage, suggesting that SYT1 may exert a repressing effect on CRC occurrence and development. Our results strongly suggest that downregulation of SYT1 promotes CRC progression.

Pseudopodial protrusion and the local reorganization of the actin cytoskeleton at the leading edge are related to tumor metastasis [44,45]. A key finding of the present study was that SYT1 could suppress CRC metastasis by inhibiting the pseudopodial formation of tumor cells. We applied a variety of assays to assess the biological functions of SYT1 in CRC metastasis, including pseudopodium formation assessment, wound healing assay, and transwell assay in vitro. In addition, we used xenograft metastasis mice model to evaluate the effect of SYT1 on CRC metastasis in vivo. Both in vitro and in vivo experiments demonstrated that SYT1 repressed CRC metastasis most likely by inhibiting tumor cell pseudopodial formation and migration.

EMT is the basis of tumor cell migration and invasion [18,46]. Vimentin is a mesenchymal marker and plays an vital role in promoting cell migration and is significantly upregulated during tumor metastasis [47]. Slug is also an important factor in promoting tumor cell migration by triggering EMT [48]. We observed that SYT1 expression level affected Vimentin and Slug expressions, suggesting that SYT1 may inhibit CRC cell migration and invasion by regulating EMT. As previously reported, ERK/MAPK signaling pathway is essential for promoting cell proliferation and migration during the occurrence and development of various malignant tumors and participates in EMT regulation process related to tumor cell migration [21,22,49–52]. Slug and Vimentin are regulated by the ERK/MAPK signaling pathway [24–27]. Here, we further demonstrated that SYT1 promoted the dephosphorylation of ERK1/2 and decreased the expression of Slug and Vimentin, strongly suggesting that SYT1 overexpression inhibits the migration and invasion of CRC cells by regulating the ERK/MAPK signaling pathway. Taken together, the present study suggests that SYT1 may be a tumor suppressor in CRC. Detailed mechanisms warrant future studies.

5. Conclusions

SYT1 expression is downregulated in CRC. SYT1 overexpression suppresses CRC metastasis both in vivo and in vitro. The inhibitory effect of SYT1 overexpression on CRC metastasis is associated with reductions of CRC cell pseudopodial formation, migration, and invasion. SYT1 overexpression can induce ERK1/2 dephosphorylation and result in inhibition of EMT, thereby resulting in suppression of CRC cell migration and invasion. Figure 9 provides a schematic outlining the signaling by which SYT1 suppresses CRC metastasis. Our findings provide new insights into CRC development and indicate the potential of SYT1 as a biomarker and potential therapeutic target for CRC.

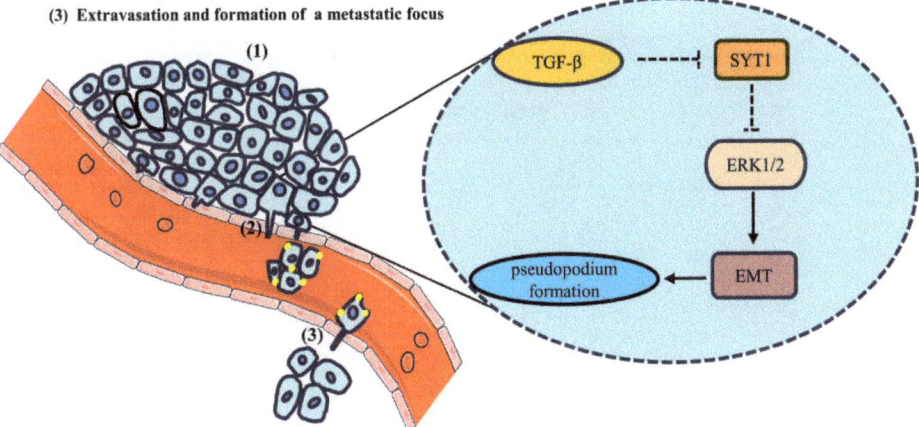

Figure 9. Schematic outlines the signaling by which SYT1 represses CRC cell metastasis.

Supplementary Materials: The following supporting information can be downloaded at: https://www.mdpi.com/article/10.3390/cancers15215282/s1, Table S1: Results of Cox regression analysis; Table S2: Sources of antibodies used in this study. Figure S1: SYT1 suppressed the progression of CRC; Figure S2: Original blots for Figures 1–8.

Author Contributions: J.C. (Jimin Cao), Y.W., and J.S. (Jianyun Shi) conceived the project. J.C. (Jimin Cao) and J.S. (Jianyun Shi) designed the experiments and analyzed data. W.L., Z.J., Y.P., and L.Z. performed the cellular experiments. J.S. (Jing Shen), R.M., W.F., L.W., and Y.F. conducted the animal experiments. J.H. and N.L. accomplished the human tissue analysis. D.W., J.S. (Jing Shen), and J.C. (Jiasong Chang) contributed to data processing. J.C. (Jimin Cao), Y.W., and J.S. (Jianyun Shi) wrote the manuscript. J.C. (Jimin Cao) and Y.W. guided and reviewed the manuscript. All authors have read and agreed to the published version of the manuscript.

Funding: This study was supported by the Shanxi Province Science Foundation for Youths (201901D211316), and partially by the National Natural Science Foundation of China (82002063, 82170523, and 82202622), the Shanxi "1331" Project Quality and Efficiency Improvement Plan (1331KFC), and a grant from the Shanxi Medical University of Shanxi Province in China (XD1809).

Institutional Review Board Statement: Animal experimentation was performed in accordance with the standards of the National Institute of Health Guide for the Care and Use of Laboratory Animals, and further approved by the Animal Ethics Committee of the Shanxi Medical University (Taiyuan, China) (approval no.: SYDL2023-085).

Informed Consent Statement: Informed consent was obtained from all subjects involved in the study.

Data Availability Statement: The data presented in this study are available on request from the corresponding author.

Acknowledgments: We thank the Shanxi Provincial Academy of Traditional Chinese Medicine for the help in human tissue collection.

Conflicts of Interest: The authors declare no conflict of interest.

Abbreviations

AMP: adenosine monophosphate; CRC: colorectal cancer; DAB: diaminobenzidine; EMT: epithelial–mesenchymal cell transformation; HPA: The Human Protein Atlas Project; MAPK: mitogen-activated protein kinase; PCNA: proliferating cell nuclear antigen; SD: standard deviation; SYT: synaptotagmin; TCGA: The Cancer Genome Atlas; TGF-β: transforming growth factor β.

References

1. Rabeneck, L.; Chiu, H.-M.; Senore, C. International Perspective on the Burden of Colorectal Cancer and Public Health Effects. *Gastroenterology* **2020**, *158*, 447–452. [CrossRef] [PubMed]
2. Hossain, M.S.; Karuniawati, H.; Jairoun, A.A.; Urbi, Z.; Ooi, J.; John, A.; Lim, Y.C.; Kibria, K.M.K.; Mohiuddin, A.K.M.; Ming, L.C.; et al. Colorectal Cancer: A Review of Carcinogenesis, Global Epidemiology, Current Challenges, Risk Factors, Preventive and Treatment Strategies. *Cancers* **2022**, *14*, 1732. [CrossRef] [PubMed]
3. Keum, N.; Giovannucci, E. Global burden of colorectal cancer: Emerging trends, risk factors and prevention strategies. *Nat. Gastroenterol. Hepatol.* **2019**, *16*, 713–732. [CrossRef] [PubMed]
4. Sawicki, T.; Ruszkowska, M.; Danielewicz, A.; Niedzwiedzka, E.; Arlukowicz, T.; Przybylowicz, K.E. A Review of Colorectal Cancer in Terms of Epidemiology, Risk Factors, Development, Symptoms and Diagnosis. *Cancers* **2021**, *13*, 2025. [CrossRef]
5. Jiang, H.; Du, J.; Gu, J.; Jin, L.; Pu, Y.; Fei, B. A 65-gene signature for prognostic prediction in colon adenocarcinoma. *Int. J. Mol. Med.* **2018**, *41*, 2021–2027. [CrossRef]
6. Zhu, X.L.; Qi, S.T.; Liu, J.; Chen, L.; Zhang, C.; Yang, S.W.; Ouyang, Y.C.; Hou, Y.; Schatten, H.; Song, Y.L.; et al. Synaptotagmin1 is required for spindle stability and metaphase-to-anaphase transition in mouse oocytes. *Cell Cycle* **2012**, *11*, 818–826. [CrossRef]
7. Zhu, X.L.; Li, S.F.; Zhang, X.Q.; Xu, H.; Luo, Y.Q.; Yi, Y.H.; Lv, L.J.; Zhang, C.H.; Wang, Z.B.; Ouyang, Y.C.; et al. Synaptotagmin1 regulates cortical granule exocytosis during mouse oocyte activation. *Zygote* **2019**, *28*, 97–102. [CrossRef]
8. Musch, M.W.; Arvans, D.L.; Walsh-Reitz, M.M.; Uchiyama, K.; Fukuda, M.; Chang, E.B. Synaptotagmin I binds intestinal epithelial NHE3 and mediates cAMP- and Ca2+-induced endocytosis by recruitment of AP2 and clathrin. *Am. J. Physiol. Gastrointest. Liver Physiol.* **2007**, *292*, G1549–G1558. [CrossRef]
9. Musch, M.W.; Arvans, D.L.; Wang, Y.; Nakagawa, Y.; Solomaha, E.; Chang, E.B. Cyclic AMP-mediated endocytosis of intestinal epithelial NHE3 requires binding to synaptotagmin 1. *Am. J. Physiol. Gastrointest. Liver Physiol.* **2010**, *298*, G203–G211. [CrossRef]
10. Nord, H.; Hartmann, C.; Andersson, R.; Menzel, U.; Pfeifer, S.; Piotrowski, A.; Bogdan, A.; Kloc, W.; Sandgren, J.; Olofsson, T.; et al. Characterization of novel and complex genomic aberrations in glioblastoma using a 32K BAC array. *Neuro Oncol.* **2009**, *11*, 803–818. [CrossRef]
11. Liu, G.; Zeng, X.; Wu, B.; Zhao, J.; Pan, Y. RNA-Seq analysis of peripheral blood mononuclear cells reveals unique transcriptional signatures associated with radiotherapy response of nasopharyngeal carcinoma and prognosis of head and neck cancer. *Cancer Biol. Ther.* **2020**, *21*, 139–146. [CrossRef] [PubMed]
12. Yang, J.; Hou, Z.; Wang, C.; Wang, H.; Zhang, H. Gene expression profiles reveal key genes for early diagnosis and treatment of adamantinomatous craniopharyngioma. *Cancer Biol. Ther.* **2018**, *25*, 227–239. [CrossRef] [PubMed]
13. Shi, J.; Ma, X.; Su, Y.; Song, Y.; Tian, Y.; Yuan, S.; Zhang, X.; Yang, D.; Zhang, H.; Shuai, J.; et al. MiR-31 Mediates Inflammatory Signaling to Promote Re-Epithelialization during Skin Wound Healing. *J. Investig. Dermatol.* **2018**, *138*, 2253–2263. [CrossRef]
14. Aikemu, B.; Shao, Y.; Yang, G.; Ma, J.; Zhang, S.; Yang, X.; Hong, H.; Yesseyeva, G.; Huang, L.; Jia, H.; et al. NDRG1 regulates Filopodia-induced Colorectal Cancer invasiveness via modulating CDC42 activity. *Int. J. Biol. Sci.* **2021**, *17*, 1716–1730. [CrossRef] [PubMed]
15. Shankar, J.; Messenberg, A.; Chan, J.; Underhill, T.M.; Foster, L.J.; Nabi, I.R. Pseudopodial actin dynamics control epithelial-mesenchymal transition in metastatic cancer cells. *Cancer Res.* **2010**, *70*, 3780–3790. [CrossRef] [PubMed]
16. Vu, T.; Datta, P. Regulation of EMT in Colorectal Cancer: A Culprit in Metastasis. *Cancers* **2017**, *9*, 171. [CrossRef]

17. Trelford, C.B.; Dagnino, L.; Di Guglielmo, G.M. Transforming growth factor-β in tumour development. *Front. Mol. Biosci.* **2022**, *9*, 991612. [CrossRef]
18. Brabletz, S.; Schuhwerk, H.; Brabletz, T.; Stemmler, M.P. Dynamic EMT: A multi-tool for tumor progression. *EMBO J.* **2021**, *40*, e108647. [CrossRef]
19. Yilmaz, M.; Christofori, G. EMT, the cytoskeleton, and cancer cell invasion. *Cancer Metastasis Rev.* **2009**, *28*, 15–33. [CrossRef]
20. Polyak, K.; Weinberg, R.A. Transitions between epithelial and mesenchymal states: Acquisition of malignant and stem cell traits. *Nat. Rev. Cancer* **2009**, *9*, 265–273. [CrossRef]
21. Guo, Y.J.; Pan, W.W.; Liu, S.B.; Shen, Z.F.; Xu, Y.; Hu, L.L. ERK/MAPK signalling pathway and tumorigenesis. *Exp. Ther. Med.* **2020**, *19*, 1997–2007. [CrossRef] [PubMed]
22. Sheng, W.; Chen, C.; Dong, M.; Wang, G.; Zhou, J.; Song, H.; Li, Y.; Zhang, J.; Ding, S. Calreticulin promotes EGF-induced EMT in pancreatic cancer cells via Integrin/EGFR-ERK/MAPK signaling pathway. *Cell Death Dis.* **2017**, *8*, e3147. [CrossRef] [PubMed]
23. Zhang, P.; Liu, H.; Xia, F.; Zhang, Q.W.; Zhang, Y.Y.; Zhao, Q.; Chao, Z.H.; Jiang, Z.W.; Jiang, C.C. Epithelial-mesenchymal transition is necessary for acquired resistance to cisplatin and increases the metastatic potential of nasopharyngeal carcinoma cells. *Int. J. Mol. Med.* **2014**, *33*, 151–159. [CrossRef] [PubMed]
24. Cai, S.; Li, N.; Bai, X.; Liu, L.; Banerjee, A.; Lavudi, K.; Zhang, X.; Zhao, J.; Venere, M.; Duan, W.; et al. ERK inactivation enhances stemness of NSCLC cells via promoting Slug-mediated epithelial-to-mesenchymal transition. *Theranostics* **2022**, *12*, 7051–7066. [CrossRef] [PubMed]
25. Pradhan, N.; Parbin, S.; Kar, S.; Das, L.; Kirtana, R.; Suma Seshadri, G.; Sengupta, D.; Deb, M.; Kausar, C.; Patra, S.K. Epigenetic silencing of genes enhanced by collective role of reactive oxygen species and MAPK signaling downstream ERK/Snail axis: Ectopic application of hydrogen peroxide repress CDH1 gene by enhanced DNA methyltransferase activity in human breast cancer. *Biochim. Biophys. Acta Mol. Basis Dis.* **2019**, *1865*, 1651–1665. [CrossRef]
26. Zeng, K.; Chen, X.; Xu, M.; Liu, X.; Li, C.; Xu, X.; Pan, B.; Qin, J.; He, B.; Pan, Y.; et al. LRIG3 represses cell motility by inhibiting slug via inactivating ERK signaling in human colorectal cancer. *IUBMB Life* **2020**, *72*, 1393–1403. [CrossRef]
27. Gong, Z.; Gao, X.; Yang, Q.; Lun, J.; Xiao, H.; Zhong, J.; Cao, H. Phosphorylation of ERK-Dependent NF-kappaB Triggers NLRP3 Inflammasome Mediated by Vimentin in EV71-Infected Glioblastoma Cells. *Molecules* **2022**, *27*, 4190. [CrossRef]
28. Hou, Y.; Zhou, B.; Ni, M.; Wang, M.; Ding, L.; Li, Y.; Liu, Y.; Zhang, W.; Li, G.; Wang, J.; et al. Nonwoven-based gelatin/polycaprolactone membrane loaded with ERK inhibitor U0126 for treatment of tendon defects. *Stem. Cell Res. Ther.* **2022**, *13*, 5. [CrossRef]
29. Huang, L.; Chen, S.; Fan, H.; Ji, D.; Chen, C.; Sheng, W. GINS2 promotes EMT in pancreatic cancer via specifically stimulating ERK/MAPK signaling. *Cancer Gene Ther.* **2021**, *28*, 839–849. [CrossRef]
30. Guren, M.G. The global challenge of colorectal cancer. *Lancet Gastroenterol. Hepatol.* **2019**, *4*, 894–895. [CrossRef]
31. Li, N.; Zhou, Z.; Zhang, L.; Tang, H.; Chen, X.; Zhou, H. High expression of TTC21A predict poor prognosis of colorectal cancer and influence the immune infiltrating level. *Transl. Cancer Res.* **2022**, *11*, 981–992. [CrossRef] [PubMed]
32. Piawah, S.; Venook, A.P. Targeted therapy for colorectal cancer metastases: A review of current methods of molecularly targeted therapy and the use of tumor biomarkers in the treatment of metastatic colorectal cancer. *Cancer* **2019**, *125*, 4139–4147. [CrossRef] [PubMed]
33. Li, Q.; Zhang, S.; Hu, M.; Xu, M.; Jiang, X. Silencing of synaptotagmin 13 inhibits tumor growth through suppressing proliferation and promoting apoptosis of colorectal cancer cells. *Int. J. Mol. Med.* **2020**, *45*, 234–244. [CrossRef] [PubMed]
34. Chen, T.H.; Chang, S.W.; Huang, C.C.; Wang, K.L.; Yeh, K.T.; Liu, C.N.; Lee, H.; Lin, C.C.; Cheng, Y.W. The prognostic significance of APC gene mutation and miR-21 expression in advanced-stage colorectal cancer. *Colorectal Dis.* **2013**, *15*, 1367–1374. [CrossRef] [PubMed]
35. Falchook, G.S.; Kurzrock, R. VEGF and dual-EGFR inhibition in colorectal cancer. *Cell Cycle* **2015**, *14*, 1129–1130. [CrossRef] [PubMed]
36. Koh, T.W.; Bellen, H.J. Synaptotagmin I, a Ca2+ sensor for neurotransmitter release. *Trends Neurosci.* **2003**, *26*, 413–422. [CrossRef]
37. Vinet, A.F.; Fukuda, M.; Descoteaux, A. The exocytosis regulator synaptotagmin V controls phagocytosis in macrophages. *J. Immunol.* **2008**, *181*, 5289–5295. [CrossRef]
38. Han, Q.; Zou, D.; Lv, F.; Wang, S.; Yang, C.; Song, J.; Wen, Z.; Zhang, Y. High SYT7 expression is associated with poor prognosis in human non-small cell lung carcinoma. *Pathol. Res. Pract.* **2020**, *216*, 153101. [CrossRef]
39. Kanda, M.; Tanaka, H.; Shimizu, D.; Miwa, T.; Umeda, S.; Tanaka, C.; Kobayashi, D.; Hattori, N.; Suenaga, M.; Hayashi, M.; et al. SYT7 acts as a driver of hepatic metastasis formation of gastric cancer cells. *Oncogene* **2018**, *37*, 5355–5366. [CrossRef]
40. Wang, K.; Xiao, H.; Zhang, J.; Zhu, D. Synaptotagmin7 is Overexpressed in Colorectal Cancer and Regulates Colorectal Cancer Cell Proliferation. *J. Cancer* **2018**, *9*, 2349–2356. [CrossRef]
41. Liu, X.; Li, C.; Yang, Y.; Liu, X.; Li, R.; Zhang, M.; Yin, Y.; Qu, Y. Synaptotagmin 7 in twist-related protein 1-mediated epithelial-Mesenchymal transition of non-small cell lung cancer. *EBioMedicine* **2019**, *46*, 42–53. [CrossRef] [PubMed]
42. Fu, Z.; Liang, X.; Shi, L.; Tang, L.; Chen, D.; Liu, A.; Shao, C. SYT8 promotes pancreatic cancer progression via the TNNI2/ERRalpha/SIRT1 signaling pathway. *Cell Death Discov.* **2021**, *7*, 390. [CrossRef] [PubMed]
43. Zhang, Y.D.; Zhong, R.; Liu, J.Q.; Sun, Z.X.; Wang, T.; Liu, J.T. Role of synaptotagmin 13 (SYT13) in promoting breast cancer and signaling pathways. *Clin. Transl. Oncol.* **2023**, *25*, 1629–1640. [CrossRef] [PubMed]

44. Iwaya, K.; Norio, K.; Mukai, K. Coexpression of Arp2 and WAVE2 predicts poor outcome in invasive breast carcinoma. *Mod. Pathol.* **2007**, *20*, 339–343. [CrossRef]
45. Iwaya, K.; Oikawa, K.; Semba, S.; Tsuchiya, B.; Mukai, Y.; Otsubo, T.; Nagao, T.; Izumi, M.; Kuroda, M.; Domoto, H.; et al. Correlation between liver metastasis of the colocalization of actin-related protein 2 and 3 complex and WAVE2 in colorectal carcinoma. *Cancer Sci.* **2007**, *98*, 992–999. [CrossRef]
46. Marconi, G.D.; Fonticoli, L.; Rajan, T.S.; Pierdomenico, S.D.; Trubiani, O.; Pizzicannella, J.; Diomede, F. Epithelial-Mesenchymal Transition (EMT): The Type-2 EMT in Wound Healing, Tissue Regeneration and Organ Fibrosis. *Cells* **2021**, *10*, 1587. [CrossRef]
47. Battaglia, R.A.; Delic, S.; Herrmann, H.; Snider, N.T. Vimentin on the move: New developments in cell migration. *F1000Research* **2018**, *7*, F1000 Faculty Rev-1796. [CrossRef]
48. Recouvreux, M.V.; Moldenhauer, M.R.; Galenkamp, K.M.O.; Jung, M.; James, B.; Zhang, Y.; Lowy, A.; Bagchi, A.; Commisso, C. Glutamine depletion regulates Slug to promote EMT and metastasis in pancreatic cancer. *J. Exp. Med.* **2020**, *217*, e20200388. [CrossRef]
49. Liu, F.; Yang, X.; Geng, M.; Huang, M. Targeting ERK, an Achilles' Heel of the MAPK pathway, in cancer therapy. *Acta Pharm. Sin. B* **2018**, *8*, 552–562. [CrossRef]
50. Shi, Y.; Sun, H. Down-regulation of lncRNA LINC00152 Suppresses Gastric Cancer Cell Migration and Invasion Through Inhibition of the ERK/MAPK Signaling Pathway. *OncoTargets Ther.* **2020**, *13*, 2115–2124. [CrossRef]
51. Zhang, F.; Ni, Z.J.; Ye, L.; Zhang, Y.Y.; Thakur, K.; Cespedes-Acuna, C.L.; Han, J.; Zhang, J.G.; Wei, Z.J. Asparanin A inhibits cell migration and invasion in human endometrial cancer via Ras/ERK/MAPK pathway. *Food Chem. Toxicol.* **2021**, *150*, 112036. [CrossRef] [PubMed]
52. Sheng, W.; Shi, X.; Lin, Y.; Tang, J.; Jia, C.; Cao, R.; Sun, J.; Wang, G.; Zhou, L.; Dong, M. Musashi2 promotes EGF-induced EMT in pancreatic cancer via ZEB1-ERK/MAPK signaling. *J. Exp. Clin. Cancer Res.* **2020**, *39*, 16. [CrossRef] [PubMed]

Disclaimer/Publisher's Note: The statements, opinions and data contained in all publications are solely those of the individual author(s) and contributor(s) and not of MDPI and/or the editor(s). MDPI and/or the editor(s) disclaim responsibility for any injury to people or property resulting from any ideas, methods, instructions or products referred to in the content.

Article

Possibility of Using Conventional Computed Tomography Features and Histogram Texture Analysis Parameters as Imaging Biomarkers for Preoperative Prediction of High-Risk Gastrointestinal Stromal Tumors of the Stomach

Milica Mitrovic Jovanovic [1,2], Aleksandra Djuric Stefanovic [1,2], Dimitrije Sarac [1], Jelena Kovac [1,2], Aleksandra Jankovic [1,2], Dusan J. Saponjski [1,2], Boris Tadic [3,4], Milena Kostadinovic [5], Milan Veselinovic [4,6], Vladimir Sljukic [4,6], Ognjan Skrobic [4,6], Marjan Micev [7], Dragan Masulovic [1,2], Predrag Pesko [4,6] and Keramatollah Ebrahimi [4,6],*

[1] Center for Radiology and Magnetic Resonance Imaging, University Clinical Centre of Serbia, Pasterova No. 2, 11000 Belgrade, Serbia; milica.mitrovic-jovanovic@med.bg.ac.rs (M.M.J.)

[2] Department for Radiology, Faculty of Medicine, University of Belgrade, Dr Subotica No. 8, 11000 Belgrade, Serbia

[3] Department for HBP Surgery, Clinic for Digestive Surgery, University Clinical Centre of Serbia, Koste Todorovica Street, No. 6, 11000 Belgrade, Serbia

[4] Department for Surgery, Faculty of Medicine, University of Belgrade, Dr Subotica No. 8, 11000 Belgrade, Serbia

[5] Center for Physical Medicine and Rehabilitation, University Clinical Centre of Serbia, Pasterova Street, No. 2, 11000 Beograd, Serbia

[6] Department of Stomach and Esophageal Surgery, Clinic for Digestive Surgery, University Clinical Centre of Serbia, Koste Todorovica Street No. 6, 11000 Belgrade, Serbia

[7] Department for Pathology, Clinic for Digestive Surgery, University Clinical Centre of Serbia, Koste Todorovica Street, No. 6, 11000 Belgrade, Serbia

* Correspondence: keramatollah.ebrahimi@med.bg.ac.rs; Tel.: +381-642448595

Simple Summary: Gastrointestinal stromal tumors are the most common mesenchymal tumors that can have a malignant character. Definitive diagnosis is obtained by pathohistological and immunohistochemical analysis of the resected tumor. Preoperative stratification of metastatic risk using non-invasive imaging methods would be of great importance in the selection of patients with high-risk GIST and the application of neoadjuvant target therapy. This could enable tumor shrinkage, avoiding multivisceral resections and reducing the risk of tumor rupture. It also could provide better long-term outcomes, including increased overall survival rates, by optimizing surgical resection and systemic control of the disease. Evaluation of the morphological characteristics of the tumor obtained by computed tomography examination as well as the histogram parameters of the textural analysis of tumor tissue may improve the preoperative prediction of the metastatic risk of GIST. Texture analysis is part of the growing field of radiomics, with significant contributions to oncology so far.

Abstract: Background: The objective of this study is to determine the morphological computed tomography features of the tumor and texture analysis parameters, which may be a useful diagnostic tool for the preoperative prediction of high-risk gastrointestinal stromal tumors (HR GISTs). Methods: This is a prospective cohort study that was carried out in the period from 2019 to 2022. The study included 79 patients who underwent CT examination, texture analysis, surgical resection of a lesion that was suspicious for GIST as well as pathohistological and immunohistochemical analysis. Results: Textural analysis pointed out min norm ($p = 0.032$) as a histogram parameter that significantly differed between HR and LR GISTs, while min norm ($p = 0.007$), skewness ($p = 0.035$) and kurtosis ($p = 0.003$) showed significant differences between high-grade and low-grade tumors. Univariate regression analysis identified tumor diameter, margin appearance, growth pattern, lesion shape, structure, mucosal continuity, enlarged peri- and intra-tumoral feeding or draining vessel (EFDV) and max norm as significant predictive factors for HR GISTs. Interrupted mucosa ($p < 0.001$) and presence of EFDV ($p < 0.001$) were obtained by multivariate regression analysis as independent predictive

factors of high-risk GISTs with an AUC of 0.878 (CI: 0.797–0.959), sensitivity of 94%, specificity of 77% and accuracy of 88%. Conclusion: This result shows that morphological CT features of GIST are of great importance in the prediction of non-invasive preoperative metastatic risk. The incorporation of texture analysis into basic imaging protocols may further improve the preoperative assessment of risk stratification.

Keywords: gastrointestinal stromal tumor (GIST); multidetector computed tomography (MDCT); texture analysis; metastatic risk

1. Introduction

Gastrointestinal stromal tumors (GISTs) are relatively rare mesenchymal tumors with a potentially malignant and aggressive behavior [1]. They can occur anywhere in the digestive tract, but are most commonly localized in the stomach and small intestine [2]. These tumors tend to spread and metastasize. As they are often detected at an advanced stage, they can pose a serious challenge for management. Surgery is the only curative treatment, and recently, the minimally invasive approach has proven to be feasible and safe [3].

Although these tumors do not have the same biological behavior, any GIST should be considered potentially malignant [3]. The location and size of the tumor are important factors that determine the modality of GIST treatment. Regardless of the size of the tumor, complete removal in challenging locations sometimes requires extensive, risky and mutilating surgery associated with functional disability or morbidity. In such cases, especially when the tumor is small, clinical guidelines generally recommend only follow-up [4,5].

The fact that a GIST is small does not exclude the possibility that its proliferative activity may be aggressive [6]. Tumor biopsy is the best way to obtain tissue samples for subsequent pathological diagnosis. However, it is associated with possible complications such as tumor rupture or bleeding, and the results are often inconclusive. Most GIST guidelines for the surveillance of small lesions recommend initial follow-up by EUS. The Japanese guidelines point out that GISTs may be potentially aggressive if the tumors show growth features ulceration or irregular margins at follow-up [5]. The NCCN sarcoma guidelines also recommend that a small tumor that has high-risk features should be removed, while others that do not have these features can be followed by EUS [7]. However, manu studies have shown that extrinsic or exophytic tumor growth may be missed by these examinations. In such situations, new or additional biomarkers would make an important contribution to deciding on the optimal treatment modality.

Given the potential benefits of and research on neoadjuvant therapy for GIST, preoperative risk stratification may be of particular importance. Preoperative administration of tyrosine kinase inhibitors could enable tumor shrinkage and reduce the risk of tumor rupture [8]. It also could provide better long-term outcomes, including increased overall survival rates, by optimizing the systemic control of the disease [9]. Two comparative systems are most commonly used in clinical practice: the Armed Forces Institute of Pathology (Miettinen's) criteria and the National Institutes of Health (NIH) classification [10,11]. Both systems use the mitotic index as an important factor in the assessment of tumor aggressiveness. According to the Miettinen criteria, the risk of recurrence or metastasis in 2 to 5 cm gastric GISTs mainly depends on their mitotic activity. Studies have shown that grading based on mitotic count is not accurate in regular biopsy [12]. Considering the above facts, the neoadjuvant strategy for GISTs may complicate the selection of suitable patients.

The most commonly used diagnostic modality for the diagnosis and staging of these tumors is computed tomography (CT). Many studies have demonstrated a correlation of certain morphologic CT features of GISTs with a high metastatic risk, especially the diameter of the lesion [13–16]. CT texture analysis (CTTA) is a relatively new postprocessing imaging tool used to assess the heterogeneity of tumor tissue [17]. It has proven to be very useful in differentiating diagnosis, stratifying the grade and risk of different tumors, assessing prognosis and predicting the response to the implemented therapy [17–20].

The aim of this study is to evaluate the diagnostic value of the morphologic parameters of conventional CT diagnostics and the histogram parameters of texture analysis in the non-invasive, preoperative assessment of the metastatic potential of GISTs in correlation with macropathological and pathohistological findings, especially with the mitotic index, as a gold standard. Preoperative diagnosis of high-risk GISTs could facilitate decisions on further treatment protocol in patients in whom tumor localization requires multivisceral or extensive surgery. The use of neoadjuvant therapy with tyrosine kinase inhibitors in patients with HR GISTs would lead to tumor dimension reduction, thus effectively improving the resection rate of surgery. Further, it would also reduce the risk of tumor recurrence and lead to better prognosis of the disease [9].

2. Materials and Methods

2.1. Patients

Seventy-nine patients who underwent a CT diagnostic protocol followed by surgery during the period from 2019 to 2022 were included in this prospective research. Criteria for inclusion in the study were as follows: (1) clinically suspected GIST as mainly submucosal gastric lesion; (2) abdominal CT exam according to a dual-phase protocol; (3) no more than 20 days between CT examination and surgery. Criteria for exclusion from the study were as follows: (1) extra-gastric localization of GIST, (2) histopathological findings suggestive of other stomach tumors, (3) patients whose CT exam was of poor technical quality without the possibility of further processing and (4) more than 20 days from the performed CT examination to the surgical resection of the tumor. Gastric resections such as total and subtotal gastrectomy or wedge resection or tumor enucleation were performed in all patients with histopathological and immunohistochemical analysis of the resected tumor. Further, the tumors were staged using the American Joint Committee on Tumor/Lymph Node/Carcinoma Metastases (TNM) classification (8th edition) [21]. According to the TNM supplemented with AFIP classification, patients were divided into two groups, low-risk (LR) and high-risk (HR) [22].

Our research was permitted by the Ethical Committee of the Faculty of Medicine, University of Belgrade, and written informed consent was obtained from all patients.

2.2. Abdominal CT Examination

CT diagnosis was performed on a multidetector CT (MDCT) machine with 64 rows of detectors (Aquuilion One, Toshiba Medical Systems, Ottawa, Japan). Immediately before the examination, the patients were given 250–500 mL of water "per os" (as a negative contrast), in order to adequately distend the stomach.

Abdominal CT examination was performed as standard after iv. administration of 60–100 mL of iodinated contrast (1–1.5 mL/kg body weight), in the arterial and portal venous phase.

2.3. Abdominal CT Scan Analysis

The following morphological characteristics of the tumor were analyzed:

1. Maximum diameter: the largest diameter of the tumor in mm (Figure 1);
2. Appearance of mucosa: intact/continuous and discontinuous (Figures 1 and 2);
3. Tumor structure: solid–necrotic and cystic changes (Figures 1–4);
4. Tumor shape: regular or irregular (Figure 1);
5. Tumor localization in relation to the region of the stomach: corpus, antrum and pylorus (Figures 1, 2 and 5);
6. Growth mode: exophytic/mixed and endophytic (Figure 1);
7. Level of opacification of the solid part of the tumor: weak and intense (Figures 3 and 6);
8. The presence of visible enlarged vascular structures draining/feeding the tumor (EFDV "enlarged feeding or draining vessel") (Figure 6);

9. Margin appearance: well-defined and ill-defined (Figures 1 and 7).

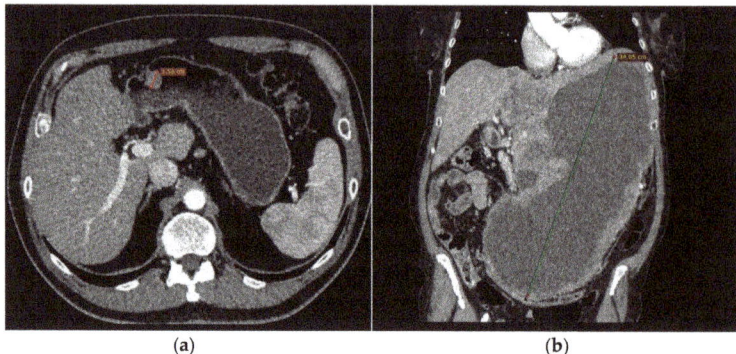

Figure 1. The lowest tumor diameter was 15 mm (LR GIST) in the pyloric stomach region (**a**) and the largest lesion measured 340 mm (HR GIST) (**b**). LR GIST shows a predominantly round shape, well-defined margins and a homogenous, solid appearance and intact mucosa (**a**). Notable difference in tumor structure with massive cystic degeneration, irregular shape and exophytic growth pattern in HR tumor (**b**).

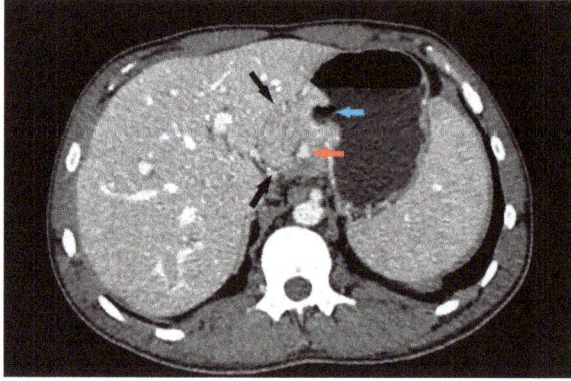

Figure 2. Axial CT scan shows exophytic tumor growth into gastro-hepatic ligament (black arrows) with presence of intralesional vascular structures (red arrow) and mucosal defect: umbilication (blue arrow). Tumor involves corpus region of the stomach.

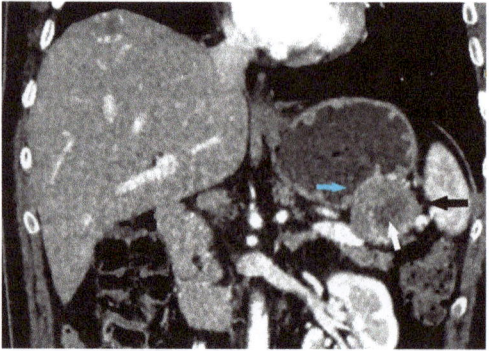

Figure 3. CT scan coronal view clearly depicts oval, clearly demarcated lesion (black arrow) covered by intact mucosa (blue arrow) with partially necrotic structure (white arrow) and weak postcontrast opacification of solid part of the lesion.

Figure 4. Abdominal CT exam coronal view shows irregular tumor shape with presence of cystic structural changes (white arrow). The tumor corresponds to HR GIST.

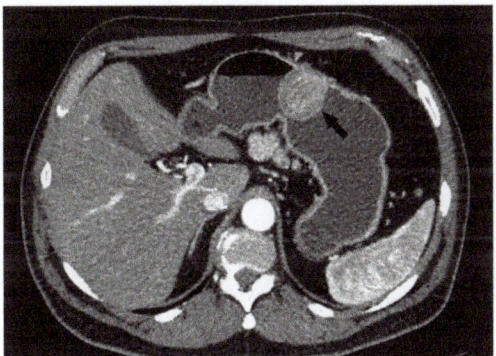

Figure 5. Endophytic growth of oval, solid GIST covered by intact mucosa (black arrow) in the antral region of the stomach.

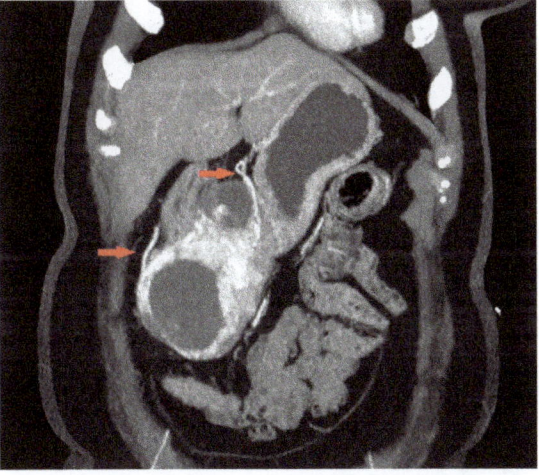

Figure 6. CT coronal view shows irregular, partially cystic tumor with hyperdense solid component and presence of peri- and intra-tumoral vascular vessels (red arrows) (EFDV).

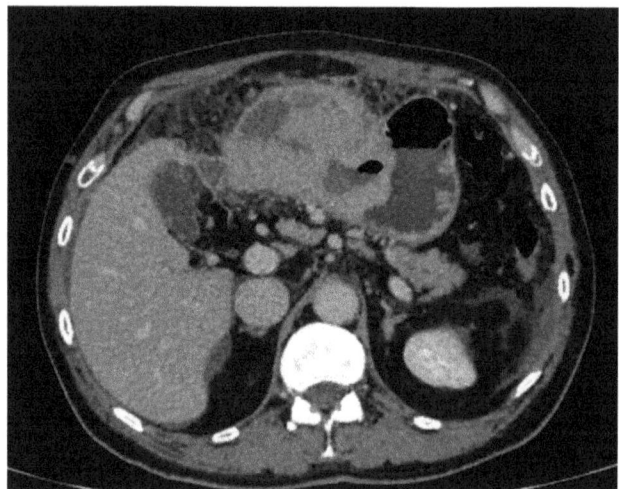

Figure 7. CT scan demonstrates massive necrotic tumor with ill-defined margins and disrupted mucosa with exulceration (HR GIST).

2.4. CT Texture Analysis

Texture analysis was performed with the software MaZda (Version 4.6 for Windows, Institute of Electronics, Technical University of Lodz, Poland). The solid part of the tumor was segmented into three consecutive sections in the portal venous phase. A healthy structure was also marked; in our case, it was a normal gastric wall. The values of the first-order texture, i.e., histogram parameters, were automatically obtained and were as follows: the normalized frequency of pixels of the lowest intensity ("min norm") and the highest intensity ("max norm"), mean intensity ("mean intensity") and standard deviation ("variance"), as well as "skewness ", i.e., asymmetry, and "kurtosis", i.e., the peak/flatness of the histogram. The mentioned values were obtained for each of the three sections, while their mean values for the GISTs and stomach wall were used for further statistical data processing (Figures 8 and 9).

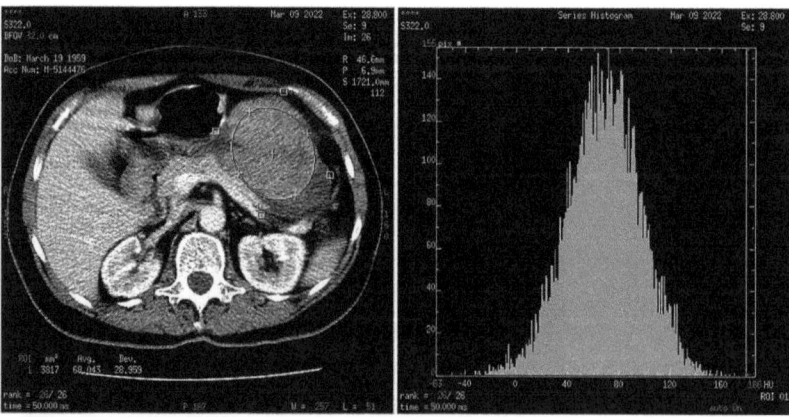

Figure 8. Segmented tumor (green ROI) with histogram.

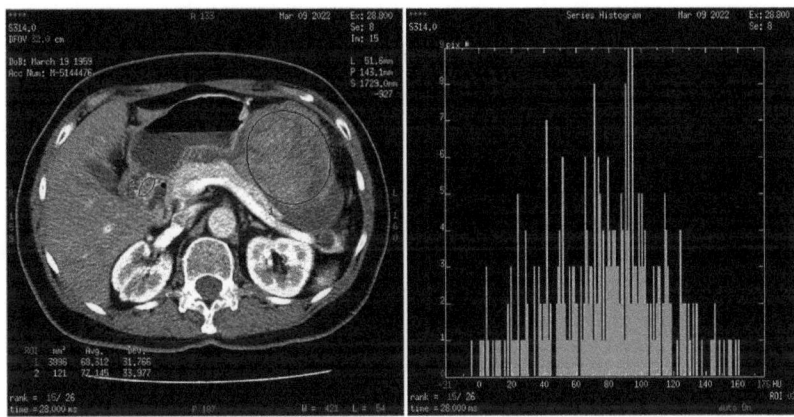

Figure 9. Segmented healthy stomach wall (green ROI) with histogram. Segmented tumor (purple ROI).

2.5. Pathological Analysis and Risk Stratification of Gastric GISTs

The main therapeutic option for localized GISTs is surgery. Wedge resection and subtotal and total gastrectomy are the most frequently used surgical procedures. The resected tumor needs a complete pathohistological evaluation according to established protocols of fixation in 10% formaldehyde and incorporation in paraffin and hematoxylin and eosin. The presence of spindle or epithelioid cells or both is necessary for GIST definition as well as positive immunohistochemical staining for C-KIT or DOG-1. The TNM classification is the standard for risk stratification [21]. Miettinen et al. have established a classification system, AFIP classification, where tumor diameter, mitotic index and localization are the most significant prognostic factors [10]. In addition to metastatic risk, the grade of these tumors is determined by the value of the mitotic index, with a cut-off of five or fewer mitoses visualized per 5 mm^2 or per 50 HPF. Based on these classifications, GISTs are further categorized into four different stages according to mitotic index and the size and presence of metastases in the lymph nodes, liver and peritoneum. We subclassified GIST patients into high-risk (HR GIST) (high-risk and intermediate-risk) and LR GIST (low-risk) groups.

2.6. Statistical Analysis

Normality of distribution of numerical data was assessed by the Kolmogorov–Smirnov test. Mean ± standard deviation (SD) or median value with range were presented depending on the distribution. The chi-square test or Fisher's exact probability test were used to assess differences in morphological features between HR and LR GISTs and, for quantitative parameters, t-test of independent samples or the Mann–Whitney test were used depending on the normality of distribution. t-test for paired samples or Wilcoxon's test for equivalent pairs were used in testing the differences in histogram texture parameters of gastric GISTs in comparison to the normal gastric wall. Univariate and multivariate binary logistic regression analysis was used to identify the morphologic characteristics and histogram texture parameters that are significant predictors of HR GISTs as well as to build a preoperative predictive model suggesting HR GISTs that was further tested by ROC analysis. The level of statistical significance was set at 0.05, while all statistical analyses were performed using SPSS software (Version 17.0 for Windows; SPSS, Chicago, IL, USA).

3. Results

The study included 79 patients with gastric GISTs (45 male, 34 female, with mean age 65 ± 11). HR GISTs were confirmed in 36 patients and LR GISTs in 43. In terms of age, there was no significant difference between the LR and HR groups (64 ± 12 vs. 62 ± 10,

$p = 0.772$), nor was there a difference in terms of gender (26 vs. 19 men and 17 vs. 17 women, $p = 0.472$).

3.1. Tumor Diameter in HR and LR Group

The smallest tumor diameter in the LR group was 15 mm and the largest was 150 mm (mean 56 ± 25 mm), while the range in the HR group was from 40 mm to 340 mm (mean 131 ± 58 mm), $p < 0.001$ (Figure 10).

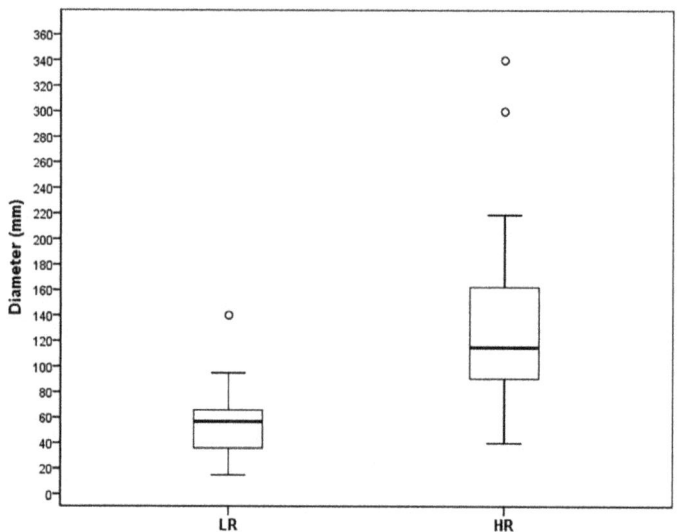

Figure 10. Box plot of tumor diameter (mm) in LR and HR group.

ROC analysis showed that a cut-off diameter of 95 mm most accurately predicted HR GISTs (AUC 0.863; CI 0.772–0.954), with a sensitivity of 98% and specificity of 75% (Figure 11).

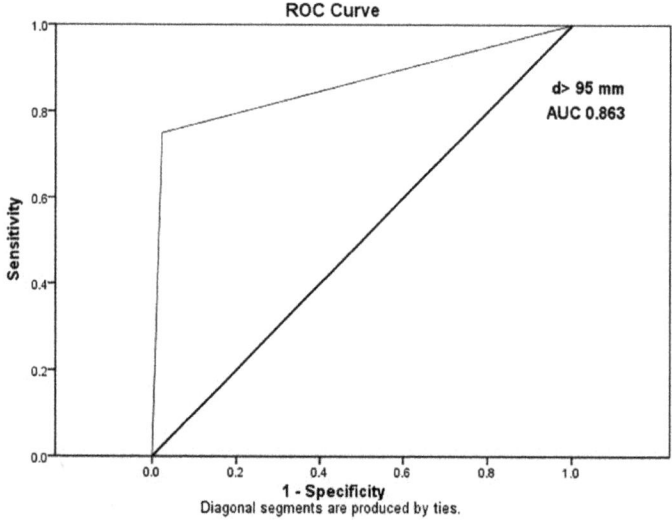

Figure 11. ROC curve representing a diameter cut-off value of 95 mm between the LR and HR group with an AUC of 0.863 (CI 0.772–0.954).

Only one of forty-three LR GISTs was larger than 95 mm, but nine of thirty-six HR GISTs (25%) were smaller than 95 mm in maximal diameter.

3.2. Classic CT Features in HR and LR Group

A comparison of classic CT features of gastric GISTs in the HR and LR groups is presented in Table 1.

Table 1. Comparison of classic CT features of gastric GISTs in the HR and LR groups.

CT Characteristics of Gastric GISTs		LR GISTs ($n = 43$)	HR GISTs ($n = 36$)	p Values
Localization	Body	13	21	
	Antrum	23	8	0.014
	Pylorus	7	7	
Margins	1—well defined	42	28	0.006
	2—ill defined	1	8	
Growth pattern	1—exophytic/mixed	32	35	0.005
	2—endophytic	11	1	
Tumor enhancement	0—low	29	33	0.009
	1—high	14	3	
Shape	1—regular (round)	38	11	<0.001
	2—irregular	5	25	
Structure	1—solid/necrotic	34	10	<0.001
	2—cystic	9	26	
Mucosa	1—continuous	36	13	<0.001
	2—discontinuous (rupture)	7	23	
EFDV *	0—absent	37	10	<0.001
	1—present	6	26	

* Enlarged feeding or draining vessels.

3.3. Histogram Parameters in HR and LR Group

The differences in textural parameters between HR and LR GISTs are shown in Table 2.

Table 2. Histogram parameters in the LR and HR groups. *: statistically significant parameter. Bold: a significant result.

Histogram Parameters	LR GISTs ($n = 43$)	HR GISTs ($n = 36$)	p
Min norm	32,866.776 (32,816.739–33,866.283)	32,851.065 (32,815.016–33,875.819)	**0.032** *
Max norm	612.419 (230.676–1572.068)	524.927 (177.284–835.740)	0.052
Mean	−0.058 (−3.570–0.213)	−0.001 (−0.428–0.304)	0.093
Variance	−0.113 (−0.560–8.145)	−0.096 (−0.557–2.248)	0.806
Skewness	32,812.667 (32,709.333–33,815.667)	32,800.166 (32,764.000–33,813.667)	0.182
Kurtosis	32,837 (32,788.000–33,841.333)	32,822.166 (32,785.667–33,838.667)	0.058

3.4. Histogram Parameters in HR and LR Group

The differences in textural parameters between HR and LR GISTs are shown in Table 3.

Table 3. Histogram parameters in the LR and HR GIST groups. *: statistically signifikant parameter; **: highly statistically significant parameter.

Histogram Parameters	LR GISTs (MI ≤ 5) (n = 52)	HR GISTs (MI > 5) (n = 27)	p
Min norm	32,867.360 (32,816.740–33,875.819)	32,845.770 (32,815.060–33,844.490)	0.007 **
Max norm	610.350 (230.700–1572.100)	516.245 (177.280–835.730)	0.051
Mean	−0.051 (−3.570–0.304)	0.007 (−0.428–0.290)	0.089
Variance	−0.106 (−0.560–8.145)	−0.113 (−0.557–2.248)	0.836
Skewness	32,815.160 (32,709.300–33,815.600)	32,797.000 (32,764.000–33,813.000)	0.035 *
Kurtosis	32,838.000 (32,788.000–33,841.000)	32,818.670 (32,785.660–33,827.330)	0.009 **

3.5. Univariate and Multivariate Predictive Models

Univariate regression analysis confirmed tumor diameter, margin appearance, growth pattern, lesion shape, structure, mucosal continuity, presence of EFDV and the textural parameter max norm as significant predictive factors for HR GISTs (Table 4) (Figure 12).

Table 4. Classical CT morphological and histogram predictive factors for HR GISTs obtained by univariate regression analysis.

		Variables in the Equation					
		B	S.E.	Wald	df	Sig.	Exp (B)
Step 1	Diameter (mm)	0.013	0.013	0.936	1	0.333	1.013
	Margins	−0.120	1.502	0.006	1	0.936	0.887
	Growth pattern	−2.425	1.570	2.386	1	0.122	0.088
	Shape	1.566	0.961	2.653	1	0.103	4.786
	Structure	0.554	0.987	0.315	1	0.575	1.740
	Mucosa	2.219	0.942	5.551	1	0.018	9.199
	EFDV	2.067	0.961	4.628	1	0.031	7.903
	Max norm	−0.001	0.002	0.398	1	0.528	0.999
	Constant	−4.751	2.994	2.518	1	0.113	0.009

Multivariate regression analysis identified interrupted mucosa ($p < 0.001$) and presence of EFDV ($p < 0.001$) as independent predictive CT features for HR GISTs (Table 5).

Table 5. Significant predictive parameters of HR GISTs by multivariate logistic regression analysis.

	Model	Unstandardized Coefficients B	Std. Error	Standardized Coefficients Beta	t	Sig.
1	(Constant)	−0.222	0.127		−1.742	0.086
	Mucosa 1—continuous 2—ruptured	0.346	0.092	0.337	3.779	0.000
	EFD 0—absent 1—present	0.493	0.091	0.486	5.445	0.000

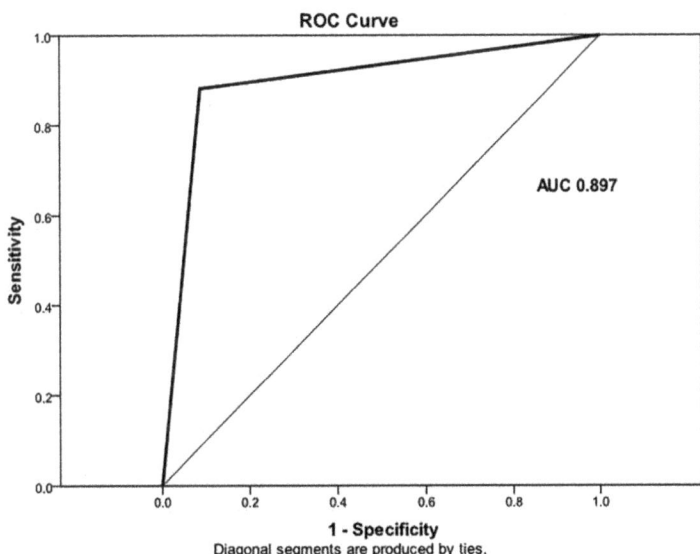

Figure 12. ROC curve shows the following linear regression model including diameter, margins, growth, shape, structure, mucosa, EFDV and max norm predicts HR GISTs with an AUC of 0.897 (CI: 0.817–0.976) with a sensitivity of 83.3%, specificity of 90.7% and accuracy of 87.3%.

ROC analysis showed the multivariate linear regression model with extracted mucosa appearance and presence of EFDV achieved an AUC of 0.878 (CI: 0.797–0.959) with a sensitivity of 94%, specificity of 77% and accuracy of 88% in the prediction of HR GISTs (Figure 13).

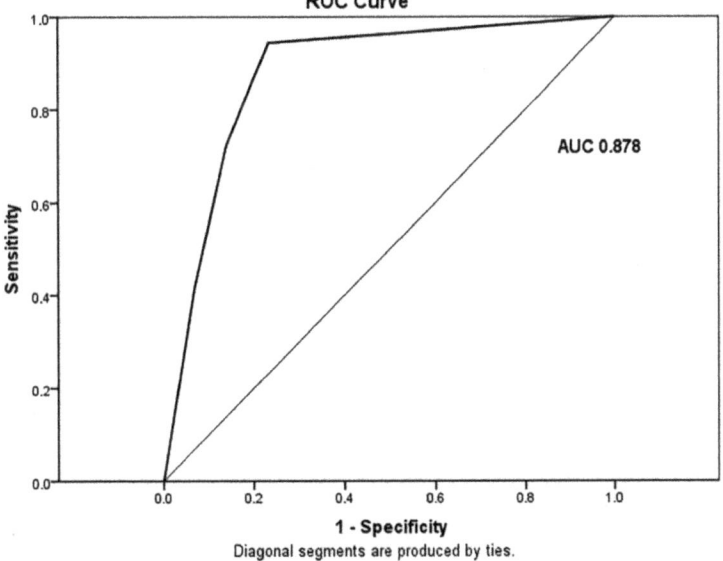

Figure 13. ROC curve of multivariate regression model with two independent predictors for HR GISTs (P (0–1)) = 0.346 × mucosa (1—continuous/2—interrupted) + 0.493 × EFDV (1—present/0—absent)–0.222). AUC = 0.878 (CI: 0.797–0.959), sensitivity 94%, specificity 77% and accuracy 88%.

4. Discussion

Our study confirmed the great importance of morphological CT characteristics of GISTs, which proved to be significant predictive factors in the risk stratification of these tumors. Parameters such as diameter, localization, margins, growth pattern, structure, intensity of postcontrast tumor opacification, shape, continuity of the mucosa and the presence of EFDV showed statistical significance in the prediction of HR GISTs (Table 1). In our research, risk assessment was based on the TNM and AFIP calcification systems, where the diameter of the lesion is an important factor in predicting the metastatic risk of these tumors. A cut-off value of 5 cm has been established within many classifications, and lesions below 5 cm are considered benign variants of this tumor [23]. We also concluded that tumor diameter is a very important predictive factor of high metastatic potential in these tumors, with a cut-off value of 9.5 cm between the LR and HR groups. However, 25% of HR tumors in our series were smaller than 95 mm.

According to our results, the most common localization of GISTs was the area of the corpus and antrum, which coincides with the predominance of Cajal cells in the stomach wall in this area, which is consistent with the results of other studies [24]. Ill-defined tumor margins showed a high statistical significance in predicting HR GISTs in many previous studies [14,16]. Similarly, in our study, univariate regression analysis revealed that this parameter was a significant predictive factor in metastatic risk stratification. In the current study, growth patterns were observed to be both exophytic and endophytic, but also a combination of both variants. Thirty-five patients with proven HR GISTs showed an exophytic and mixed growth pattern. This growth pattern was proven to be a highly statistically significant parameter regarding HR GISTs, which was also shown in other studies as well. Peng et al. used multivariate regression analysis and identified exophytic growth, irregular shape and discontinuous gastric mucosa covering the tumor as significant and independent predictors of HR GISTs [25]. In a series of 129 patients, Zhou et al. analyzed the morphological characteristics of tumors and their regression model extracted tumor diameter, mixed tumor growth and the presence of EFDV as independent predictive factors of high-risk GISTs [19]. When the intensity of postcontrast opacification was analyzed, the largest number of patients (62) showed lower postcontrast enhancement in the portal venous phase of the examination. Among them, 33 patients had a HR GIST. In the present study, regular tumor shape (oval and round shape) was found mostly in LR GISTs, while an irregular, lobulated CT tumor presentation correlated with a higher metastatic risk. An irregular tumor shape is certainly a very important and statistically significant parameter in the prediction of HR GISTs. In our study, the linear regression model included this morphological CT feature as a predictive factor for HR GISTs (AUC 0.897). In previous studies, irregular tumor shape was exclusively characteristic of HR GISTs [18,26]. The structure of tumors can vary from homogeneous and predominantly solid to heterogeneous due to the appearance of intralesional necrosis and cystic degeneration. CT examination clearly shows the mentioned structural differences. Solid and partially necrotic lesions are predominant within the LR group, while cystically degraded tumors were mostly high-risk. Contrary to our results, previous studies have shown that necrotic tumors are associated with a higher mitotic index and metastatic risk. A high MI reflects more intense tissue proliferation, which results in structural degradation and the appearance of intratumor hemorrhage, necrosis and cystic degeneration. Therefore, it is likely that the necrosis and heterogeneity of a tumor observed by visual inspection could be associated with an increased number of mitoses. Larger lesions tend to have a heterogeneous structure and correspond to high-risk tumors. In a study by Grazzini et al., necrosis was shown to be an independent predictive factor for HR GISTs [15]. GISTs are tumors of submucosal localization. In smaller lesions (low-risk GISTs), the mucosa is usually smooth and continuous. Interruption of the continuity of the mucosa leads to the formation of ulcers or umbilications, which are often the source of bleeding. In our study, discontinuous mucosa is a statistically significant factor in the prediction of HR GISTs, in line with the results of previous studies [20]. The multivariate linear regression model in the present study included mucosa appearance and the presence of EFDV as significant predictive parameters of HR GISTs (AUC 0.878). Presence of EFDV was an independent predictor of HR GISTs. This parameter is a reliable index for evaluating the malignancy of these tumors, which can be explained by the fact that accentuated neovascularization is crucial in tumor proliferation and the occurrence of distant hematogenous metastases. This result is consistent with the results of other studies underlining this parameter as an important predictor of high-risk tumors [24,27]. CTTA of tumor heterogeneity showed a significant contribution in the characterization of lesions, such as distinguishing benign from malignant tumors or indicating more biologically aggressive lesions. This technique has also shown progress in the initial evaluation of tumors before treatment and in evaluating the therapeutic response for some types of tumors as well [19]. Although there are encouraging data suggesting that CTTA is a promising imaging biomarker, one should not forget the significant variability in methods and examined parameters and in association with biological correlates. Before this advanced CT diagnostic method can be considered for global clinical practice implementation, the standardization of tumor segmentation and measurement techniques, as well as postprocessing, is necessary to identify the most important textural parameters. The continuation of research, external verification of histopathological correlates and a specified, uniform formulation of reports are also of great importance for the further application of this method [19]. Tumors are generally heterogeneous lesions not only at the cellular level, but also genetically and phenotypically, with spatial heterogeneity of cell density, angiogenesis and necrosis. This tissue heterogeneity is an important factor that has an impact on prognosis and treatment, bearing in mind that more intense structural degradation of the lesion and its heterogeneity can be associated with very malignant and aggressive tumor behavior with increased resistance to treatment [25]. CTTA is only one segment of the growing and very promising field of radiomics, which involves the extraction, complex analysis and interpretation of quantitative parameters obtained from diagnostic images. In our study, histogram parameters were analyzed in 79 patients and max norm showed statistical significance in terms of predicting HR GISTs. In contrast, a study by Choi et al. including 145 GIST patients showed that the main predictive factors for HR GISTs were kurtosis and MPP (mean positive pixels) [17]. In the same study, in the HR GIST group, lower mean, SD and MPP values were observed, while the kurtosis parameter was significantly higher. Moreover, higher values of skewness and kurtosis were characteristic of lesions with a high mitotic index. Based on subjective assessment, lesion characteristics such as lower density, necrosis and mucosal ulceration were identified as predictive factors for HR GISTs [17]. It should be kept in mind that histogram parameters may have a different significance depending on the type of tumor, the type of imaging performed as well as the analytical method. A high mitotic index in GISTs reflects rapid tissue proliferation that leads to a heterogeneous structure and necrotic and cystic degradation of lesions, so it can be concluded that visual confirmation of damaged tumor tissue may suggest a higher mitotic index. Previous studies have shown a correlation of larger diameter (>11 cm), tumor heterogeneity and presence of necrosis with a higher mitotic index and metastatic risk [14,27]. In addition, necrotic lesions showed low mean and MPP values and higher values for kurtosis, which is consistent with the results of our research. This can be explained by the low attenuation caused by tissue necrosis and increased heterogeneity of the tumor structure [28]. In a study by Liu et al., in terms of predicting the metastatic risk of GISTs, it was shown that the peak value on the histogram (maximum frequency) has the greatest superiority in comparison with other parameters of texture analysis, which is in concordance with our results [29]. Another study by the same authors indicated a correlation between CT texture parameters and immunohistochemical biomarkers such as E-cadherin, Ki67, VEGFR2 and EGFR in 139 patients with gastric cancer [30].

Our study has several limitations. The sample of patients was relatively small and our research did not include a follow-up of the included patients. Certainly, a prospective study with a larger cohort is needed in further research to confirm the findings of this study and to incorporate the analyzed diagnostic method into daily clinical practice.

5. Conclusions

Our study resulted in a regression model that identified mucosal discontinuity and the presence of EFDV features as independent and significant predictors of HR GISTs, which leads us to the conclusion that morphological CT features have the greatest value in the non-invasive, preoperative prediction of metastatic risk of gastric GISTs. A significant statistical significance was shown regarding the functional parameter max norm within the textural analysis of these tumors. The incorporation of advanced CT techniques into the basic diagnostic protocol can further benefit the preoperative assessment of risk stratification in GISTs. Preoperative risk stratification is of great significance to evaluate the risk of tumor recurrence and guide treatment planning before and after surgery. This improves the management of treatment, especially in terms of the application of neoadjuvant therapy, which further enables tumor shrinkage, reduces tumor ruptures, increases overall survival rates and optimizes surgical resection and systemic control of the disease. Our model may serve as a diagnostic tool for the non-invasive prediction of HR GISTs to support personalized treatment strategies.

Author Contributions: Conceptualization, M.M.J. and A.D.S.; methodology, A.D.S. and K.E.; software, B.T.; validation, D.M., M.V. and P.P.; formal analysis, A.J.; resources, O.S. and D.J.S.; data curation, M.K.; writing—original draft preparation, M.M.J. and A.D.S.; writing—review and editing, J.K., V.S. and B.T.; supervision, D.S.; project administration, M.M. and D.J.S. All authors have read and agreed to the published version of the manuscript.

Funding: This research received no external funding.

Institutional Review Board Statement: This study was conducted in accordance with the Declaration of Helsinki and approved by the Institutional Review Board of the University Clinical Center of Serbia (protocol code 14/4, approved 25 January 2019).

Informed Consent Statement: Informed consent was obtained from all subjects involved in this study.

Data Availability Statement: Data are contained within the article.

Conflicts of Interest: The authors declare no conflict of interest.

References

1. Wang, Q.; Huang, Z.P.; Zhu, Y.; Fu, F.; Tian, L. Contribution of Interstitial Cells of Cajal to Gastrointestinal Stromal Tumor Risk. *Med. Sci. Monit.* **2021**, *27*, e929575. [CrossRef] [PubMed]
2. Blay, J.-Y.; Kang, Y.-K.; Nishida, T.; von Mehren, M. Gastrointestinal stromal tumours. *Nat. Rev. Dis. Primers* **2021**, *7*, 22. [CrossRef] [PubMed]
3. Casali, P.G.; Blay, J.Y.; Abecassis, N.; Bajpai, J.; Bauer, S.; Biagini, R.; Bielack, S.; Bonvalot, S.; Boukovinas, I.; Bovee, J.; et al. Gastrointestinal stromal tumours: ESMO-EURACAN-GENTURIS Clinical Practice Guidelines for diagnosis, treatment and follow-up. *Ann. Oncol.* **2022**, *33*, 20–33. [CrossRef]
4. Nishida, T.; Goto, O.; Raut, C.P.; Yahagi, N. Diagnostic and treatment strategy for small gastrointestinal stromal tumors. *Cancer* **2016**, *122*, 3110–3118. [CrossRef] [PubMed]
5. Nishida, T.; Hirota, S.; Yanagisawa, A.; Sugino, Y.; Minami, M.; Yamamura, Y.; Otani, Y.; Shimada, Y.; Takahashi, F.; Kubota, T.; et al. Clinical practice guidelines for gastrointestinal stromal tumor (GIST) in Japan: English version. *Int. J. Clin. Oncol.* **2008**, *13*, 416–430. [CrossRef] [PubMed]
6. Sawaki, A.; Mizuno, N.; Takahashi, K.; Nakamura, T.; Tajika, M.; Kawai, H.; Isaka, T.; Imaoka, H.; Okamoto, Y.; Aoki, M.; et al. Long-Term Follow up of Patients with Small Gastrointestinal Stromal Tumors in the Stomach Using Endoscopic Ultrasonography-Guided Fine-Needle Aspiration Biopsy. *Dig. Endosc.* **2005**, *18*, 40–44. [CrossRef]
7. von Mehren, M.; Randall, R.L.; Benjamin, R.S.; Boles, S.; Bui, M.M.; Ganjoo, K.N.; George, S.; Gonzalez, R.J.; Heslin, M.J.; Kane, J.M.; et al. Soft Tissue Sarcoma, Version 2.2018, NCCN Clinical Practice Guidelines in Oncology. *J. Natl. Compr. Cancer Netw.* **2018**, *16*, 536–563. [CrossRef] [PubMed]
8. Rodrigues, J.; Campanati, R.G.; Nolasco, F.; Bernardes, A.M.; Sanches, S.R.A.; Savassi-Rocha, P.R. Pre-Operative Gastric Gist Downsizing: The Importance of Neoadjuvant Therapy. *Arq. Bras. Cir. Dig.* **2019**, *32*, e1427. [CrossRef]
9. Ishikawa, T.; Kanda, T.; Kameyama, H.; Wakai, T. Neoadjuvant therapy for gastrointestinal stromal tumor. *Transl. Gastroenterol. Hepatol* **2018**, *3*, 3. [CrossRef]
10. Miettinen, M.; Lasota, J. Gastrointestinal stromal tumors: Pathology and prognosis at different sites. *Semin. Diagn. Pathol.* **2006**, *23*, 70–83. [CrossRef]

11. Fletcher, C.D.; Berman, J.J.; Corless, C.; Gorstein, F.; Lasota, J.; Longley, B.J.; Miettinen, M.; O'Leary, T.J.; Remotti, H.; Rubin, B.P.; et al. Diagnosis of gastrointestinal stromal tumors: A consensus approach. *Hum. Pathol.* **2002**, *33*, 459–465. [CrossRef] [PubMed]
12. Demetri, G.D.; Benjamin, R.S.; Blanke, C.D.; Blay, J.Y.; Casali, P.; Choi, H.; Corless, C.L.; Debiec-Rychter, M.; DeMatteo, R.P.; Ettinger, D.S.; et al. NCCN Task Force report: Management of patients with gastrointestinal stromal tumor (GIST)--update of the NCCN clinical practice guidelines. *J. Natl. Compr. Cancer Netw.* **2007**, *5* (Suppl. S2), S1–S29. [CrossRef]
13. Tateishi, U.; Hasegawa, T.; Satake, M.; Moriyama, N. Gastrointestinal stromal tumor. Correlation of computed tomography findings with tumor grade and mortality. *J. Comput. Assist. Tomogr.* **2003**, *27*, 792–798. [CrossRef] [PubMed]
14. Kim, H.C.; Lee, J.M.; Kim, K.W.; Park, S.H.; Kim, S.H.; Lee, J.Y.; Han, J.K.; Choi, B.I. Gastrointestinal stromal tumors of the stomach: CT findings and prediction of malignancy. *AJR Am. J. Roentgenol.* **2004**, *183*, 893–898. [CrossRef] [PubMed]
15. Grazzini, G.; Guerri, S.; Cozzi, D.; Danti, G.; Gasperoni, S.; Pradella, S.; Miele, V. Gastrointestinal stromal tumors: Relationship between preoperative CT features and pathologic risk stratification. *Tumori J.* **2021**, *107*, 556–563. [CrossRef] [PubMed]
16. Choi, H.; Charnsangavej, C.; Faria, S.C.; Macapinlac, H.; Burgess, M.A.; Patel, S.R.; Chen, L.L.; Podoloff, D.A.; Benjamin, R.S. Correlation of computed tomography and positron emission tomography in patients with metastatic gastrointestinal stromal tumor treated at a single institution with imatinib mesylate: Proposal of new computed tomography response criteria. *J. Clin. Oncol. Off. J. Am. Soc. Clin. Oncol.* **2007**, *25*, 1753–1759. [CrossRef] [PubMed]
17. Choi, I.Y.; Yeom, S.K.; Cha, J.; Cha, S.H.; Lee, S.H.; Chung, H.H.; Lee, C.M.; Choi, J. Feasibility of using computed tomography texture analysis parameters as imaging biomarkers for predicting risk grade of gastrointestinal stromal tumors: Comparison with visual inspection. *Abdom. Radiol. (NY)* **2019**, *44*, 2346–2356. [CrossRef]
18. Lubner, M.G.; Smith, A.D.; Sandrasegaran, K.; Sahani, D.V.; Pickhardt, P.J. CT Texture Analysis: Definitions, Applications, Biologic Correlates, and Challenges. *Radiographics* **2017**, *37*, 1483–1503. [CrossRef]
19. Zhou, C.; Duan, X.; Zhang, X.; Hu, H.; Wang, D.; Shen, J. Predictive features of CT for risk stratifications in patients with primary gastrointestinal stromal tumour. *Eur. Radiol.* **2016**, *26*, 3086–3093. [CrossRef]
20. Lubner, M.G.; Stabo, N.; Lubner, S.J.; del Rio, A.M.; Song, C.; Halberg, R.B.; Pickhardt, P.J. CT textural analysis of hepatic metastatic colorectal cancer: Pre-treatment tumor heterogeneity correlates with pathology and clinical outcomes. *Abdom. Imaging* **2015**, *40*, 2331–2337. [CrossRef]
21. Amin, M.B.; Edge, S.B.; Greene, F.L.; Byrd, D.R.; Brookland, R.K.; Washington, M.K.; Gershenwald, J.E.; Compton, C.C.; Hess, K.R.; Sullivan, D.C. *AJCC Cancer Staging Manual*; Springer: Berlin/Heidelberg, Germany, 2017; Volume 1024.
22. Rutkowski, P.; Wozniak, A.; Debiec-Rychter, M.; Kakol, M.; Dziewirski, W.; Zdzienicki, M.; Ptaszynski, K.; Jurkowska, M.; Limon, J.; Siedlecki, J.A. Clinical utility of the new American Joint Committee on Cancer staging system for gastrointestinal stromal tumors: Current overall survival after primary tumor resection. *Cancer* **2011**, *117*, 4916–4924. [CrossRef] [PubMed]
23. Jo, V.Y.; Fletcher, C.D. WHO classification of soft tissue tumours: An update based on the 2013 (4th) edition. *Pathology* **2014**, *46*, 95–104. [CrossRef] [PubMed]
24. Burkill, G.J.; Badran, M.; Al-Muderis, O.; Meirion Thomas, J.; Judson, I.R.; Fisher, C.; Moskovic, E.C. Malignant gastrointestinal stromal tumor: Distribution, imaging features, and pattern of metastatic spread. *Radiology* **2003**, *226*, 527–532. [CrossRef] [PubMed]
25. Davnall, F.; Yip, C.S.; Ljungqvist, G.; Selmi, M.; Ng, F.; Sanghera, B.; Ganeshan, B.; Miles, K.A.; Cook, G.J.; Goh, V. Assessment of tumor heterogeneity: An emerging imaging tool for clinical practice? *Insights Imaging* **2012**, *3*, 573–589. [CrossRef] [PubMed]
26. Bashir, U.; Siddique, M.M.; McLean, E.; Goh, V.; Cook, G.J. Imaging Heterogeneity in Lung Cancer: Techniques, Applications, and Challenges. *AJR Am. J. Roentgenol.* **2016**, *207*, 534–543. [CrossRef]
27. Maldonado, F.J.; Sheedy, S.P.; Iyer, V.R.; Hansel, S.L.; Bruining, D.H.; McCollough, C.H.; Harmsen, W.S.; Barlow, J.M.; Fletcher, J.G. Reproducible imaging features of biologically aggressive gastrointestinal stromal tumors of the small bowel. *Abdom. Radiol.* **2018**, *43*, 1567–1574. [CrossRef]
28. Iannarelli, A.; Sacconi, B.; Tomei, F.; Anile, M.; Longo, F.; Bezzi, M.; Napoli, A.; Saba, L.; Anzidei, M.; D'Ovidio, G.; et al. Analysis of CT features and quantitative texture analysis in patients with thymic tumors: Correlation with grading and staging. *Radiol. Med.* **2018**, *123*, 345–350. [CrossRef]
29. Liu, S.; Pan, X.; Liu, R.; Zheng, H.; Chen, L.; Guan, W.; Wang, H.; Sun, Y.; Tang, L.; Guan, Y.; et al. Texture analysis of CT images in predicting malignancy risk of gastrointestinal stromal tumours. *Clin. Radiol.* **2018**, *73*, 266–274. [CrossRef]
30. Liu, S.; Shi, H.; Ji, C.; Guan, W.; Chen, L.; Sun, Y.; Tang, L.; Guan, Y.; Li, W.; Ge, Y.; et al. CT textural analysis of gastric cancer: Correlations with immunohistochemical biomarkers. *Sci. Rep.* **2018**, *8*, 11844. [CrossRef]

Disclaimer/Publisher's Note: The statements, opinions and data contained in all publications are solely those of the individual author(s) and contributor(s) and not of MDPI and/or the editor(s). MDPI and/or the editor(s) disclaim responsibility for any injury to people or property resulting from any ideas, methods, instructions or products referred to in the content.

Review

Photodynamic Therapy and Immunological View in Gastrointestinal Tumors

David Aebisher [1,*], Paweł Woźnicki [2], Klaudia Dynarowicz [3], Aleksandra Kawczyk-Krupka [4], Grzegorz Cieślar [4] and Dorota Bartusik-Aebisher [5]

1. Department of Photomedicine and Physical Chemistry, Medical College of the University of Rzeszów, 35-959 Rzeszów, Poland
2. Students English Division Science Club, Medical College of the University of Rzeszów, 35-959 Rzeszów, Poland
3. Center for Innovative Research in Medical and Natural Sciences, Medical College of the University of Rzeszów, 35-310 Rzeszów, Poland; kdynarowicz@ur.edu.pl
4. Department of Internal Medicine, Angiology and Physical Medicine, Center for Laser Diagnostics and Therapy, Medical University of Silesia, Batorego 15 Street, 41-902 Bytom, Poland; akawczyk@sum.edu.pl (A.K.-K.); cieslar1@tlen.pl (G.C.)
5. Department of Biochemistry and General Chemistry, Medical College of the University of Rzeszów, 35-959 Rzeszów, Poland; dbartusikaebisher@ur.edu.pl
* Correspondence: daebisher@ur.edu.pl

Citation: Aebisher, D.; Woźnicki, P.; Dynarowicz, K.; Kawczyk-Krupka, A.; Cieślar, G.; Bartusik-Aebisher, D. Photodynamic Therapy and Immunological View in Gastrointestinal Tumors. *Cancers* **2024**, *16*, 66. https://doi.org/10.3390/cancers16010066

Academic Editor: Shihori Tanabe

Received: 29 October 2023
Revised: 13 December 2023
Accepted: 20 December 2023
Published: 22 December 2023

Copyright: © 2023 by the authors. Licensee MDPI, Basel, Switzerland. This article is an open access article distributed under the terms and conditions of the Creative Commons Attribution (CC BY) license (https://creativecommons.org/licenses/by/4.0/).

Simple Summary: Many clinical cases of gastrointestinal tumors exist that require the use of high-precision technology for eradication due to their proximity to vital anatomical sites. These sites within the gastrointestinal system are often inaccessible or unsafe for treatment by traditional surgical procedures. Therefore, we reviewed the current literature on the potential of photodynamic therapy (PDT) and associated immunological anti-tumor mechanisms in gastrointestinal tumors. Since its discovery, PDT has emerged as a powerful method for the treatment of skin and esophageal cancers. Traditionally, PDT uses intravenously injected photosensitizers to generate cytotoxic singlet oxygen upon local illumination. Prodrug delivery strategies have shown promise, but the selectivity of the photosensitizer drug in diseased tissue could be improved. Thus, there is a critical need for treatment strategies that enable photodynamic action site-specifically for enhanced tumor destruction.

Abstract: Gastrointestinal cancers are a specific group of oncological diseases in which the location and nature of growth are of key importance for clinical symptoms and prognosis. At the same time, as research shows, they pose a serious threat to a patient's life, especially at an advanced stage of development. The type of therapy used depends on the anatomical location of the cancer, its type, and the degree of progression. One of the modern forms of therapy used to treat gastrointestinal cancers is PDT, which has been approved for the treatment of esophageal cancer in the United States. Despite the increasingly rapid clinical use of this treatment method, the exact immunological mechanisms it induces in cancer cells has not yet been fully elucidated. This article presents a review of the current understanding of the mode of action of photodynamic therapy on cells of various gastrointestinal cancers with an emphasis on colorectal cancer. The types of cell death induced by PDT include apoptosis, necrosis, and pyroptosis. Anticancer effects are also a result of the destruction of tumor vasculature and activation of the immune system. Many reports exist that concern the mechanism of apoptosis induction, of which the mitochondrial pathway is most often emphasized. Photodynamic therapy may also have a beneficial effect on such aspects of cancer as the ability to develop metastases or contribute to reducing resistance to known pharmacological agents.

Keywords: gastrointestinal cancers; photodynamic therapy; anticancer effect

1. Introduction

1.1. Gastric Cancers—Morbidity

Cancers are one of the leading causes of death worldwide. One of the more diverse groups of cancers are those located in the gastrointestinal tract. Currently, there is a significant increase in new cases of this disease [1]. An example of an aggressive cancer of the gastrointestinal tract is esophageal cancer (EC), which is more often diagnosed in men, and is a significant cause of cancer-related mortality worldwide, accounting for 16,910 new cases and 15,910 deaths in the United States in 2016 [2,3]. Nonspecific symptoms can delay a patient's examination by a physician resulting in an inoperable tumor stage and the presence of metastases in more than 50% of patients, giving EC a poor prognosis [3,4]. More than 95% of new cases of esophageal cancer are adenocarcinoma, which is more common in developed countries, and squamous cell carcinoma which is prevalent in non-industrialized countries. Smoking, obesity, and gastroesophageal reflux disease predispose a person to the development of adenocarcinoma, while achalasia, alcohol consumption, and smoking are risk factors for squamous cell carcinoma [3]. The third most common cause of cancer deaths worldwide is gastric cancer (GC). Common societal risk factors for this disease include high salt intake, a diet low in fruits and vegetables, and H. pylori infection [5]. Gastric cancer should be treated in a multidisciplinary manner. Surgical resection is the primary treatment method with demonstrated treatment efficacy and this is being expanded with adjuvant and neoadjuvant therapies for the treatment of locally advanced lesions. In patients found to have metastases, therapy has unsatisfactory results, with a median survival of about 1 year [6]. Colorectal cancer (CRC), which is the third most common cancer worldwide, is estimated to occur in more patients than cancers of the upper gastrointestinal tract [7]. Its occurrence, mainly in developed Western countries is increasing annually, and it is the fourth most common cause of death among cancer patients [8]. Lifestyles, such as an unhealthy diet, smoking, and alcohol consumption, as well as chronic diseases, age, and environmental factors, predispose a person to the development of this cancer [8,9]. Two pathways lead to the development of colorectal cancer. The first is the multi-step adenoma-carcinoma sequence of mutations of the APC gene, and the second is the development of serrated adenoma to cancer, in which the genetic defect responsible has not yet been determined. Early-stage pre-cancerous adenomatous polyps, as well as advanced cancer, can be asymptomatic, worsening the prognosis, making early diagnosis difficult, and warranting screening in people over 50. Localized colorectal cancer should not be associated with a poor prognosis; however, most cancers are diagnosed at a locally advanced tumor stage or with lymph node metastasis, which is responsible for an unfavorable prognosis. Up to 20% of patients have metastases, most commonly in the liver [10].

1.2. Photodynamic Therapy—One of the Treatment Methods

The significant increase in new cancer cases worldwide creates the need to discover new and effective therapeutic methods. The search for innovative forms of therapy, including photodynamic therapy (PDT), is ongoing. Photodynamic therapy exploits photodynamic action initiated by the excitation of photosensitizers (PSs) with light and the subsequent interaction of the excited PSs with oxygen to produce reactive oxygen species (ROS) including singlet oxygen ($^1O_2$) [11,12]. Figure 1 presents the mechanism of $^1O_2$ generation upon excitation of a PS.

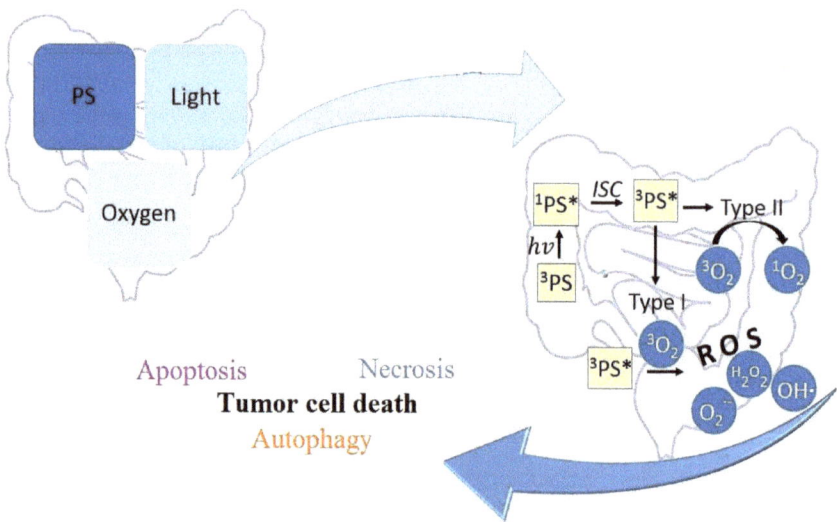

Figure 1. The mechanism of PDT. Most PDT therapies used in clinical settings are based on three components: oxygen, PS, and light. After the application of these three elements, a number of reactions are initiated in the tissue. PS in the ground state becomes excited to singlet state under the influence of light of a specific wavelength. A photosensitizer in the excited singlet state may end up in the excited triplet state as a result of intersystem crossing. A photosensitizer in the excited triplet state can generate reactive oxygen species by electron transfer to an oxygen molecule (Type I). The main components of ROS are superoxide anion, hydroxyl radical, and hydrogen peroxide. In turn, in type (II), energy is transferred from the photosensitizer in the excited triplet state to oxygen in the triplet state, generating singlet oxygen. Both processes have the ability to eliminate cancer cells. Molecular biology describes three main and fundamental mechanisms of cell death: apoptosis, necrosis, and autophagy. PS—photosensitizer, ^{3}PS—photosensitizer in the ground state, ^{1}PS*—excited singlet state, hv—specific wavelength of light, ^{3}PS*—excited triplet state, *ISC*—intersystem crossing, ROS—reactive oxygen species, O_2^-—superoxide anion, OH—hydroxyl radical, $^3O_2$—triplet state oxygen, $^1O_2$—singlet oxygen.

Cell signal transmission depends on the healthy amount of ROS, but when that level is excessively increased, it can result in irreparable cellular damage. ROS produced through PDT oxidizes biological macromolecules in tumor cells, including DNA/RNA and protein, leading to tumor cell death which is known as apoptosis, necrosis, and autophagy. Additionally, ROS in tumor tissue might harm microvascular structures and result in immunogenic cell death. The main signs of apoptosis are shrinking cells and the appearance of vesicles in the cell membrane. Typically, apoptotic cells are surrounded by healthy-looking neighboring cells.

Apoptosis is characterized by several microscopically detectable changes, which include the most striking condensation of chromatin into well-defined granular masses along the nuclear envelope, cell shrinkage, twisting of cell and nucleus contours, and fragmentation of the nucleus. Eventually, the cell disintegrates into membrane-bound apoptotic bodies that contain, among other things, nuclear debris and that are rapidly removed by neighboring macrophages. During this process, the cell membrane and the membrane surrounding the apoptotic fragments maintain their integrity. In addition, lysosomes remain intact, and therefore, lysosomal enzymes are not released into surrounding tissues. Consequently, there is no accompanying inflammation in apoptosis.

Necrosis is a non-programmed process, defined as accidental cell death caused by physical or chemical damage. It is characterized by pyknotic nuclei, cytoplasmic swelling, and progressive breakdown of cytoplasmic membranes, leading to cell fragmentation and the release of material into the extracellular space.

Autophagy is a progressive course of degradation and restoration of cytoplasmic parts. Moreover, it is important for maintaining cell homeostasis and development. It is a physiological cycle in which the cytosol and whole organelles become surrounded by a double layer of vacuoles, known as the autophagosome. As a result of the fusion of the lysosome with the autophagosome, the autophagosome becomes damaged. Studies have confirmed that PDT can induce pathways of apoptosis, autophagy, mitotic catastrophe, and necrosis. The phototoxic effect of PDT leads to photodamage due to irreversible degradation of cell membranes and organelles. Induction of multiple cell death pathways is considered to be a useful feature of PDT because it enhances the photo-killing of cancer cells resistant to a particular cell death pathway. At the level of molecular biology, PDT induces concentration-dependent cell death mechanisms, physicochemical properties, subcellular localization of PSs, oxygen concentration, and the appropriate wavelength and intensity of light. Cell-type-specific properties can affect the mode and extent of cell death. Where the PS enters the cell depends on the chemical properties of each compound. Hydrophobic molecules can diffuse rapidly into plasma membranes, whereas more polar molecules tend to be internalized by endocytosis or assisted transport by lipids and serum proteins. Most PSs are found in organelle membranes, but in general, the cellular localization of PSs includes the endoplasmic reticulum (ER), mitochondria, Golgi complex, lysosomes, and cell membrane. Photogenerated ROS are very short-lived and have limited diffusion distance in biological systems. Therefore, the subcellular location of a PS is often the site where the generated ROS will cause more damage through the activation of cell death mechanisms. Mitochondria, the main intracellular target of PDT, plays a critical role in apoptosis by controlling the release of crucial factors involved in this process.

1.3. Overall Cellular Response to PDT

Numerous studies have demonstrated the effectiveness of PDT as a cancer therapy and also in the treatment of many non-oncological diseases, e.g., dermatoses [13–16]. In the presence of oxygen, the reactions of photo-excited PSs in tumor tissue result in direct cell death through various pathways, the induction of an inflammatory response, and, especially in the case of vascular photosensitizers, the destruction of blood vessels supplying the tumor [17–21]. The phenomenon of photon absorption is key to the excitation of photosensitizers (PSs) to a higher energy level and the formation of a triplet state, which is responsible for energy transfer or electron transfer leading to the production of ROS including $^1O_2$ [19]. Generating $^1O_2$ in concentrations sufficient to destroy a hypoxic tumor is a major technological challenge. Most of the clinical experience with gastrointestinal PDT involves patients who are considered to be at risk of poor surgical outcomes, and follow-up reports are limited [22]. Despite the demonstrated efficacy of PDT and knowledge of its underlying mechanism of action, elucidation of the exact mechanisms by which it leads to cell death is ongoing. The best documented cytotoxic effects of PDT on organelles are associated with photodamage to mitochondria and lysosomes (Figure 2).

A type of non-apoptotic cell death is ferroptosis [23]. Its basic characteristic is the gradual loss of mitochondria in response to the administered therapy [24]. In turn, inflammasomes are multi-proteins that contribute to the activation of the inflammatory process and, consequently, to cell death, called pyroptosis [25]. PSs should not accumulate in cell nuclei to prevent the formation of resistant cells [26]. It has been shown that at the cellular level, PDT-induced cell death subroutines may or may not be random [27]. Accidental cell death is an uncontrolled form of death characterized by the gradual loss of cell membrane integrity and swelling of cell organelles [28]. Regulated cell death (RCD) is triggered by the activation of one or more signal transduction modules, such that it can be modulated in some sense pharmacologically or genetically. There are also PDT-associated RCD subroutines that involve apoptosis and various mechanisms of regulated necrosis, including necroptosis and autophagy-dependent cell death [29].

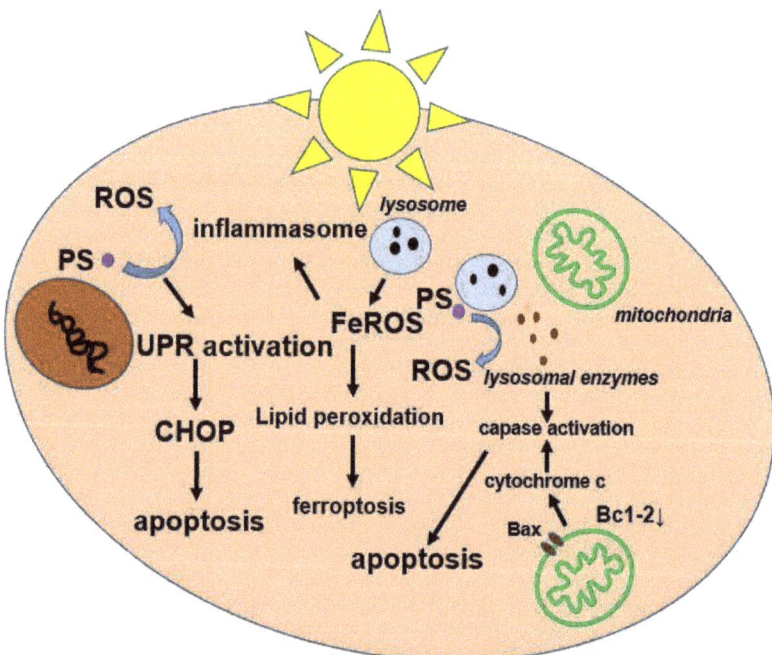

Figure 2. Diagram of activation of processes in cells as a result of PDT activity. Initially, PS penetrates the cancer cells and accumulates in the mitochondria. Upon activation of PS with laser light of a specific wavelength, ROS photogeneration and destruction of the apoptotic protein Bcl-2 occur, which causes the permeabilization of the outer membrane of the mitochondria. As a consequence, cytochrome c is released from the mitochondria into the cytosol, enhancing the apoptotic signal by activating caspases. Most often, PS is found in the plasma membrane, endoplasmic reticulum, mitochondrion, or lysosome. Depending on its location, when activated with light, it can directly damage the plasma membrane causing unregulated necrosis or lead to one or more mechanisms of regulated cell death. UPR: unfolded protein response; Fe: iron; ROS: reactive oxygen species; CHOP: pro-apoptotic transcription factor.

1.4. Merits and Defects of PDT

1.4.1. Merits

For the treatment of gastrointestinal cancers, PDT has found application mainly in the treatment of lesions located in the esophagus. In addition, it has been shown in studies that PDT is indicated not only in the treatment of already-formed cancer but also in Barrett's esophagus [30]. The origins of PDT used to treat esophageal cancer include the palliative treatment of patients with obstructive esophageal cancer [31]. Moreover, PDT is being used to treat superficial esophageal cancers characterized by difficulty in endoscopic treatment, and this indication has already been approved for treatment in Japan [32]. In patients in whom local radiotherapy has not achieved the intended therapeutic goals, and in whom treatment by other means may be insufficiently effective, PDT using second-generation photosensitizers is indicated [33]. An example of the strength of PDT is that the cure of early mucosal disease is possible after a single endoscopic procedure [34]. Photodynamic therapy using the photosensitizer sodium porfimer (Photofrin®) was approved in the United States in 1995 for use in patients with advanced esophageal cancer [32]. Photodynamic therapy has a favorable side-effect profile, is less invasive, and minimizes systemic toxicity, making it well tolerated by patients [35,36]. Moreover, PDT does not impair fertility and does not affect pregnancy [37]. Enhancing the systemic immune response against cancer may increase the effectiveness of PDT as well as act synergistically with other forms of therapy [38].

1.4.2. Defects

The use of PDT is limited to the treatment of flat superficial lesions that are usually accessible endoscopically due to the limited tissue penetration depth of light [39]. At least partially, this problem can be solved by implantable devices or lasers in the near-infrared range, enabling tissue penetration up to 3 cm [40,41]. Another aspect is that the usefulness of PDT is also limited by hypoxia which is typical of many tumors that limit photodynamic action [42,43]. Additionally, reducing tumor oxygenation may promote proliferation and metastasis [43]. Standard guidelines for PDT treatment protocols are still not available, which makes the selection of parameters difficult and affects the quality of treatment [44].

Recent research has focused on improving the effectiveness of PDT. Zaigang Zhou et al. synthesized nanoparticles capable of dually disrupting the PD-1/PD-L1 axis and reversing tumor hypoxia [45]. In turn, Liao W. et al. described the synthesis of a nanogel with the ability to increase the production of ROS in cancer cells [46]. Recent reports have also demonstrated the potential of using PDT based on synthetic hypericin in the treatment of early stages of early-stage cutaneous t-cell lymphoma (Mycosis Fungoides) [47]. It was also found that 5-ALA PDT achieves high effectiveness in the treatment of low-grade squamous intraepithelial lesions with high-risk HPV infection and that the effectiveness of 5-ALA PDT in the treatment of actinic keratosis is increased by microneedling and cryotherapy [48,49]. The aim of this study is to review studies on the treatment of gastroenterological diseases with PDT and its immunological effects.

2. Materials and Methods

The literature search, which focused on the immunological mechanisms induced by PDT in the treatment of gastrointestinal cancers, was conducted using articles from PubMed, ScienceDirect, Web of Science, and Google Scholar from 1990 to September 2023. The authors of this review worked according to an agreed framework, selecting articles based on their title, language, abstract, and access. Duplicate works have been removed. The review included papers describing the immunological view of photodynamic therapy in the treatment of gastrointestinal cancers, such as esophageal, stomach, and colon cancer.

3. Literature Review

3.1. Esophageal Cancer

The anti-tumor effect of PDT in esophageal cancer is due to a combination of direct cell damage, destruction of tumor blood vessels, and activation of the immune response [50]. However, the exact mechanisms of action of PDT have not yet been precisely researched and established. The mechanism of photosensitizer accumulation in cancer cells is also insufficiently understood.

One study suggests that in the case of Photofrin-II, the mechanism responsible for the accumulation of the photosensitizer in cancer cells is the direct uptake of this compound by the cells, while others negate these conclusions [51,52]. It has been shown that after administration of 5-alpha-aminolevulinic acid, porphyrins accumulate in greater amounts in Barrett's epithelium and esophageal adenocarcinoma, which results from an imbalance between the activity of porphobilinogen deaminase and ferrochelatase enzymes [53].

Photodynamic therapy causes cell death by apoptosis and necrosis and induces autophagy and pyroptosis of esophageal cancer cells [54–58]. In a study by Shi Y. et al., PDT using sinoporphyrin sodium (DVDMs-PDT) was shown to induce apoptosis and autophagy of Eca-109 esophageal cancer cells [54]. By inducing the formation of reactive oxygen species in Eca-109, PDT led to the activation of p38MAPK and JNK kinases and HO-1 heme oxygenase proteins responsible for cellular responses to stress [54,59–61]. In Eca-109 cells, apoptosis is also induced by ALA-PDT, stopping the cell cycle in the G0/G1 phase and increasing the level of the pro-apoptotic Bax protein while decreasing the anti-apoptotic Bcl-2 [62,63]. Despite the observed increased levels of apoptosis and caspase-3 activity in esophageal adenocarcinoma cells, PDT using Photofrin-II has not been shown to be responsible for these differences [55].

Necrosis of Eca-109 esophageal cancer cells was induced by PDT with hematoporphyrin, while significantly increasing the level of malondialdehyde (a product of peroxidation of omega-6 fatty acids) without an increase in the expression of caspase-3, a key proenzyme in the apoptosis process [64,65].

By inhibiting the last enzyme involved in glycolysis, pyruvate kinase (PMK-2), and consequently activating caspase-8 and caspase-9, ultimately leading to the release of gasdermin E (GSDME), PDT can induce pyroptosis of esophageal squamous cell carcinoma cells [66]. A reduction in PKM-2 activity was also observed when examining the effect of ALA-PDT on the Warburg effect. It was shown that in esophageal cancer cells, glucose uptake was inhibited within 4 h after ALA-PDT; however, after 24 h, a significant increase in the expression of this enzyme and glucose uptake was observed [67]. ALA-PDT enhances the effect of the EGFR inhibitor AG1478 and the PI34 inhibitor LY294002, significantly reducing the expression of EGFR/PI3K and PI3K/AKT proteins, leading to a synergistic reduction in the growth and migration ability of Eca-109 esophageal cancer cells in vitro [68].

Photodynamic therapy increased NF-κB activity and HIF-1α and VEGF gene expression in vitro and *in vivo*, which may maintain their proliferation, protect against apoptosis, and promote tumor development. Dihydroartemisinin (DHA) may enhance the effect of PDT on esophageal cancer cells [69–71]. Zhou et al. examined the mechanism of action of DHA and showed that it was at least partially due to the deactivation of NF-kB [72].

3.2. Stomach Cancer

There are a very limited number of reports on the mechanism of action of PDT on gastric cancer cells.

One study showed that the mechanism underlying gastric-cancer-specific porphyrin accumulation is closely related to both nitric oxide (NO) and heme carrier protein-1 (HCP-1). Moreover, NO has been found to inactivate ferrochelatase, and thus, intracellular porphyrin levels in cells are increased following administration of a NO donor after 5-aminolevulinic acid treatment, and HCP-1 transports not only heme but also other porphyrins. Since NO stabilizes hypoxia-inducible factor (HIF)-1α, causing up-regulation of heme biosynthesis, HCP-1 expression may be increased by stabilizing HIF-1α, which affects the efficiency of porphyrin accumulation by cancer cells [73].

One study tested the dependence of the type of gastric cancer cell death induced by PDT using the chlorin-based photosensitizer DH-II-24 on dose level. It was shown that through intracellular free radical production and an increase in intracellular Ca^{2+} ion levels, low-dose PDT (LDP) led to apoptosis, while high-dose PDT (HDP) induced a massive and prolonged increase in intracellular Ca^{2+} ion levels and was thus responsible for inducing necrosis. Moreover, LDP activated caspase-3 [74].

It was observed that 5-ALA-PDT applied to human gastric cancer xenografts in vivo caused the apoptosis and necrosis of cells, and in histological examination, most of the tumor blood vessels were hyperemic [75]. It has been shown that PDT via the photosensitizer Photofrin in the MKN45 gastric cancer cell line within 15 min leads to an increase in the activity of caspase-3 and caspase-9 and chromatin condensation. The reduction in rhodamine 123 uptake begins after 30 min and induces mitochondrial damage and apoptosis after 60 min [76]. Moreover, due to its ability to activate the immune system, PDT has a specific effect on metastatic lesions [77]. The effect of PDT on gastric adenocarcinoma cells was studied in patients receiving immune checkpoint inhibitors. Immune cell infiltration increased in tumors after PDT, which is associated with the up-regulation of the B2M gene, which is lost in tumor cells. TCR analysis revealed specific clonal expansion after PDT in cytotoxic T cells but constriction in Treg cells [78,79].

3.3. Colon Cancer

The largest number of reports on the effects of PDT on gastrointestinal cancer concern colorectal cancer treatment. However, the exact sequence of reactions occurring after PDT has not yet been fully explained [80]. It has been established that PDT leads to the direct killing of cancer cells by $^1O_2$ and the indirect killing of cells through damage to blood vessels and the induced immune response [81]. The effectiveness of PDT itself depends on the concentration of the photosensitizer in the cell, but it has been shown that precise intracellular localization has an additional impact on the way in which the therapy causes damage. Moreover, the degree of differentiation of cancer cells also affects the effectiveness of therapy. It was shown that well-differentiated tumor cells had a better response to PDT using protoporphyrin IX (PpIX) than less differentiated cells [82]. Research results indicate that the internalization of a photosensitizer may be the result of partitioning, pinocytosis, and endocytosis, and the target place of its accumulation in the cell is different for different photosensitizers [83,84]. In the case of PpIX, it was found that the tumor-preferential accumulation of this compound is influenced by the difference in activity between porphobilinogen deaminase and ferrochelatase [85]. Photodynamic therapy causes the death of colorectal cancer cells by apoptosis and necrosis [86–105].

One of the most important mechanisms of apoptosis triggered by PDT appears to be the mitochondrial pathway. PDT, using hexaminolevulinaine as a photosensitizer, leads to the loss of mitochondrial membrane potential, the release of cytochrome c from mitochondria into the cytosol, and the rapid activation of caspase-9 and caspase-3 and consequently to the apoptosis of 320 DM colon cancer cells [97]. Identical observations were made in the case of PDT with silicon (IV) phthalocyanine [91]. It has been shown that the calcium signal plays an important role in the apoptosis of SW480 cells induced by PDT with the pre-photosensitizer 5-ALA [78]. However, the role of this signal may also contribute to the failure of PDT, as it induces the activation of the ERK pathway, which plays a key role in the survival and development of cancer cells. Calcium ions released from the endoplasmic reticulum were found to result in an increase in the expression level of the chaperone protein GRP78, which in many cancer models, both in vitro and *in vivo*, confers a growth advantage and drug resistance to solid tumors [87,88,91,92].

Another study highlighting the involvement of the mitochondrial apoptosis pathway is the study by Guoqing Ouyang et al. who showed that PDT with PpIX led to an increase in the expression of the pro-apoptotic protein bax and caspase-3 while decreasing the expression of the anti-apoptotic bcl-2 [96]. It was shown that cell lines with cytosolic or mitochondrial localization of PpIX were characterized by a loss of mitochondrial transmembrane potential, which led to growth arrest [82]. In turn, in the case of PDT with pyropheophorbide methylester (PPME) that accumulates in the endoplasmic reticulum/Golgi apparatus and lysosomes, it was not demonstrated that transmitters such as calcium ions, Bid proteins, Bap31, phosphorylated Bcl-2, and caspase-12 were involved in triggering the release of cytochrome c from mitochondria when provoking apoptosis [75]. The loss of mitochondrial functionality and therefore apoptosis was also induced by PDT using [Ir-b]Cl and [Rh-b]Cl complexes and PpIX attached to triphenylphosphonium (TPP), which has the ability to target mitochondria [90,100]. Moreover, it is assumed that the leakage of lysosomal protease into the cytosol may also be involved in the induction of apoptosis [99].

The effect of PDT on gene expression, which can contribute to resistance, also appears to be important. A study by H. Abrahamse et al. tested the effect of photodynamic therapy on the expression of pro-apoptotic and anti-apoptotic genes in DLD-1 and Caco-2 colon cancer cells. In the case of DLD-1 cells, with increased tumorigenicity, apoptosis was observed with the up-regulation of 3 genes and down-regulation of 20 genes, and these cells were found to have an increased risk of resistance. Caco-2 cells responded better to PDT, and the up-regulation of 16 genes and down-regulation of 22 were observed in these cells [94]. As mentioned earlier, PDT can also cause necrosis of colorectal cancer cells; however, there are no precise reports on the specific mechanism by which this cell death occurs. One study on HT29 colon adenocarcinoma cells showed that the predominant type of cell death provoked by PDT with Foscan® was not apoptosis but necrosis and that changes in mitochondrial membrane potential and cytochrome c release were responsible for cell photoinactivation. HT29 multicellular spheroids loaded with Foscan® showed significantly higher anti-tumor activity at equivalent light doses and the lowest fluence applied. At the lowest fluence rate, and at fluences of moderate levels, 65% cell death was observed via apoptosis. It was also found that the level of caspase-3 activation was not affected by the use of higher fluence values (at identical levels of photocytotoxicity) [106]. It is known that membrane-bound PpIX induces loss of membrane integrity and subsequent necrosis and that 21-selenaporphyrin probably induces necrosis through the endothelial cells of newly formed tumor vessels [82,104]. Necrosis may also be induced by other photosensitizers, e.g., Soranjidol and Rubiadin [103].

So far, it has been established that cellular interactions in the tumor microenvironment also participate in the induction of cancer cell death. It has been shown that due to their plasticity, macrophages residing in or recruited from the tumor can enhance tumor development by promoting tumor cell migration and endothelial stimulation. The increased cytotoxicity of PDT mediated by the production of nitric oxide, interleukin-6 (IL-6), and tumor necrosis factor alpha was in turn achieved in the presence of non-resident macrophages with a strong anti-tumor phenotype (TNF-α) [107]. In contrast, a study by A. Jalili et al. [102] determined the anti-tumor efficacy of combining PDT with the administration of immature dendritic cells. They found that inactivation of C-26 colon cancer cells after PDT was followed by necrosis and apoptosis. Moreover, there was also an increase in the expression of HSP72/73, HSP90, HSP27, HSP60, HO-1, and GRP78 proteins [101]. It was observed that immature dendritic cells cultured with cancer cells after PDT exhibited the ability to engulf dead cancer cells, acquired functional maturation characteristics, and produced significant amounts of interleukin-12 (IL 12), thereby enhancing the activity of macrophages, NK cells, and monocytes. Moreover, these cells also stimulated the cytotoxic activity of NK cells and T lymphocytes and stimulated their influx into lymph nodes [101,102].

It has been shown that PDT can also lead to the systemic induction of anti-tumor immune responses. In order to test the potential mechanism of this phenomenon, the effect of vascular PDT on β-galactosidase antigen-expressing colon adenocarcinomas BALB/c, CT26WT, and CT26. CL25 was studied. It was found that the efficacy of the therapy depended on the level of β-galactosidase expression, as complete cure occurred only in antigen-positive tumors. The destruction of distant metastases was also observed in 70% of the mice tested. It was found that T cells in these mice were able to recognize the epitope derived from the beta-galactosidase antigen and specifically destroy the cancerous antigen-positive cells. In the remaining 30% of mice, the tumor antigen was lost and the metastatic lesions were not cured [108]. The effect of PDT on the ability of cells to migrate and metastasize seems to be significant. It is known that PDT using low concentrations (5 μM) of hyperforin and aristophorin not only inhibits cell cycle progression and induces apoptosis but also reduces the expression of metalloproteinases-2/-9 and cell adhesion potential [89]. Similar observations were found in the case of PDT therapy using m-THPC, which also reduces the colony formation and migration ability of SW480 and SW620 colorectal cancer cells [95].

The possible mechanisms of this effect were investigated during PDT involving the photosensitizer chlorin-e6 (Ce6-PDT). It was shown that the therapy led to the inhibition of proliferation, almost complete disappearance of pseudopodia, a decrease in the migration ability of SW480 cells, and an increase in the expression of F-actin, α-tubulin, β-tubulin, vimentin, and E-cadherin. It is assumed that the possible inhibition of cancer cell migration was due to the increased expression of E-cadherin, the loss of which is often observed during metastases, causing the disappearance of pseudopodia and destruction of the cytoskeleton [109,110]. In another study, it was shown that under the influence of Ce6-PDT, the healing and migration rate of SW620 cells was significantly reduced, the pseudopodia of the cells were reduced or disappeared, the original microfilament structure was destroyed, and the expression of F-actin was significantly reduced. The Rac1/PAK1/LIMK1/cofilin signaling pathway, which is one of the main pathways through which Rho GTPases regulate microfilaments, was down-regulated by Ce6-PDT [111].

Another aspect of PDT's action is the ability to reduce the resistance of colorectal cancer cells. As shown by M. Luo et al., PDT with the photosensitizer chlorin-e6 can inhibit oxaliplatin (L-OHP)-induced autophagy while promoting apoptosis and increasing the expression of procaspase-3 protein, while the combination of Ce-6PDT with L-OHP led to the same effects and an increase in the expression of proapoptotic Bcl-2 and reduced the migration capacity of SW620 colorectal cancer cells [105].

An important aspect of PDT is the possibility of developing tumor resistance to this type of therapy, resulting from cellular responses to stress, hypoxia, or the heterogeneity of PS uptake by individual tumor cells [112–114]. It is known that in response to hypoxia, cells can induce HIF-1α-mediated autophagy, leading to increased colon cancer cell survival and reduced cell death after PDT. By binding to hypoxia-responsive elements in the VMP1 promoter, stabilization of HIF-1α has been shown to significantly increase the VMP1-related autophagy process [115]. An important factor involved in tumorigenesis is hypoxia-inducible factor-1 alpha (HIF-1α), which may also contribute to the development of PDT resistance [116,117]. Investigating the effects of PDT using Me-ALA (a pro-drug of PS PpIX) on human colon cancer spheroids, it was discovered that the PDT resistance phenotype was due to the highly regulated transcriptional activity of hypoxia-inducible factor-1α (HIF-1α). Abolishing the RNA interference (RNAi) of HIF-1α reduced the degree of resistance to PDT, while inhibition of the MEK/ERK pathway and removal of ROS abolished the regulation of HIF-1α by PDT [117]. It is known that elevated levels of Hsp27 may play an important role in colorectal cancer cell resistance (Figure 3), as phosphorylation of this protein plays an important role in cytoprotection.

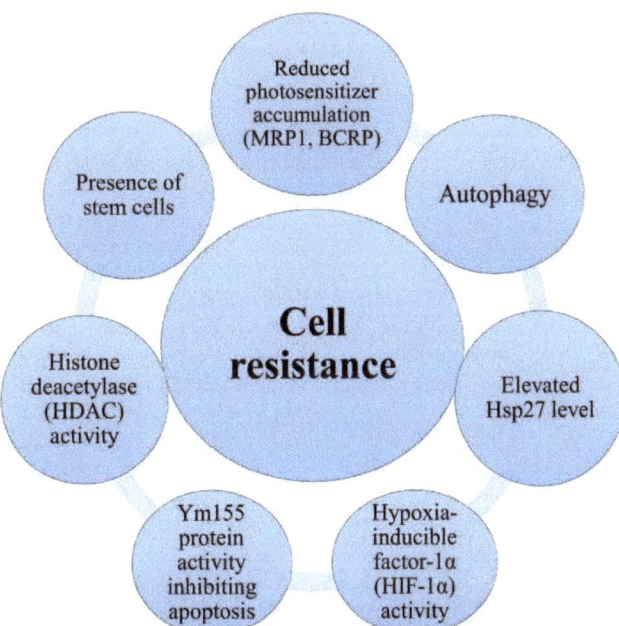

Figure 3. Mechanisms of cell resistance. The basic mechanisms of cellular resistance are presence of stem cells; reduced photosensitizer accumulation (P-gp, MRP1, BCRP); autophagy; elevated Hsp27 level; hypoxia-inducible factor-1α (HIF-1α) activity; Ym155 protein activity inhibiting apoptosis; histone deacetylase (HDAC) activity.

Studying the effects of Photofrin-PDT on HT29-P14 colon cancer cells, it was found that pathways leading to Hsp27 phosphorylation may contribute to cell resistance to photooxidative damage [118].

By examining the effect of YM155, a small molecule inhibitor of survivin expression in HT-29 colorectal adenocarcinoma cells resistant to dynamic phototherapy with hypericin, it was shown that proteins that inhibit apoptosis play a key role in cancer progression and therapeutic resistance [119]. Further, the interaction of hypericin with the mechanisms of elimination of anticancer drugs by cancer cells is unclear. It is known that they are complex. In HT-29 colon cancer cells treated with hypericin, increased activity of multidrug resistance-related protein 1 (MRP1) and breast cancer resistance protein (BCRP) was observed. In contrast, administration of cytochrome P450 enzyme inhibitors led to an increased content of this photosensitizer. Hypericin content in these cells is also known to decrease glycoprotein-p [120]. On the other hand, examining the contribution of the mechanism of export by p-glycoprotein, it was shown that the use of verapamil, a p-glycoprotein antagonist, can reverse the resistance of HRT-18 colorectal cancer cells to PDT with hematoporphyrin, which suggests a significant role of p-glycoprotein in reducing sensitivity to treatment [121]. One study examined the effect of histone deacetylase (HDAC) inhibitors on the development of resistance to PDT with hypericin by colorectal cancer cells. Two chemical classes of histone deacetylase (HDAC) inhibitors have been studied in combination with HY-PDT: the hydroxamic acids Saha and Trichostatin A, and the short-chain fatty acids valproic acid and sodium phenylbutyrate (NaPB). Combining HDAC inhibitors with HY-PDT significantly attenuated the renewed resistance of cancer cells to treatment. The manner, selectivity, and potency of HDAC inhibition depended on the specific inhibitor. To sum up, histone deacetylase may be one of the causes of cell resistance to PDT (Figure 3) [122].

Another study showed that a total of 1096 long noncoding lncRNAs were present in HCT116 colon cancer cells treated with PDT. Resistance to PDT was determined by the interaction between Long Noncoding RNA LIFR-AS1, the miR-29a gene, and the TNF Alpha Induced Protein 3 (TNFAIP3) gene. The resistance of HCT116 cells to PDT was due to the role of LIFR-AS1, as it serves as a competitive endogenous RNA for miR-29a, inhibiting its expression and increasing TNFAIP3 expression [123]. Epigenetic changes are known to account for drug resistance in colorectal cancer [124]. At the same time, they are reversible. The regulation of polycomb proteins (PcG), which have the ability to epigenetically silence genes, polycomb group RING finger protein 4 (BMI1) and Enhancer of zeste homolog 2 (EZH2), and the associated cancer progression are potential therapeutic targets. A study of resistance to PDT with hypericin by M. N. Sardoiwala et al. showed that Protein phosphatase 2 mediated the degradation of BMI1 and that inhibition of HMI1 and EZH2 contributed to improved treatment outcomes [125].

Stem cells are believed to be resistant to PDT, which may be another reason for the lack of therapeutic efficacy. Through their ability to self-renew cyclically with a long duration of one cycle, they increase resistance to treatment, which contributes to PDT failure and an increased risk of recurrence [126]. A major concern is the ability of cancer cells to acquire resistance to drug treatment. The sensitivity of colorectal cancer cells to treatment may be enhanced by PDT. It was shown that PDT increased the efficacy of L-oxaliplatin (L-OHP) treatment. A multilevel mechanism for this phenomenon has been established, involving the decreased efflux of L-OHP (dependent on multidrug resistance-associated protein 1 (MRP-2)), inhibition of glutathione S-transferase activity and intracellular glutathione, increased DNA double-strand breaks, and decreased expression of DNA excision repair protein (ERCC-1) along with DNA repair endonuclease XPF, involved in the nucleotide excision repair pathway [127]. Photodynamic therapy works synergistically with drugs that block Programmed death-ligand 1 (PD-L1), which may increase the effectiveness of treatment. A study by Z. Yuan et al. showed that the combination can inhibit primary and distant tumor growth, as well as contribute to long-term host immune memory, which prevents cancer recurrence. The mechanism of this interaction has been shown to induce cell death and stimulate a systemic immune response, which can be further promoted by PD-L1 blockade [128]. It is known that the efficacy of PDT of HT-29 colon cancer cells can be enhanced by stimulating apoptosis by administering the specific 5-lipoxygenase inhibitor MK-886. Further analysis of individual ROS groups revealed the effect of increasing MK-886 concentration on peroxide accumulation, which was accompanied by a decrease in the level of hydrogen peroxide in cells. A clonogenicity test revealed impaired colony formation when both agents were combined compared to MK-886 or PDT alone [129]. Figure 3 shows the mechanisms of cell resistance.

Photodynamic therapy does not always lead to complete cure [39]. This phenomenon involves mutations related to the inhibition of apoptosis, drug–drug interactions, increased drug efflux, reduced photosensitizer concentration and light exposure, and local hypoxia [39,130–134]. Much research has been undertaken to develop a new generation of nanomaterial-based photosensitizers that could address this problem [39]. Emerging evidence indicates that overcoming the resistance of cancer cells can be achieved by using photosensitizers with the regulation of ROS production, targeting organelles, nanosubstituted photoactive drugs, and PS delivery nanosystems and combining different types of therapies [131]. Pramual et al. created a new hybrid molecule and demonstrated that it had the potential to deliver a photosensitizer or chemotherapeutic drug for the treatment of multidrug-resistant lung cancer cells [134]. In turn, Qian-Li Ma et al. found that the combination of an ATM inhibitor with PDT has the ability to inhibit the DNA damage response and increase the effectiveness of therapy against PDT-resistant lung cancer cells [110]. Moreover, Deken et al. showed that nanoparticles can induce the regression of tumors overexpressing HER2 during one treatment session, which may be used in the treatment of trastuzumab-resistant cancers [135]. Zhijian Luo et al. created molecules that bind to annexin 1, which improved the cellular uptake of drugs and, consequently, increased cytotoxicity against multidrug-resistant breast cancer cells [136]. As shown by Zhong et al., a properly constructed nanoparticle with palitaxel intended for combined chemo-photodynamic therapy can break the resistance of lung cancer cells to this drug [137].

3.4. Interaction of PDT with Gastrointestinal Tumor Cells

Modern research methods and advanced drug complexes are being developed to observe and understand the immunological processes occurring in tumor metabolism. One of the main aspects of the analysis is the characterization of the immune response (Table 1), mainly the process of programmed cell death. An example of advanced research assessing the immune response of a tumor is the research conducted by Liu et al. The therapeutic method involving an increase in the infiltration of T lymphocytes has completely revolutionized the therapeutic technique of cancer. Although many metabolic processes are known and investigated, the mechanisms of the tumor's immune response to PDT remain undiscovered. Additionally, there is still uncertainty about the safety of applied photosensitizers, drugs that target selected cell organelles (i.e., mitochondria). Work by Liu et al. describes an innovatively designed drug that is safe and effective both in vivo and *in vitro*. Drug-assisted PDT has the ability to inhibit tumor growth. Additionally, it alleviates the phenomenon of hypoxia, i.e., tumor hypoxia, by generating a higher number of ROS. The designed multifunctional drug and PDT enable it to influence tumor metabolism and its immune system [138]. Table 1 shows the mechanisms of interaction of PDT with gastrointestinal tumor cells (from the type of accumulation of photosensitizer through the mechanism of destruction to the type of response).

Table 1. Mechanisms of interaction of PDT with gastrointestinal tumor cells.

	Mechanism of Interaction of PDT with Gastrointestinal Tumor Cells	
Esophageal cancer	Accumulation of photosensitizer	Imbalance between the activity of porphobilinogen deaminase and ferrochelatase enzymes (5-ALA)
	Mechanism of cell damage	Direct cell damage
		Destruction of tumor blood vessels
		Activation of the immune response
	Type of response and cell death	Apoptosis
		Necrosis
		Pyroptosis
		Autophagy
Gastric cancer	Accumulation of photosensitizer	Dependent on nitric oxide (NO) and heme carrier protein-1 (HCP-1)
	Mechanism of cell damage	Direct cell damage
		Activation of the immune response
	Type of response and cell death	Apoptosis
		Necrosis
Colorectal cancer	Accumulation of photosensitizer	Partitioning
		Pinocytosis
		Endocytosis
		Difference in activity between porphobilinogen deaminase and ferrochelatase (PPIX)
	Mechanism of cell damage	Direct cell damage
		Destruction of tumor blood vessels
		Activation of the immune response
	Type of response and cell death	Apoptosis
		Necrosis

3.5. Clinical Challenges

In the field of PDT in cancer treatment, valuable insights are provided by the dual perspective of photosensitizers undergoing clinical trials and those already in clinical use. Clinical trials are a source of innovation, presenting a diverse range of photosensitizers of different generations. These trials highlight ongoing efforts to improve and expand the potential of PDT. In particular, third-generation photosensitizers demonstrate increased tumor specificity, improved tissue penetration, and reduced side effects, representing significant progress. Challenges such as poor water solubility and aggregation remain, highlighting the complexity of developing effective photosensitizers.

Certainly, one of the main limitations and challenges of conducting PDT in a clinical setting is the difficult process of monitoring the entire treatment process.

Additionally, uneven and varied distribution of therapy components (such as light and oxygen) may result in numerous side effects. Currently, various types of simulations are practiced to improve PDT in clinical conditions at every stage of treatment (from the application of a photosensitizer to the exposure process and follow-up observations) [139].

Currently, interstitial PDT supported by chemotherapy and immunotherapy is also practiced, introducing a number of combination options in the treatment of gastroenterological diseases. More often, pilot studies are carried out as initial verification and the initial stage of clinical trials.

Despite the high effectiveness of first- and second-generation photosensitizers, new solutions are still being sought. An example of improving treatment results is the use of nanotechnology, i.e., third-generation photosensitizers (Table 2). Currently, ongoing research and the latest literature reports on the use of PDT in gastroenterological diseases give hope for improving the effectiveness, sensitivity, and specificity of treatment. Table 2 shows a review of third-generation photosensitizers in gastroenterological cancers.

Table 2. A review of third-generation photosensitizers in gastroenterological cancers.

Type of Disease	A Type of Third-generation Photosensitizer	Wavelength of Laser Light (nm)	Immunological Effect	References
Colon cancer	porphyrin grafted lipid (PGL) nanoparticles	650	The results confirmed that the designed nanoplatform effectively eliminates differences in oxygen content, which positively affects the process of generating singlet oxygen and the process of weakening COX-2 expression.	[140]
Colon cancer	liposome encapsulating phosphoinositide 3-kinase gamma (PI3Kγ) inhibitor IPI-549 and chlorin e6	660	The proposed therapy significantly limited the development and growth of the tumor by positively affecting the physiology of dendritic cells and T lymphocytes.	[141]
Colorectal cancer	CD133-Pyro	670	The study showed that the designed composite increases ROS production and induces cell death.	[142]
Colorectal cancer	Sinoporphyrin sodium (DVDMS)	635	The therapy induced programmed cell death, among others, by generating the caspase pathway in CX-1 cells.	[143]

Third-generation photosensitizers (Table 2) and their effectiveness are another challenge in transforming laboratory and preclinical research into clinical trials. The combination of nanotechnology and the process of developing various types of nanoparticles supporting PDT is a promising tool but not free from obstacles and challenges. Many nanocomplexes being developed are in the process of improvement to be safely used in clinical trials. The nanomaterials used are not always free of toxicity, which is why they are subject to control and testing. Due to the fact that some of the research is conducted at an early stage, we can look forward to the future with hope for the development of the PDT technique in the treatment of gastroenterological cancers. We hope that the results of future studies will allow us to improve the effectiveness of clinical trials using PDT as much as possible [144].

4. Conclusions

Cancer treatment using PDT poses many challenges. One of them is the possibility of cancer cells becoming resistant to this type of therapy. This article presents evidence that mechanisms such as the removal of photosensitizer from cancer cells, induction of autophagy in response to damage, natural increased resistance of tumor stem cells, and, finally, increased presence of various cytoprotective proteins are involved in this process. Interactions between tumor cells and other cells are also an important aspect, as they may contribute to weakening the effect of PDT and even to accelerating tumor development. Further research is necessary to determine the exact mechanisms of action of dynamic phototherapy on gastrointestinal cancer cells, taking into account the type of photosensitizers, the classification of cancers, and their stage of advancement. Understanding the precise impact of PDT on the treatment of this disease may help discover new photosensitizers and their transport mechanism or determine the appropriate, most effective therapeutic regimens. It also seems promising to investigate the mechanisms by which PDT can lead to the activation of the immune system and, as a result, to the treatment of metastases. Moreover, based on this review, it can be concluded that a thorough examination of the mechanisms responsible for cellular resistance to PDT may contribute to the discovery of new therapeutic agents that can inhibit this resistance. In summary, the immunological mechanisms of the action of PDT on gastrointestinal cancer cells are still insufficiently understood, and their detailed examination may contribute to increasing the effectiveness of this therapy. Solutions to certain challenges and application problems emerging in clinical trials are still being sought. The solution turns out to be not only nanotechnology and its possibilities but also designed drugs targeting selected cell organelles. The therapeutic process of cancer is very complex, and the biological and immunological mechanisms initiated as a result of PDT are still not clear and understandable. It is satisfactory that such a difficult and important topic as the immunological aspects of PDT is constantly being explored and addressed.

Author Contributions: Conceptualization, D.A., P.W., K.D., A.K.-K., G.C. and D.B.-A.; methodology, D.A., P.W., K.D. and D.B.-A.; software, D.A., P.W., K.D. and D.B.-A.; validation, D.A., P.W., K.D. and D.B.-A.; formal analysis, D.A., P.W., K.D., A.K.-K., G.C. and D.B.-A.; investigation, D.A., P.W., K.D., A.K.-K., G.C. and D.B.-A.; resources, D.A., P.W., K.D., A.K.-K., G.C. and D.B.-A.; data curation, D.A., P.W., K.D., A.K.-K., G.C. and D.B.-A.; writing—original draft preparation, D.A., P.W., K.D. and D.B.-A.; writing—review and editing, D.A., P.W., K.D., A.K.-K., G.C. and D.B.-A.; visualization, D.A., P.W., K.D., A.K.-K., G.C. and D.B.-A.; supervision, D.A.; project administration, D.A.; funding acquisition, A.K.-K., G.C. and D.A. All authors have read and agreed to the published version of the manuscript.

Funding: This research received no external funding.

Data Availability Statement: The data can be shared up on request.

Conflicts of Interest: The authors declare no conflict of interest.

References

1. Tong, Y.; Gao, H.; Qi, Q.; Liu, X.; Li, J.; Gao, J.; Li, P.; Wang, Y.; Du, L.; Wang, C. High fat diet, gut microbiome and gastrointestinal cancer. *Theranostics* **2021**, *11*, 5889–5910. [CrossRef] [PubMed]
2. Abbas, G.; Krasna, M. Overview of esophageal cancer. *Ann. Cardiothorac. Surg.* **2017**, *6*, 131–136. [CrossRef] [PubMed]
3. Short, M.W.; Burgers, K.G.; Fry, V.T. Esophageal Cancer. *Am. Fam. Physician* **2017**, *95*, 22–28. [PubMed]
4. Enzinger, P.C.; Mayer, R.J. Esophageal cancer. *N. Engl. J. Med.* **2003**, *349*, 2241–2252. [CrossRef] [PubMed]
5. Smyth, E.C.; Nilsson, M.; Grabsch, H.I.; van Grieken, N.C.; Lordick, F. Gastric cancer. *Lancet* **2020**, *396*, 635–648. [CrossRef] [PubMed]
6. Van Cutsem, E.; Sagaert, X.; Topal, B.; Haustermans, K.; Prenen, H. Gastric cancer. *Lancet* **2016**, *388*, 2654–2664. [CrossRef] [PubMed]
7. Catalano, V.; Labianca, R.; Beretta, G.D.; Gatta, G.; de Braud, F.; Van Cutsem, E. Gastric cancer. *Crit. Rev. Oncol. Hematol.* **2009**, *71*, 127–164. [CrossRef] [PubMed]
8. Mármol, I.; Sánchez-de-Diego, C.; Pradilla Dieste, A.; Cerrada, E.; Rodriguez Yoldi, M.J. Colorectal Carcinoma: A General Overview and Future Perspectives in Colorectal Cancer. *Int. J. Mol. Sci.* **2017**, *18*, 197. [CrossRef]

9. Haraldsdottir, S.; Einarsdottir, H.M.; Smaradottir, A.; Gunnlaugsson, A.; Halfdanarson, T.R. Colorectal cancer—Review. *Laeknabladid* **2014**, *100*, 75–82.
10. Cappell, M.S. Pathophysiology, clinical presentation, and management of colon cancer. *Gastroenterol. Clin. N. Am.* **2008**, *37*, 1–24. [CrossRef]
11. Rkein, A.M.; Ozog, D.M. Photodynamic therapy. *Dermatol. Clin.* **2014**, *32*, 415–425. [CrossRef] [PubMed]
12. Bartusik-Aebisher, D.; Serafin, I.; Dynarowicz, K.; Aebisher, D. Photodynamic therapy and associated targeting methods for treatment of brain cancer. *Front. Pharmacol.* **2023**, *14*, 1250699. [CrossRef] [PubMed]
13. Chiba, K.; Aihara, Y.; Oda, Y.; Fukui, A.; Tsuzuk, S.; Saito, T.; Nitta, M.; Muragaki, Y.; Kawamata, T. Photodynamic therapy for malignant brain tumors in children and young adolescents. *Front. Oncol.* **2022**, *12*, 957267. [CrossRef] [PubMed]
14. Aldosari, L.I.N.; Hassan, S.A.B.; Alshadidi, A.A.F.; Rangaiah, G.C.; Divakar, D.D. Short-term influence of antimicrobial photodynamic therapy as an adjuvant to mechanical debridement in reducing soft-tissue inflammation and subgingival yeasts colonization in patients with peri-implant mucositis. *Photodiagn. Photodyn. Ther.* **2023**, *42*, 103320. [CrossRef] [PubMed]
15. Gil-Pallares, P.; Navarro-Bielsa, A.; Almenara-Blasco, M.; Gracia-Cazaña, T.; Gilaberte, Y. Photodynamic Therapy, a successful treatment for granular parakeratosis. *Photodiagn. Photodyn. Ther.* **2023**, *42*, 103562. [CrossRef] [PubMed]
16. Alexiades-Armenakas, M. Laser-mediated photodynamic therapy. *Clin. Dermatol.* **2006**, *24*, 16–25. [CrossRef] [PubMed]
17. Kwiatkowski, S.; Knap, B.; Przystupski, D.; Saczko, J.; Kędzierska, E.; Knap-Czop, K.; Kotlińska, J.; Michel, O.; Kotowski, K.; Kulbacka, J. Photodynamic therapy—Mechanisms, photosensitizers and combinations. *Biomed. Pharmacother.* **2018**, *106*, 1098–1107. [CrossRef] [PubMed]
18. Agostinis, P.; Berg, K.; Cengel, K.A.; Foster, T.H.; Girotti, A.W.; Gollnick, S.O.; Hahn, S.M.; Hamblin, M.R.; Juzeniene, A.; Kessel, D.; et al. Photodynamic therapy of cancer: An update. *CA Cancer J. Clin.* **2011**, *61*, 250–281. [CrossRef]
19. Juarranz, A.; Jaén, P.; Sanz-Rodríguez, F.; Cuevas, J.; González, S. Photodynamic therapy of cancer. Basic principles and applications. *Clin. Transl. Oncol.* **2008**, *10*, 148–154. [CrossRef]
20. Kübler, A.C. Photodynamic therapy. *Med. Laser Appl.* **2005**, *20*, 37–45. [CrossRef]
21. Triesscheijn, M.; Baas, P.; Schellens, J.H.; Stewart, F.A. Photodynamic therapy in oncology. *Oncologist* **2006**, *11*, 1034–1044. [CrossRef] [PubMed]
22. Webber, J.; Herman, M.; Kessel, D.; Fromm, D. Photodynamic treatment of neoplastic lesions of the gastrointestinal tract. Recent advances in techniques and results. *Langenbecks Arch. Surg.* **2000**, *385*, 299–304. [CrossRef] [PubMed]
23. Dixon, S.J.; Lemberg, K.M.; Lamprecht, M.R.; Skouta, R.; Zaitsev, E.M.; Gleason, C.E.; Patel, D.N.; Bauer, A.J.; Cantley, A.M.; Yang, W.S.; et al. Ferroptosis: An iron-dependent form of nonapoptotic cell death. *Cell* **2012**, *149*, 1060–1072. [CrossRef] [PubMed]
24. Li, J.; Cao, F.; Yin, H.L.; Huang, Z.J.; Lin, Z.T.; Mao, N.; Sun, B.; Wang, G. Ferroptosis: Past, present and future. *Cell Death Dis.* **2020**, *11*, 88. [CrossRef] [PubMed]
25. Pan, Y.; Cai, W.; Huang, J.; Cheng, A.; Wang, M.; Yin, Z.; Jia, R. Pyroptosis in development, inflammation and disease. *Front. Immunol.* **2022**, *13*, 991044. [CrossRef]
26. Slastnikova, T.A.; Rosenkranz, A.A.; Lupanova, T.N.; Gulak, P.V.; Gnuchev, N.V.; Sobolev, A.S. Study of efficiency of the modular nanotransporter for targeted delivery of photosensitizers to melanoma cell nuclei in vivo. *Dokl. Biochem. Biophys.* **2012**, *446*, 235–237. [CrossRef]
27. Maharjan, P.S.; Bhattarai, H.K. Singlet Oxygen, Photodynamic Therapy, and Mechanisms of Cancer Cell Death. *J. Oncol.* **2022**, *2022*, 7211485. [CrossRef]
28. Tang, D.; Kang, R.; Berghe, T.V.; Vandenabeele, P.; Kroemer, G. The molecular machinery of regulated cell death. *Cell Res.* **2019**, *29*, 347–364. [CrossRef]
29. Cui, J.; Zhao, S.; Li, Y.; Zhang, D.; Wang, B.; Xie, J.; Wang, J. Regulated cell death: Discovery, features and implications for neurodegenerative diseases. *Cell Commun. Signal.* **2021**, *19*, 120. [CrossRef]
30. Bartusik-Aebisher, D.; Osuchowski, M.; Adamczyk, M.; Stopa, J.; Cieślar, G.; Kawczyk-Krupka, A.; Aebisher, D. Advancements in photodynamic therapy of esophageal cancer. *Front. Oncol.* **2022**, *12*, 1024576. [CrossRef]
31. Yano, T.; Hatogai, K.; Morimoto, H.; Yoda, Y.; Kaneko, K. Photodynamic therapy for esophageal cancer. *Ann. Transl. Med.* **2014**, *2*, 29.
32. Qumseya, B.J.; David, W.; Wolfsen, H.C. Photodynamic Therapy for Barrett's Esophagus and Esophageal Carcinoma. *Clin. Endosc.* **2013**, *46*, 30–37. [CrossRef] [PubMed]
33. Yano, T.; Wang, K.K. Photodynamic Therapy for Gastrointestinal Cancer. *Photochem. Photobiol.* **2020**, *96*, 517–523. [CrossRef] [PubMed]
34. Barr, H. Photodynamic therapy in gastrointestinal cancer: A realistic option? *Drugs Aging* **2000**, *16*, 81–86. [CrossRef] [PubMed]
35. Li, X.; Lee, S.; Yoon, J. Supramolecular photosensitizers rejuvenate photodynamic therapy. *Chem. Soc. Rev.* **2018**, *47*, 1174–1188. [CrossRef] [PubMed]
36. Niculescu, A.G.; Grumezescu, A.M. Photodynamic Therapy—An Up-to-Date Review. *Appl. Sci.* **2021**, *11*, 3626. [CrossRef]
37. Chizenga, E.P.; Chandran, R.; Abrahamse, H. Photodynamic therapy of cervical cancer by eradication of cervical cancer cells and cervical cancer stem cells. *Oncotarget* **2019**, *10*, 4380–4396. [CrossRef]
38. Xu, J.; Gao, J.; Wei, Q. Combination of photodynamic therapy with radiotherapy for cancer treatment. *J. Nanomater.* **2016**, *2016*, 8507924. [CrossRef]

39. Bhattacharya, D.; Mukhopadhyay, M.; Shivam, K.; Tripathy, S.; Patra, R.; Pramanik, A. Recent developments in photodynamic therapy and its application against multidrug resistant cancers. *Biomed. Mater.* **2023**, *18*, 062005. [CrossRef]
40. Bhanja, D.; Wilding, H.; Baroz, A.; Trifoi, M.; Shenoy, G.; Slagle-Webb, B.; Hayes, D.; Soudagar, Y.; Connor, J.; Mansouri, A. Photodynamic Therapy for Glioblastoma: Illuminating the Path toward Clinical Applicability. *Cancers* **2023**, *15*, 3427. [CrossRef]
41. Bartusik-Aebisher, D.; Woźnicki, P.; Dynarowicz, K.; Aebisher, D. Photosensitizers for Photodynamic Therapy of Brain Cancers—A Review. *Brain Sci.* **2023**, *13*, 1299. [CrossRef] [PubMed]
42. Gunaydin, G.; Gedik, M.E.; Ayan, S. Photodynamic Therapy-Current Limitations and Novel Approaches. *Front. Chem.* **2021**, *9*, 691697. [CrossRef] [PubMed]
43. Wan, Y.; Fu, L.H.; Li, C.; Lin, J.; Huang, P. Conquering the Hypoxia Limitation for Photodynamic Therapy. *Adv. Mater.* **2021**, *33*, e2103978. [CrossRef] [PubMed]
44. Quirk, B.J.; Brandal, G.; Donlon, S.; Vera, J.C.; Mang, T.S.; Foy, A.B.; Lew, S.M.; Girotti, A.W.; Jogal, S.; LaViolette, P.S.; et al. Photodynamic therapy (PDT) for malignant brain tumors—Where do we stand? *Photodiagn. Photodyn. Ther.* **2015**, *12*, 530–544. [CrossRef] [PubMed]
45. Zhou, Z.; Liu, Y.; Song, W.; Jiang, X.; Deng, Z.; Xiong, W.; Shen, J. Metabolic reprogramming mediated PD-L1 depression and hypoxia reversion to reactivate tumor therapy. *J. Control Release* **2022**, *352*, 793–812. [CrossRef]
46. Liao, W.; Xiao, S.; Yang, J.; Shi, X.; Zheng, Y. Multifunctional nanogel based on carboxymethyl cellulose interfering with cellular redox homeostasis enhances phycocyanobilin photodynamic therapy. *Carbohydr. Polym.* **2024**, *323*, 121416. [CrossRef]
47. Kim, E.J.; Mangold, A.R.; DeSimone, J.A.; Wong, H.K.; Seminario-Vidal, L.; Guitart, J.; Appel, J.; Geskin, L.; Lain, E.; Korman, N.J.; et al. Efficacy and Safety of Topical Hypericin Photodynamic Therapy for Early-Stage Cutaneous T-Cell Lymphoma (Mycosis Fungoides): The FLASH Phase 3 Randomized Clinical Trial. *JAMA Dermatol.* **2022**, *158*, 1031–1039. [CrossRef]
48. Niu, J.; Cheng, M.; Hong, Z.; Ling, J.; Di, W.; Gu, L.; Qiu, L. The effect of 5-Aminolaevulinic Acid Photodynamic Therapy versus CO_2 laser in the Treatment of Cervical Low-grade Squamous Intraepithelial Lesions with High-Risk HPV Infection: A non-randomized, controlled pilot study. *Photodiagn. Photodyn. Ther.* **2021**, *36*, 102548. [CrossRef]
49. Qiao, S.; Tang, H.; Xia, J.; Ding, M.; Qiao, S.; Niu, Y.; Jiang, G. Efficacy and safety of microneedling, fractional CO_2 laser, and cryotherapy combined with 5-aminolevulinic acid photodynamic therapy in the treatment of actinic keratosis: A multicenter prospective randomized controlled study. *Photodiagn. Photodyn. Ther.* **2023**, *43*, 103700. [CrossRef]
50. Filip, A.G.; Clichici, S.; Daicoviciu, D.; Olteanu, D.; Mureşan, A.; Dreve, S. Photodynamic therapy--indications and limits in malignant tumors treatment. *Rom. J. Intern. Med.* **2008**, *46*, 285–293.
51. Gao, S.; Liang, S.; Ding, K.; Qu, Z.; Wang, Y.; Feng, X. Specific cellular accumulation of photofrin-II in EC cells promotes photodynamic treatment efficacy in esophageal cancer. *Photodiagn. Photodyn. Ther.* **2016**, *14*, 27–33. [CrossRef] [PubMed]
52. Gao, S.G.; Wang, L.D.; Feng, X.S.; Qu, Z.F.; Shan, T.Y.; Xie, X.H. Absorption and elimination of photofrin-II in human immortalization esophageal epithelial cell line SHEE and its malignant transformation cell line SHEEC. *Chin. J. Cancer* **2009**, *28*, 1248–1254. [CrossRef] [PubMed]
53. Hinnen, P.; de Rooij, F.W.; van Velthuysen, M.L.; Edixhoven, A.; van Hillegersberg, R.; Tilanus, H.W.; Wilson, J.H.; Siersema, P.D. Biochemical basis of 5-aminolaevulinic acid-induced protoporphyrin IX accumulation: A study in patients with (pre)malignant lesions of the oesophagus. *Br. J. Cancer* **1998**, *78*, 679–682. [CrossRef] [PubMed]
54. Shi, Y.; Zhang, B.; Feng, X.; Qu, F.; Wang, S.; Wu, L.; Wang, X.; Liu, Q.; Wang, P.; Zhang, K. Apoptosis and autophagy induced by DVDMs-PDT on human esophageal cancer Eca-109 cells. *Photodiagn. Photodyn. Ther.* **2018**, *24*, 198–205. [CrossRef] [PubMed]
55. Chen, X.; Zhao, P.; Chen, F.; Li, L.; Luo, R. Effect and mechanism of 5-aminolevulinic acid-mediated photodynamic therapy in esophageal cancer. *Lasers Med. Sci.* **2011**, *26*, 69–78. [CrossRef] [PubMed]
56. McGarrity, T.J.; Peiffer, L.P.; Granville, D.J.; Carthy, C.M.; Levy, J.G.; Khandelwal, M.; Hunt, D.W. Apoptosis associated with esophageal adenocarcinoma: Influence of photodynamic therapy. *Cancer Lett.* **2001**, *163*, 33–41. [CrossRef]
57. Chen, X.H.; Luo, R.C.; Li, L.B.; Ding, X.M.; Lv, C.W.; Zhou, X.P.; Yan, X. Mechanism of photodynamic therapy against human esophageal carcinoma xenografts in nude mice. *J. South. Med. Univ.* **2009**, *29*, 2222–2224.
58. Li, L.; Song, D.; Qi, L.; Jiang, M.; Wu, Y.; Gan, J.; Cao, K.; Li, Y.; Bai, Y.; Zheng, T. Photodynamic therapy induces human esophageal carcinoma cell pyroptosis by targeting the PKM2/caspase-8/caspase-3/GSDME axis. *Cancer Lett.* **2021**, *520*, 143–159. [CrossRef]
59. Xue, Q.; Wang, P.; Wang, X.; Zhang, K.; Liu, Q. Targeted inhibition of p38MAPK-enhanced autophagy in SW620 cells resistant to photodynamic therapy-induced apoptosis. *Lasers Med. Sci.* **2015**, *30*, 1967–1975. [CrossRef]
60. Ip, Y.T.; Davis, R.J. Signal transduction by the c-Jun N-terminal kinase (JNK)—From inflammation to development. *Curr. Opin. Cell Biol.* **1998**, *10*, 205–219. [CrossRef]
61. Sun, Y.; Liu, W.Z.; Liu, T.; Feng, X.; Yang, N.; Zhou, H.F. Signaling pathway of MAPK/ERK in cell proliferation, differentiation, migration, senescence and apoptosis. *J. Recept. Signal Transduct. Res.* **2015**, *35*, 600–604. [CrossRef] [PubMed]
62. Gutiérrez-Puente, Y.; Zapata-Benavides, P.; Tari, A.M.; López-Berestein, G. Bcl-2-related antisense therapy. *Semin. Oncol.* **2002**, *29*, 71–76. [CrossRef] [PubMed]
63. Weng, C.; Li, Y.; Xu, D.; Shi, Y.; Tang, H. Specific cleavage of Mcl-1 by caspase-3 in tumor necrosis factor-related apoptosis-inducing ligand (TRAIL)-induced apoptosis in Jurkat leukemia T cells. *J. Biol. Chem.* **2005**, *280*, 10491–11500. [CrossRef] [PubMed]
64. Ayala, A.; Muñoz, M.F.; Argüelles, S. Lipid peroxidation: Production, metabolism, and signaling mechanisms of malondialdehyde and 4-hydroxy-2-nonenal. *Oxid. Med. Cell Longev.* **2014**, *2014*, 360438. [CrossRef] [PubMed]

65. Asadi, M.; Taghizadeh, S.; Kaviani, E.; Vakili, O.; Taheri-Anganeh, M.; Tahamtan, M.; Savardashtaki, A. Caspase-3: Structure, function, and biotechnological aspects. *Biotechnol. Appl. Biochem.* **2022**, *69*, 1633–1645. [CrossRef] [PubMed]
66. Vaupel, P.; Harrison, L. Tumor hypoxia: Causative factors, compensatory mechanisms, and cellular response. *Oncologist* **2004**, *9*, 4–9. [CrossRef]
67. Gan, J.; Li, S.; Meng, Y.; Liao, Y.; Jiang, M.; Qi, L.; Li, Y.; Bai, Y. The influence of photodynamic therapy on the Warburg effect in esophageal cancer cells. *Lasers Med. Sci.* **2020**, *35*, 1741–1750. [CrossRef]
68. Zhang, X.; Cai, L.; He, J.; Li, X.; Li, L.; Chen, X.; Lan, P. Influence and mechanism of 5-aminolevulinic acid-photodynamic therapy on the metastasis of esophageal carcinoma. *Photodiagn. Photodyn. Ther.* **2017**, *20*, 78–85. [CrossRef]
69. Li, Y.; Sui, H.; Jiang, C.; Li, S.; Han, Y.; Huang, P.; Du, X.; Du, J.; Bai, Y. Dihydroartemisinin Increases the Sensitivity of Photodynamic Therapy Via NF-κB/HIF-1α/VEGF Pathway in Esophageal Cancer Cell in vitro and in vivo. *Cell Physiol. Biochem.* **2018**, *48*, 2035–2045. [CrossRef]
70. Carmeliet, P. VEGF as a key mediator of angiogenesis in cancer. *Oncology* **2005**, *69*, 4–10. [CrossRef]
71. Vlahopoulos, S.A. Aberrant control of NF-κB in cancer permits transcriptional and phenotypic plasticity, to curtail dependence on host tissue: Molecular mode. *Cancer Biol. Med.* **2017**, *14*, 254–270. [CrossRef] [PubMed]
72. Zhou, J.H.; Du, X.X.; Jia, D.X.; Wu, C.L.; Huang, P.; Han, Y.; Sui, H.; Wei, X.L.; Liu, L.; Yuan, H.H.; et al. Dihydroartemisinin accentuates the anti-tumor effects of photodynamic therapy via inactivation of NF-κB in Eca109 and Ec9706 esophageal cancer cells. *Cell Physiol. Biochem.* **2014**, *33*, 1527–1536.
73. Kurokawa, H.; Ito, H.; Terasaki, M.; Matano, D.; Taninaka, A.; Shigekawa, H.; Matsui, H. Nitric oxide regulates the expression of heme carrier protein-1 via hypoxia inducible factor-1α stabilization. *PLoS ONE* **2019**, *14*, e0222074. [CrossRef]
74. Yoo, J.O.; Lim, Y.C.; Kim, Y.M.; Ha, K.S. Differential cytotoxic responses to low- and high-dose photodynamic therapy in human gastric and bladder cancer cells. *J. Cell Biochem.* **2011**, *112*, 3061–3071. [CrossRef] [PubMed]
75. Zhou, G.J.; Huang, Z.H.; Yu, J.L.; Li, Z.; Ding, L.S. Effect of 5-aminolevulinic acid-mediated photodynamic therapy on human gastric cancer xenografts in nude mice in vivo. *Chin. J. Gastrointest. Surg.* **2008**, *11*, 580–583.
76. Takahira, K.; Sano, M.; Arai, H.; Hanai, H. Apoptosis of gastric cancer cell line MKN45 by photodynamic treatment with photofrin. *Lasers Med. Sci.* **2004**, *19*, 89–94. [CrossRef] [PubMed]
77. Wang, H.; Ewetse, M.P.; Ma, C.; Pu, W.; Xu, B.; He, P.; Wang, Y.; Zhu, J.; Chen, H. The "Light Knife" for Gastric Cancer: Photodynamic Therapy. *Pharmaceutics* **2022**, *15*, 101. [CrossRef] [PubMed]
78. Yu, Y.; Xu, B.; Xiang, L.; Ding, T.; Wang, N.; Yu, R.; Gu, B.; Gao, L.; Maswikiti, E.P.; Wang, Y.; et al. Photodynamic therapy improves the outcome of immune checkpoint inhibitors via remodelling anti-tumour immunity in patients with gastric cancer. *Gastric Cancer* **2023**, *26*, 798–813. [CrossRef]
79. Zhao, Y.; Cao, Y.; Chen, Y.; Wu, L.; Hang, H.; Jiang, C.; Zhou, X. B2M gene expression shapes the immune landscape of lung adenocarcinoma and determines the response to immunotherapy. *Immunology* **2021**, *164*, 507–523. [CrossRef]
80. Xue, Q.; Wang, X.; Wang, P.; Zhang, K.; Liu, Q. Role of p38MAPK in apoptosis and autophagy responses to photodynamic therapy with Chlorin e6. *Photodiagn. Photodyn. Ther.* **2015**, *12*, 84–91. [CrossRef]
81. Peng, C.L.; Lin, H.C.; Chiang, W.L.; Shih, Y.H.; Chiang, P.F.; Luo, T.Y.; Cheng, C.C.; Shieh, M.J. Anti-angiogenic treatment (Bevacizumab) improves the responsiveness of photodynamic therapy in colorectal cancer. *Photodiagn. Photodyn. Ther.* **2018**, *23*, 111–118. [CrossRef] [PubMed]
82. Krieg, R.C.; Messmann, H.; Schlottmann, K.; Endlicher, E.; Seeger, S.; Schölmerich, J.; Knuechel, R. Intracellular localization is a cofactor for the phototoxicity of protoporphyrin IX in the gastrointestinal tract: In vitro study. *Photochem. Photobiol.* **2003**, *78*, 393–399. [CrossRef] [PubMed]
83. Siboni, G.; Weitman, H.; Freeman, D.; Mazur, Y.; Malik, Z.; Ehrenberg, B. The correlation between hydrophilicity of hypericins and helianthrone: Internalization mechanisms, subcellular distribution and photodynamic action in colon carcinoma cells. *Photochem. Photobiol. Sci.* **2002**, *1*, 483–491. [CrossRef] [PubMed]
84. Orenstein, A.; Kostenich, G.; Roitman, L.; Shechtman, Y.; Kopolovic, Y.; Ehrenberg, B.; Malik, Z. A comparative study of tissue distribution and photodynamic therapy selectivity of chlorin e6, Photofrin II and ALA-induced protoporphyrin IX in a colon carcinoma model. *Br. J. Cancer* **1996**, *73*, 937–944. [CrossRef] [PubMed]
85. Krieg, R.C.; Messmann, H.; Rauch, J.; Seeger, S.; Knuechel, R. Metabolic characterization of tumor cell-specific protoporphyrin IX accumulation after exposure to 5-aminolevulinic acid in human colonic cells. *Photochem. Photobiol.* **2002**, *76*, 518–525. [CrossRef]
86. Ibarra, L.E.; Porcal, G.V.; Macor, L.P.; Ponzio, R.A.; Spada, R.M.; Lorente, C.; Chesta, C.A.; Rivarola, V.A.; Palacios, R.E. Metallated porphyrin-doped conjugated polymer nanoparticles for efficient photodynamic therapy of brain and colorectal tumor cells. *Nanomedicine* **2018**, *13*, 605–624. [CrossRef]
87. Zheng, J.H.; Shi, D.; Zhao, Y.; Chen, Z.L. Role of calcium signal in apoptosis and protective mechanism of colon cancer cell line SW480 in response to 5-aminolevulinic acid-photodynamic therapy. *Chin. J. Cancer* **2006**, *25*, 683–688.
88. Guo, Y.J.; Pan, W.W.; Liu, S.B.; Shen, Z.F.; Xu, Y.; Hu, L.L. ERK/MAPK signalling pathway and tumorigenesis. *Exp. Ther. Med.* **2020**, *19*, 1997–2007. [CrossRef]
89. Šemeláková, M.; Mikeš, J.; Jendželovský, R.; Fedoročko, P. The pro-apoptotic and anti-invasive effects of hypericin-mediated photodynamic therapy are enhanced by hyperforin or aristoforin in HT-29 colon adenocarcinoma cells. *J. Photochem. Photobiol. B.* **2012**, *117*, 115–125. [CrossRef]

90. Xu, Y.; Yao, Y.; Wang, L.; Chen, H.; Tan, N. Hyaluronic Acid Coated Liposomes Co-Delivery of Natural Cyclic Peptide RA-XII and Mitochondrial Targeted Photosensitizer for Highly Selective Precise Combined Treatment of Colon Cancer. *Int. J. Nanomed.* **2021**, *16*, 4929–4942. [CrossRef]
91. Chan, C.M.; Lo, P.C.; Yeung, S.L.; Ng, D.K.; Fong, W.P. Photodynamic activity of a glucoconjugated silicon(IV) phthalocyanine on human colon adenocarcinoma. *Cancer Biol. Ther.* **2010**, *10*, 126–134. [CrossRef] [PubMed]
92. Ye, R.; Zhang, Y.; Lee, A.S. The ER Chaperone GRP78 and Cancer. In *Protein Misfolding Disorders: A Trip into the ER*; Hetz, C., Ed.; Bentham Science Publishers: Santiago, Chile, 2009; Volume 1, pp. 47–55.
93. Gariboldi, M.B.; Ravizza, R.; Baranyai, P.; Caruso, E.; Banfi, S.; Meschini, S.; Monti, E. Photodynamic effects of novel 5,15-diaryl-tetrapyrrole derivatives on human colon carcinoma cells. *Bioorg Med. Chem.* **2009**, *17*, 2009–2016. [CrossRef] [PubMed]
94. Abrahamse, H.; Houreld, N.N. Genetic Aberrations Associated with Photodynamic Therapy in Colorectal Cancer Cells. *Int. J. Mol. Sci.* **2019**, *20*, 3254. [CrossRef] [PubMed]
95. Abdulrehman, G.; Xv, K.; Li, Y.; Kang, L. Effects of meta-tetrahydroxyphenylchlorin photodynamic therapy on isogenic colorectal cancer SW480 and SW620 cells with different metastatic potentials. *Lasers Med. Sci.* **2018**, *33*, 1581–1590. [CrossRef]
96. Ouyang, G.; Liu, Z.; Xiong, L.; Chen, X.; Li, Q.; Huang, H.; Lin, L.; Miao, X.; Ma, L.; Chen, W.; et al. Role of PpIX-based photodynamic therapy in promoting the damage and apoptosis of colorectal cancer cell and its mechanisms. *J. Cent. S. Univ. Med. Sci.* **2017**, *42*, 874–881.
97. Shahzidi, S.; Stokke, T.; Soltani, H.; Nesland, J.M.; Peng, Q. Induction of apoptosis by hexaminolevulinate-mediated photodynamic therapy in human colon carcinoma cell line 320DM. *J. Environ. Pathol. Toxicol. Oncol.* **2006**, *25*, 159–171. [CrossRef]
98. Ali, S.M.; Olivo, M. Bio-distribution and subcellular localization of Hypericin and its role in PDT induced apoptosis in cancer cells. *Int. J. Oncol.* **2002**, *21*, 531–540. [CrossRef]
99. Matroule, J.Y.; Carthy, C.M.; Granville, D.J.; Jolois, O.; Hunt, D.W.; Piette, J. Mechanism of colon cancer cell apoptosis mediated by pyropheophorbide-a methylester photosensitization. *Oncogene* **2001**, *20*, 4070–4084. [CrossRef]
100. Pérez-Arnaiz, C.; Acuña, M.I.; Busto, N.; Echevarría, I.; Martínez-Alonso, M.; Espino, G.; García, B.; Domínguez, F. Thiabendazole-based Rh(III) and Ir(III) biscyclometallated complexes with mitochondria-targeted anticancer activity and metal-sensitive photodynamic activity. *Eur. J. Med. Chem.* **2018**, *157*, 279–293. [CrossRef]
101. Wang, X.; Zhang, A.; Qiu, X.; Yang, K.; Zhou, H. The IL-12 family cytokines in fish: Molecular structure, expression profile and function. *Dev. Comp. Immunol.* **2023**, *141*, 104643. [CrossRef]
102. Jalili, A.; Makowski, M.; Switaj, T.; Nowis, D.; Wilczynski, G.M.; Wilczek, E.; Chorazy-Massalska, M.; Radzikowska, A.; Maslinski, W.; Biały, L.; et al. Effective photoimmunotherapy of murine colon carcinoma induced by the combination of photodynamic therapy and dendritic cells. *Clin. Cancer Res.* **2004**, *10*, 4498–4508. [CrossRef]
103. Cogno, I.S.; Gilardi, P.; Comini, L.; Núñez-Montoya, S.C.; Cabrera, J.L.; Rivarola, V.A. Natural photosensitizers in photodynamic therapy: In vitro activity against monolayers and spheroids of human colorectal adenocarcinoma SW480 cells. *Photodiagn. Photodyn. Ther.* **2020**, *31*, 101852. [CrossRef]
104. Marcinkowska, E.; Ziółkowski, P.; Pacholska, E.; Latos-Grazyński, L.; Chmielewski, P.; Radzikowski, C. The new sensitizing agents for photodynamic therapy: 21-selenaporphyrin and 21-thiaporphyrin. *Anticancer Res.* **1997**, *17*, 3313–3319.
105. Luo, M.; Ji, J.; Yang, K.; Li, H.; Kang, L. The role of autophagy in the treatment of colon cancer by chlorin e6 photodynamic therapy combined with oxaliplatin. *Photodiagn. Photodyn. Ther.* **2022**, *40*, 103082. [CrossRef]
106. Marchal, S.; Fadloun, A.; Maugain, E.; D'Hallewin, M.A.; Guillemin, F.; Bezdetnaya, L. Necrotic and apoptotic features of cell death in response to Foscan photosensitization of HT29 monolayer and multicell spheroids. *Biochem. Pharmacol.* **2005**, *69*, 1167–1176. [CrossRef]
107. Pansa, M.F.; Lamberti, M.J.; Cogno, I.S.; Correa, S.G.; Rumie Vittar, N.B.; Rivarola, V.A. Contribution of resident and recruited macrophages to the photodynamic intervention of colorectal tumor microenvironment. *Tumour Biol.* **2016**, *37*, 541–552. [CrossRef]
108. Mroz, P.; Szokalska, A.; Wu, M.X.; Hamblin, M.R. Photodynamic therapy of tumors can lead to development of systemic antigen-specific immune response. *PLoS ONE* **2010**, *5*, e15194. [CrossRef]
109. Na, T.Y.; Schecterson, L.; Mendonsa, A.M.; Gumbiner, B.M. The functional activity of E-cadherin controls tumor cell metastasis at multiple steps. *Proc. Natl. Acad. Sci. USA* **2020**, *117*, 5931–5937. [CrossRef]
110. Ma, H.; Yang, K.; Li, H.; Luo, M.; Wufuer, R.; Kang, L. Photodynamic effect of chlorin e6 on cytoskeleton protein of human colon cancer SW480 cells. *Photodiagn. Photodyn. Ther.* **2021**, *33*, 102201. [CrossRef]
111. Wufuer, R.; Ma, H.X.; Luo, M.Y.; Xu, K.Y.; Kang, L. Downregulation of Rac1/PAK1/LIMK1/cofilin signaling pathway in colon cancer SW620 cells treated with Chlorin e6 photodynamic therapy. *Photodiagn. Photodyn. Ther.* **2021**, *33*, 102143. [CrossRef]
112. Dobre, M.; Boscencu, R.; Neagoe, I.V.; Surcel, M.; Milanesi, E.; Manda, G. Insight into the Web of Stress Responses Triggered at Gene Expression Level by Porphyrin-PDT in HT29 Human Colon Carcinoma Cells. *Pharmaceutics* **2021**, *13*, 1032. [CrossRef]
113. Yuan, Z.; Liu, C.; Sun, Y.; Li, Y.; Wu, H.; Ma, S.; Shang, J.; Zhan, Y.; Yin, P.; Gao, F. Bufalin exacerbates Photodynamic therapy of colorectal cancer by targeting SRC-3/HIF-1α pathway. *Int. J. Pharm.* **2022**, *624*, 122018. [CrossRef]
114. West, C.M.; Moore, J.V. Mechanisms behind the resistance of spheroids to photodynamic treatment: A flow cytometry study. *Photochem. Photobiol.* **1992**, *55*, 425–430. [CrossRef]
115. Rodríguez, M.E.; Catrinacio, C.; Ropolo, A.; Rivarola, V.A.; Vaccaro, M.I. A novel HIF-1α/VMP1-autophagic pathway induces resistance to photodynamic therapy in colon cancer cells. *Photochem. Photobiol. Sci.* **2017**, *16*, 1631–1642. [CrossRef]

116. Rashid, M.; Zadeh, L.R.; Baradaran, B.; Molavi, O.; Ghesmati, Z.; Sabzichi, M.; Ramezani, F. Up-down regulation of HIF-1α in cancer progression. *Gene* **2021**, *798*, 145796. [CrossRef]
117. Lamberti, M.J.; Pansa, M.F.; Vera, R.E.; Fernández-Zapico, M.E.; Rumie Vittar, N.B.; Rivarola, V.A. Transcriptional activation of HIF-1 by a ROS-ERK axis underlies the resistance to photodynamic therapy. *PLoS ONE* **2017**, *12*, e0177801. [CrossRef]
118. Wang, H.P.; Hanlon, J.G.; Rainbow, A.J.; Espiritu, M.; Singh, G. Up-regulation of Hsp27 plays a role in the resistance of human colon carcinoma HT29 cells to photooxidative stress. *Photochem. Photobiol.* **2002**, *76*, 98–104. [CrossRef]
119. Gyurászová, K.; Mikeš, J.; Halaburková, A.; Jendželovský, R.; Fedoročko, P. YM155, a small molecule inhibitor of survivin expression, sensitizes cancer cells to hypericin-mediated photodynamic therapy. *Photochem. Photobiol. Sci.* **2016**, *15*, 812–821. [CrossRef]
120. Jendzelovský, R.; Mikeš, J.; Koval', J.; Soucek, K.; Procházková, J.; Kello, M.; Sacková, V.; Hofmanová, J.; Kozubík, A.; Fedorocko, P. Drug efflux transporters, MRP1 and BCRP, affect the outcome of hypericin-mediated photodynamic therapy in HT-29 adenocarcinoma cells. *Photochem. Photobiol. Sci.* **2009**, *8*, 1716–1723. [CrossRef]
121. Purkiss, S.F.; Grahn, M.F.; Williams, N.S. Haematoporphyrin derivative–photodynamic therapy of colorectal carcinoma, sensitized using verapamil and adriamycin. *Surg. Oncol.* **1996**, *5*, 169–175. [CrossRef]
122. Halaburková, A.; Jendželovský, R.; Koval', J.; Herceg, Z.; Fedoročko, P.; Ghantous, A. Histone deacetylase inhibitors potentiate photodynamic therapy in colon cancer cells marked by chromatin-mediated epigenetic regulation of CDKN1A. *Clin. Epigenet.* **2017**, *9*, 62. [CrossRef] [PubMed]
123. Liu, K.; Yao, H.; Wen, Y.; Zhao, H.; Zhou, N.; Lei, S.; Xiong, L. Functional role of a long non-coding RNA LIFR-AS1/miR-29a/TNFAIP3 axis in colorectal cancer resistance to pohotodynamic therapy. *Biochim. Biophys. Acta Mol. Basis Dis.* **2018**, *1864*, 2871–2880. [CrossRef] [PubMed]
124. Luo, M.; Yang, X.; Chen, H.N.; Nice, E.C.; Huang, C. Drug resistance in colorectal cancer: An epigenetic overview. *Biochim. Biophys. Acta Rev. Cancer* **2021**, *1876*, 188623. [CrossRef] [PubMed]
125. Sardoiwala, M.N.; Kushwaha, A.C.; Dev, A.; Shrimali, N.; Guchhait, P.; Karmakar, S.; Roy Choudhury, S. Hypericin-Loaded Transferrin Nanoparticles Induce PP2A-Regulated BMI1 Degradation in Colorectal Cancer-Specific Chemo-Photodynamic Therapy. *ACS Biomater. Sci. Eng.* **2020**, *6*, 3139–3153. [CrossRef] [PubMed]
126. Hodgkinson, N.; Kruger, C.A.; Abrahamse, H. Targeted photodynamic therapy as potential treatment modality for the eradication of colon cancer and colon cancer stem cells. *Tumour Biol.* **2017**, *39*, 1010428317734691. [CrossRef] [PubMed]
127. Lin, S.; Lei, K.; Du, W.; Yang, L.; Shi, H.; Gao, Y.; Yin, P.; Liang, X.; Liu, J. Enhancement of oxaliplatin sensitivity in human colorectal cancer by hypericin mediated photodynamic therapy via ROS-related mechanism. *Int. J. Biochem. Cell Biol.* **2016**, *71*, 24–34. [CrossRef] [PubMed]
128. Yuan, Z.; Fan, G.; Wu, H.; Liu, C.; Zhan, Y.; Qiu, Y.; Shou, C.; Gao, F.; Zhang, J.; Yin, P.; et al. Photodynamic therapy synergizes with PD-L1 checkpoint blockade for immunotherapy of CRC by multifunctional nanoparticles. *Mol. Ther.* **2021**, *29*, 2931–2948. [CrossRef]
129. Kleban, J.; Mikes, J.; Horváth, V.; Sacková, V.; Hofmanová, J.; Kozubík, A.; Fedorocko, P. Mechanisms involved in the cell cycle and apoptosis of HT-29 cells pre-treated with MK-886 prior to photodynamic therapy with hypericin. *J. Photochem. Photobiol. B.* **2008**, *93*, 108–118. [CrossRef]
130. Aniogo, E.C.; George, B.P.; Abrahamse, H. Molecular Effectors of Photodynamic Therapy-Mediated Resistance to Cancer Cells. *Int. J. Mol. Sci.* **2021**, *22*, 13182. [CrossRef]
131. Pucelik, B.; Sułek, A.; Barzowska, A.; Dąbrowski, J.M. Recent advances in strategies for overcoming hypoxia in photodynamic therapy of cancer. *Cancer Lett.* **2020**, *492*, 116–135. [CrossRef]
132. Aniogo, E.C.; Plackal Adimuriyil George, B.; Abrahamse, H. The role of photodynamic therapy on multidrug resistant breast cancer. *Cancer Cell Int.* **2019**, *19*, 91. [CrossRef] [PubMed]
133. Gu, B.; Wang, B.; Li, X.; Feng, Z.; Ma, C.; Gao, L.; Yu, Y.; Zhang, J.; Zheng, P.; Wang, Y.; et al. Photodynamic therapy improves the clinical efficacy of advanced colorectal cancer and recruits immune cells into the tumor immune microenvironment. *Front. Immunol.* **2022**, *13*, 1050421. [CrossRef] [PubMed]
134. Pramual, S.; Lirdprapamongkol, K.; Jouan-Hureaux, V.; Barberi-Heyob, M.; Frochot, C.; Svasti, J.; Niamsiri, N. Overcoming the diverse mechanisms of multidrug resistance in lung cancer cells by photodynamic therapy using pTHPP-loaded PLGA-lipid hybrid nanoparticles. *Eur. J. Pharm. Biopharm.* **2020**, *149*, 218–228. [CrossRef] [PubMed]
135. Deken, M.M.; Kijanka, M.M.; Beltrán Hernández, I.; Slooter, M.D.; de Bruijn, H.S.; van Diest, P.J.; van Bergen En Henegouwen, P.M.P.; Lowik, C.W.G.M.; Robinson, D.J.; Vahrmeijer, A.L.; et al. Nanobody-targeted photodynamic therapy induces significant tumor regression of trastuzumab-resistant HER2-positive breast cancer, after a single treatment session. *J. Control Release* **2020**, *323*, 269–281. [CrossRef] [PubMed]
136. Luo, Z.; Li, M.; Zhou, M.; Li, H.; Chen, Y.; Ren, X.; Dai, Y. O2-evolving and ROS-activable nanoparticles for treatment of multi-drug resistant Cancer by combination of photodynamic therapy and chemotherapy. *Nanomedicine* **2019**, *19*, 49–57. [CrossRef] [PubMed]
137. Zhong, D.; Wu, H.; Wu, Y.; Li, Y.; Yang, J.; Gong, Q.; Luo, K.; Gu, Z. Redox dual-responsive dendrimeric nanoparticles for mutually synergistic chemo-photodynamic therapy to overcome drug resistance. *J. Control Release* **2021**, *329*, 1210–1221. [CrossRef]
138. Liu, Y.; Zhou, Z.; Hou, J.; Xiong, W.; Kim, H.; Chen, J.; Zheng, C.; Jiang, X.; Yoon, J.; Shen, J. Tumor Selective Metabolic Reprogramming as a Prospective PD-L1 Depression Strategy to Reactivate Immunotherapy. *Adv Mater.* **2022**, *34*, e2206121. [CrossRef]

139. Huis In't Veld, R.V.; Heuts, J.; Ma, S.; Cruz, L.J.; Ossendorp, F.A.; Jager, M.J. Current Challenges and Opportunities of Photodynamic Therapy against Cancer. *Pharmaceutics* **2023**, *15*, 330. [CrossRef]
140. Liang, X.; Chen, M.; Bhattarai, P.; Hameed, S.; Dai, Z. Perfluorocarbon@Porphyrin Nanoparticles for Tumor Hypoxia Relief to Enhance Photodynamic Therapy against Liver Metastasis of Colon Cancer. *ACS Nano* **2020**, *14*, 13569–13583. [CrossRef]
141. Ding, D.; Zhong, H.; Liang, R.; Lan, T.; Zhu, X.; Huang, S.; Wang, Y.; Shao, J.; Shuai, X.; Wei, B. Multifunctional Nanodrug Mediates Synergistic Photodynamic Therapy and MDSCs-Targeting Immunotherapy of Colon Cancer. *Adv. Sci.* **2021**, *8*, e2100712. [CrossRef]
142. Yan, S.; Tang, D.; Hong, Z.; Wang, J.; Yao, H.; Lu, L.; Yi, H.; Fu, S.; Zheng, C.; He, G.; et al. CD133 peptide-conjugated pyropheophorbide-a as a novel photosensitizer for targeted photodynamic therapy in colorectal cancer stem cells. *Biomater. Sci.* **2021**, *9*, 2020–2031. [CrossRef] [PubMed]
143. Kong, F.; Zou, H.; Liu, X.; He, J.; Zheng, Y.; Xiong, L.; Miao, X. miR-7112-3p targets PERK to regulate the endoplasmic reticulum stress pathway and apoptosis induced by photodynamic therapy in colorectal cancer CX-1 cells. *Photodiagn. Photodyn Ther.* **2020**, *29*, 101663. [CrossRef] [PubMed]
144. Lee, D.; Kwon, S.; Jang, S.Y.; Park, E.; Lee, Y.; Koo, H. Overcoming the obstacles of current photodynamic therapy in tumors using nanoparticles. *Bioact. Mater.* **2021**, *8*, 20–34. [CrossRef] [PubMed]

Disclaimer/Publisher's Note: The statements, opinions and data contained in all publications are solely those of the individual author(s) and contributor(s) and not of MDPI and/or the editor(s). MDPI and/or the editor(s) disclaim responsibility for any injury to people or property resulting from any ideas, methods, instructions or products referred to in the content.

MDPI AG
Grosspeteranlage 5
4052 Basel
Switzerland
Tel.: +41 61 683 77 34

Cancers Editorial Office
E-mail: cancers@mdpi.com
www.mdpi.com/journal/cancers

Disclaimer/Publisher's Note: The statements, opinions and data contained in all publications are solely those of the individual author(s) and contributor(s) and not of MDPI and/or the editor(s). MDPI and/or the editor(s) disclaim responsibility for any injury to people or property resulting from any ideas, methods, instructions or products referred to in the content.

www.ingramcontent.com/pod-product-compliance
Lightning Source LLC
LaVergne TN
LVHW070405100526
838202LV00014B/1399